Decision Engineering

Series Editor

Dr Rajkumar Roy
Department of Enterprise Integration
School of Industrial and Manufacturing Science
Cranfield University
Cranfield
Bedford
MK43 0AL
UK

Other titles published in this series

Cost Engineering in Practice
John McIlwraith

IPA – Concepts and Applications in Engineering
Jerzy Pokojski

Strategic Decision Making
Navneet Bhushan and Kanwal Rai

Product Lifecycle Management
John Stark

From Product Description to Cost: A Practical Approach
Volume 1: The Parametric Approach
Pierre Foussier

Decision-Making in Engineering Design
Yotaro Hatamura

Intelligent Decision-making Support Systems: Foundations, Applications and Challenges
Jatinder N.D. Gupta, Guisseppi A. Forgionne and Manuel Mora
Publication due April 2006

Metaheuristics: A Comprehensive Guide to the Design and Implementation of Effective Optimisation Strategies
Christian Prins, Marc Sevaux and Kenneth Sörensen
Publication due December 2006

Context-aware Emotion-based Multi-agent Systems
Rajiv Khosla, Nadia Bianchi-Berthouse, Mel Seigel and Toyoaki Nishida
Publication due July 2006

Pierre Foussier

From Product Description to Cost: A Practical Approach

Volume 2: Building a Specific Model

With 171 Figures

Springer

Pierre Foussier, MBA
3f, 15, rue des Tilleuls
78960 Voisins le Bretonneux
France

British Library Cataloguing in Publication Data
Foussier, Pierre
From product description to cost: a practical approach
Volume 2: Building a specific model. - (Decision engineering)
1. Production planning - Mathematical models 2. Start-up costs
Mathematical models 3.New products - Decision-making
I. Title
658.1'552'015118
ISBN-10: 1846280427

Library of Congress Control Number: 2005937146

Decision Engineering Series ISSN 1619-5736
ISBN-10: 1-84628-042-7 e-ISBN 1-84628-043-5 Printed on acid-free paper
ISBN-13: 978-1-84628-042-9

Typesetting: Gray Publishing, Tunbridge Wells, UK
Production: LE-TEX Jelonek, Schmidt & Vöckler GbR, Leipzig, Germany

Printed in Germany

9 8 7 6 5 4 3 2 1

Springer Science+Business Media
springer.com

To my daughter Pierrine,
without whom this book could not have been written.

Preface

Volume 1 was dedicated first (Part I) to a general understanding of the cost forecasting, generally called "cost estimating", and then to the important concept of data normalization (Part II), which is a prerequisite for comparing cost data, as it was reminded that the only way the human mind found for forecasting the future was to extrapolate the results of previous "experiences", which implies the necessity to compare them. Such a comparison can only be made on comparable, or normalized, data.

Then (Part III) introduced the concept of what we called "general" models and eventually (Part IV) the use of models in the cost-estimating process. Taking into account the "risk" in cost forecasting was an important chapter of this Part IV.

By "**general models**" we mean that these models can estimate the cost of anything – at least in a given "class" of product, a class representing an industrial sector. These models, of which number is limited, are difficult to build; they represent, if they are really general, a large investment.

This volume (Volume 2) deals with the building of "**specific**" cost estimating **models** (sometimes also called "in-house models"). A specific cost model explicitly refers to a "product family", which is a set of products fulfilling the same function(s) and manufactured about the same way.

A short word at the history of science helps illustrate the fundamental differences between specific and general models.

Any science first looks at the facts and records them: in our modern language, any science starts by building databases. In the domain of astronomy the fantastic amount of data accumulated by great observers, such as Copermic, Tycho Brahé, ... deserves our admiration.

The second step – not necessarily carried out by the same persons – is to try to establish and quantify correlations between variables which, apparently, may seem different. Once a good correlation has been demonstrated between these variables – this involves what is called in this volume "data analysis" – it is rather tempting a build a mathematical relationship between these variables; these relationships do not "explain" anything; they are just a tentative to group in a few equations what we know about the facts. The nice thing about them is that they make us able to predict values of one variable when the other one(s) is (are) known, as long as the previsionist remains in the same area (we would say, in the domain of cost, in the same "product family"). These relationships are called "laws" (we would say in the cost domain cost-estimating relationships (CERs)). There are plenty of such relationships in all sciences; just remember: the three Kepler laws, the Kirchoff laws, the Van der Walls law, the law of light emission by the black body, etc. The authors of these laws did not know the regression analysis and generally work by curve fitting, but the idea is the same.

The third step is far more recent; it implies to look "below the facts" in order to understand them (which means explaining by investigation things more in depth and finding reasons for their behavior).[1] This is done by finding abstract concepts (such as the forces in the description of motion, the fields in electrodynamics, the entropy in thermodynamics, the quanta in light emission, etc.) from which the facts could be "explained". The mathematical support then becomes a must, as it is the only way the human mind can work with abstract concepts. The great names (the "giants" to cite a word used by Newton)[2] in this respect are Newton, Maxwell, Boltzman, Planck, Einstein, etc. The set of equations they developed, generally a very limited set from which all phenomena can be predicted,[3] is generally called a "theory".

In the cost domain, the abstract concept that throws a powerful light on the cost behavior is the "product structure". This concept was described by Lucien Géminard in France and maybe others. This concept, which is developed in Part III of Volume 1, helped create a general "theory" of cost behavior. But it is the only time I will use this term of "theory" in our domain and for three reasons:

1. The first reason is that human behavior is far less predictable than natural phenomena in the physical sciences. Therefore the fantastic level of precision often attained in the physical sciences cannot be obtained in the domain of cost. The word "theory" in the domain of cost could therefore be misleading and rejected, although it correctly describes the human look at the things.
2. The second reason is that – as it was said by Karl Popper – a theory can neither be considered as finished: it has always to be checked with the results of nature and just one phenomenon which does not fit with the theory seriously questions its validity: remember the experience carried out by Morley and Michelson, or the advance of Mercury perihelion. One single experience can force people to adopt another theory. But in the current language, theory is considered as the truth and, again, the common word could be misleading in the domain of cost.
3. The third reason is related to semantics: in the ordinary language, the word "theory" has two opposite meanings. First of all it is used, with great respect, to qualify the work of the giants who preceded us. But the second usage is rather dangerous: if you arrive in a meeting with a cost estimate adding that it was prepared with such or such theory, you may be sure that somebody will demolish your estimate, saying it is just a "theoretical" approach … . The word "model" is much more accepted than the word "theory" and we will use it.

As the techniques for building such models are now well understood (even if they can still be improved), preparing these models can be done by any company, and the cost analyst has just to follow the documented procedures. This does not mean that the process can be fully automated: during his/her work, the cost analyst will have to make decisions, which require a good understanding of these procedures.

---

[1] It is well known that we never "understand" nature fully, by a step-by-step analysis requiring less and less hypothesis: understanding nature really means reducing the number of the hypotheses which are necessary for predictions.

[2] If I could see farther than the other ones, it is because I was sitting on the shoulders of the giants who preceded me.

[3] This illustrates the power of both the concepts and the mathematics which use them!

The major advantage of these specific models is that they are built from the company own data (this obviously requires that the company was organized for capturing and saving its data, and this is the major constraint). Therefore:

1. The cost analyst can choose the variables, or "parameters", he/she wants to include in the model, depending on the purpose of it (for instance he/she may prefer to use functional or physical variables).
2. The credibility (and credibility is an important concept in cost forecasting!) of a cost forecast prepared by a specific model is higher that any forecast made by a general model, because the source of the forecast is clear.
3. The way the forecast was prepared is easy to explain to a decision-maker, even if only a few minutes are available.

For these reasons cost estimators are strongly encouraged to start parametric cost estimating following this path.

Using general models should come afterwards, for instance for cost estimating new products for which no comparison is possible with existing products ("first of a kind"), and therefore no specific model is available or even possible.

Understanding the procedures is the key word for creating successful specific models. For this reason all these procedures are fully described in this volume. Classical methods and new ones (such as the "Bootstrap") will be described and illustrated.

Cost estimating requires 30% of data, 30% of tools and 40% of judgment and thinking, with a minimum of 80% in total. EstimLab™ – with which most of the computations which illustrate this book have been performed – was designed to get all these 30% of tools with a minimum of effort, freeing time for collecting data and making use of judgment, which is always the most important component in cost estimating.

Paris
February 2005

Pierre Foussier

Contents

Introduction

What Are We Looking For?

Our company wants to buy (or design) a new product and we are asked by the procurement department (or the design department), for budgeting or negotiating purposes, to deliver a reliable cost estimate for this new product. This new product mass is 3.5 kg (we consider in this example one parameter only).

It happens that this product belongs to an existing product family (the concept of product family is defined in Chapter 1). This is rather fortunate and looking in our records, or in our database, we find the (normalized) cost and the mass of several products belonging to the same product family; the values are displayed in Figure 1.

These values are unfortunately rather scattered and this scattering forbids us to easily estimate the cost of the new product.

What we would like to present is a cost represented by the small square noted A. However, it is clear that, from the data we have, the cost could as well be between B and C (Figure 2).

If the data were not scattered at all, the trend (how the cost changes with the mass) would be very easy to find out: point A would be well defined and the difference between B and C would be negligible. But this is not the case and we have to cope with these values.

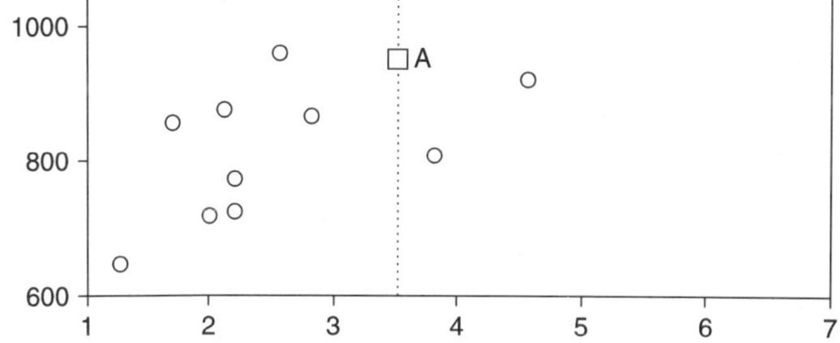

Figure 1 The available data: cost and mass of several products belonging to same product family (mass in abscissa, cost in ordinate).

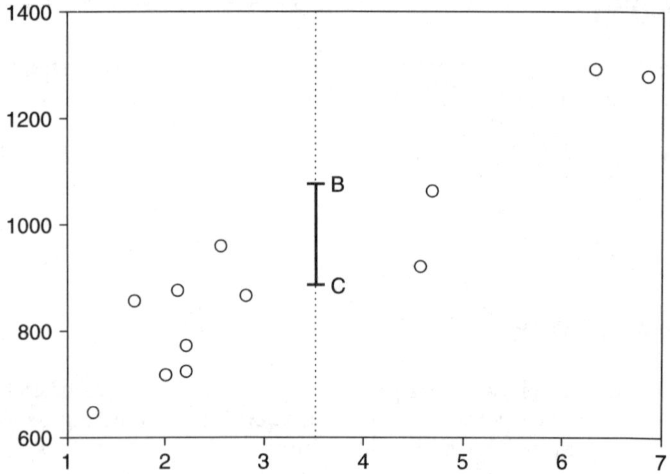

Figure 2 The available data: extreme values of cost (mass in abscissa, cost in ordinate).

Nevertheless, values B and C appear – intuitively – less likely than A, because these values are "supported" by few values of our database (only the extreme values do support them), whereas A seems to be supported by much more values. Consequently it appears natural to propose:

- First of all what can be called a "nominal cost" which can be defined as the most likely cost, in the sense it is supported by most of the data. It is represented by the small square noted A. It is the cost which could be expected if the data were not so scattered.
- The values B and C which could be considered as the "extreme" values of the cost; the costs are rather unlikely, but they are not impossible if we look at all the data of our database.

We have therefore to find out this "nominal cost", plus the extreme values: these values constitute the result, for a particular product of mass equal to 3.5 kg, of a "specific model", the word "specific" meaning that this "model" is dedicated to this product family and cannot be used for any other family.

Values A, B and C are, in this example, related to our product of which mass is 3.5 kg. But a new product belonging to this product family may have any mass. In such a case A, B and C will be defined as three curves, one for each value, as illustrated in Figure 3.

These three lines can be anything, and we will see that deciding about the shape of these lines will be one of the most important decision the cost analyst will have to make. Very often straight lines are selected, not because they are the best ones, but because their computation can easily made[4] by theorems based on the linear algebra which has been developed for centuries by mathematicians, Karl Friedrich Gauss being one of them.

[4] And also because the human mind prefers simple things each time it seems possible.

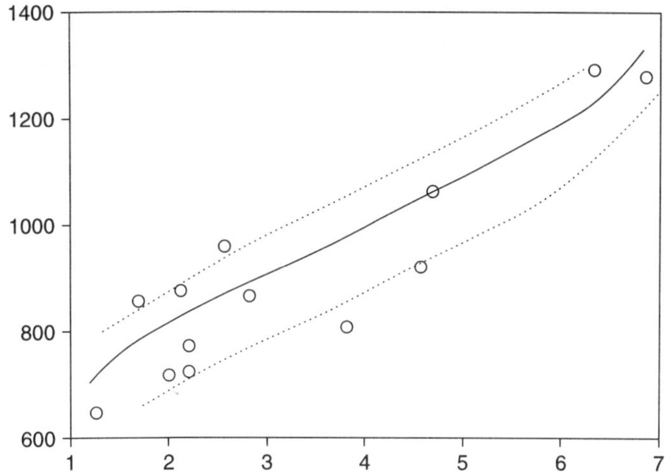

Figure 3 The three lines we are looking for.

Most often only the full line (called the "nominal cost") is computed, the two other lines being only represented by what will be called the "confidence interval", but full computations can nowadays be easily performed.

What Is a Specific Model?

A specific model is the result of these computations. It can be defined by the following way.

A specific model is a mathematical tool of which purpose is to help the cost estimator to prepare an estimate for a product belonging to a product family.

A specific model is a set of two things:

- A formula which gives the evolution of the nominal cost (this formula is generally called a "CER" which stands for "cost-estimating relationship").
- The confidence interval around the formula.

Both things really constitute the specific model (whereas many cost analysts consider only the first one); the formula will compute a "nominal cost", the distribution of the residuals around the formula will define its level of confidence.

We will use very rarely the word "CER", just to avoid the frequent confusion between a model and what is just a part of it.

How Is a Specific Model Built?

A specific model is traditionally built in two steps: first of all the formula is computed, then the distribution of the residuals (which are the deviations of our existing

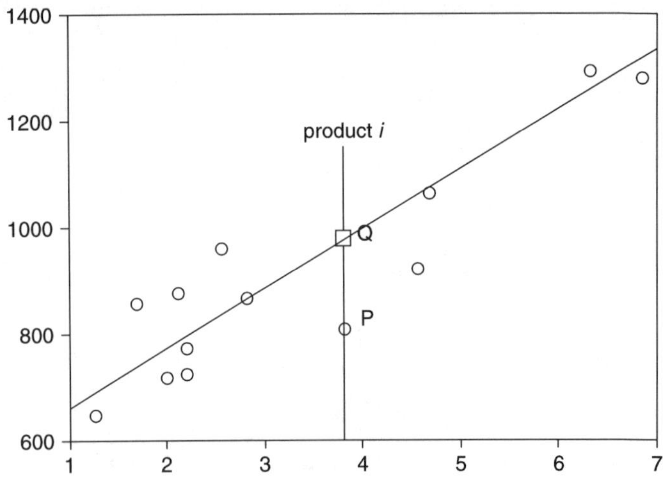

Figure 4 Adjusting the curve to the data points.

data around this formula) are globally computed. In this paragraph we just comment on the way the formula is computed.

First of all a comment about the shape of the formula: there is no procedure to automatically find out the most appropriate formula. Among several data points there is an infinite number of curves which can be drawn. Consequently, let us repeat it, **the shape of the formula is always a decision made by the cost analyst.** Most analysts prefer to use a straight line but this is a choice which has no theoretical background: we will discover that other relationships generally give better results, from the cost analyst point of view.

Once the formula shape has been decided on, **it has to be adjusted to the actual data.** For this reason the selected formula always include several "coefficients" and the cost analyst's job is to find out the coefficients which give the best fit between the curve and the data points.

What does this mean? Here also there are several ways to find out the best fit. The most common reasoning is the following one: the purpose of the formula is to replace any value such as P (see Figure 4), related to product i, by value Q given by this formula. This value Q will be considered as the "nominal cost" of this product i; *we obviously do not want this nominal cost to be too far away from the actual cost P.* This means that we would like that the deviation, or residual, $e_i = Q - P$ to be as small as possible.

As we do not want to favor any particular data point, we must find out a way such as the curve will be *as close as possible to all data points.* The natural solution would be to try minimizing the sum of all the deviations. However – and this was mentioned by Carl Friedrich Gauss – the sum of deviations is something difficult to mathematically work with, whereas the sum of the squares of the deviations is much more convenient (because minimizing such a sum of squares leads, when the selected curve is a straight line, to linear equations of which solution was discovered two centuries ago). Consequently the traditional solution is to find out the coefficients just mentioned by minimizing the sum:

$$\sum_i e_i^2$$

The general solution is logically called the "least squares method".

When the selected curve is a straight line, this procedure is known as the "linear regression". The word "linear" comes from the fact that the selected curve is a straight line, the word "regression" being explained later on.

This procedure has several mathematical properties that were studied by Gauss. However, it has also mathematical drawbacks that will appear in the following chapters. The use of computers allows now to search for more efficient curves and there is no reason to be contrived by old solutions.

These solutions will be studied in details in the following chapters.

A Preliminary Study

Before trying to work on the data in order to build a formula, it is very important to analyze the data: an algorithm – and building a formula, although in very simple situations it could be done on a piece of paper, will nowadays always use an algorithm – will always provide a result.

Some tests will be made to check the quality of a model. But, even if the results of these tests are positive, you will be confident about the result (the cost estimate you will compute from it) if you are confident in the quality of the data it is built from.

Your opinion about the data is based on the seriousness of the normalization and the analysis of these data. The first subject was dealt with in Volume 1. It is supposed from now on that this process has been carried out and that you are satisfied with the results.

You are now ready to analyze these data and to prepare a specific model.

We will do that in this volume with the same rigor as the processes we developed in the first volume.

Notations

Cost estimation handles information.

This information is generally presented into tables of figures. In order to discuss about these tables, to analyze their structure, to establish relationships between them, etc., it is convenient to use symbols to represent them. The definition of these symbols is given in this section. We try to use symbols which are – by using simple rules – easy to remember. Most of them are common with the majority of the authors; a few of them, when experience showed the symbols generally used in the literature maybe confusing, are different.

Information is relative to objects or products (or "individuals" in statistical books). It is conveyed by variables.

The sample is the set of objects for which we know the value of the variables. The population is the set of objects, of which number is supposed to be infinite, for which we want to get a cost estimate and from which the sample is supposed to be "extracted".

The "Individuals" or Products

The methods developed in this book can be applied to any set of objects. An element of a set is called an "individual" (the term reminds that statistical methods were developed for studying populations).

However, as its title mentions it, this book is principally dedicated to cost analysts and cost estimators. The subject of interest of these persons will be called "products".

The term "product" is therefore used here to designate anything we are interested in. It can be a piece of hardware, a system, a software, a tunnel, a bridge, a building, etc. Generally speaking, it is something which has to be designed and/or produced and/or maintained and/or decommissioned.

Products generally have names. In order to remain as general as possible, one will assign a letter to each one: they will be designated by capital letters such as A_1, $A_2, A_3, \ldots, A_i, \ldots, A_I$ (A for "article") the index[5] "i" being reserved for the products.

The number of products we may have to deal with simultaneously (the meaning of this adverb will become clear in the following chapters) will be called I.

A product is characterized by a set of variables.

[5] When indexes are used, it is very convenient – and easy to remember – to use a small letter for the current index and the same letter – this time in capital – for the upper limit of this index.

The Variables

A **variable** is something which can be quantified (in such a case it is called a "*quantitative*" variable, the quantity being a number such as 27, or 34.5 or even –12.78) or on which one can affect an attribute (in such a case it is called a "*qualitative*" variable; the quality being an adjective such as superior, good, poor or even a sentence, such as made by machine C, machine D, … , or even sometimes an integer);[6] the attribute can be "*objective*" (if it expresses a fact, such as the material used, or the manufacturer) or "*subjective*" if it expresses an opinion (such as little complex, very complex … the adjective "complex", or sometimes "difficult", being rather frequent for expressing an opinion about the nature of a product).

A **quantitative variable** must have a unit (such as kilogram, meter, inch, euro, dollar, or simply "quantity", generally simplified in "qty"). A limit can be imposed on the set of values it may take. It is supposed to be continuous, even it is not really, such as qty.

A **qualitative variable** refers to a limited set of attributes or modalities such as (manufactured by A or B) or (blue, green, yellow) or (very low, medium, high, very high).

A variable will always be symbolized by a capital V. This capital V will only be used to represent the variable as such (e.g. one can say V represents the product mass; sometimes one can even say – it is not really correct but generally accepted – V "is" the mass), not its value.

Several variables will have to be used. Each variable has a name, such as the mass, or the power, or the speed, … ; for practical reasons they will nevertheless be represented by the capital letter V with an index, such as $V_0, \ldots, V_j, \ldots, V_J$ the index "j" being reserved for the variables. In this book, dedicated to forecasting a value (in principle the cost, but the methods can be applied to practically anything), *two types of variables* must be distinguished:[7]

1. The "explicative" or "causal" variables, which are the variables that are known when a forecast has to be made; these variables are generally called "parameters". They will be represented by an index equal or superior to 1. Example: V_2. The number of these variables will be called J. V_0 will be used for a constant, as cost may have a constant part, which has of course also to be determined.
2. The "dependent" or "explained" variable, which is the one that we want to forecast. In order to clearly distinguish it from the causal variables its name will be called Y. There is only one such variable in any particular treatment of the data. However, it is quite possible to have different cost values for the same product: for instance you may have the cost of development, the cost of manufacturing, the cost of materials, etc. And you may be interested to find out a correlation between the development cost and the production cost. In such a case, you will have to define, for each treatment, which is the "dependent" variable, and which are the causal variables.

The Observed Values

A value is a figure (for a quantitative variable) or an attribute (for a qualitative variable) observed on a product. Values for products belonging to the sample will

---

[6] In such a case it must not be confused with a quantitative variable: we can attribute the integer "1" to the sentence "made by machine C", etc., which does not mean that such attributes may be added or multiplied.

[7] We return to this important subject – from a methodological point of view – in Chapter 1.

always be represented by small letters, capital letters being reserved for products of the population.

For the Dependent Variable

As previously mentioned, there is only one dependent variable per product. The table of the values for the I products therefore takes the form of a vector:

$$\begin{Vmatrix} y_1 \\ y_2 \\ \vdots \\ y_i \\ \vdots \\ y_I \end{Vmatrix}$$

or simply $\vec{y}$ (column vector). y_i represents the cost of product i.

Centered and Scaled Variables

The values, once centered, are noted ${}_cy_i$; once centered and scaled are noted ${}_{cs}y_i$. The definition of these values requires the knowledge of the arithmetic mean, noted $\bar{y}$, and the standard deviation, noted s_y which are defined in Chapter 2. Then:

$${}_cy_i = y_i - \bar{y} \quad {}_{cs}y_i = \frac{y_i - \bar{y}}{s_y}$$

Other notations for the sample.
Notations for the sample always use small letters:

- Arithmetic mean – or simply "mean": $\bar{y}$.
- Median: $\tilde{y}$.
- Dynamic center (the term is defined in Chapter 11) in general: $\hat{y}$, and its value for product A_i: $\hat{y}_i$.
- When data point i (one product) is eliminated from the sample: $\vec{y}_{(i)}$ represents the set of all values less y_i.

Residuals and Euclidian Distances

Residuals are an important concept in this book; it quantifies, for a given product, the "distance" between the dynamic center and the cost value. As several "distances" may be defined, one distinguishes e_{i+} defined by $e_{i+} = y_i - \hat{y}_i$, e_{i*} defined by $e_{i*} = (y_i/\hat{y}_i)$, e_{i-} defined by $e_{i*} = (y_i/\hat{y}_i) - 1$.

The Euclidian distance used in a sample is represented by Δ_i

A normalized residual is noted e_{i+}^* (defined only for additive residuals).

For the Independent or Causal Variables

An *observed* value for a variable is always represented by a small x. One can say: we measured $x = 17.5\,\text{kg}$.

An observed value always refers to a variable and a product: it is the value observed for product i when variable j is considered. In order to make this clear, an observed value, when there is more than one causal variable, must always have two indexes. These indexes will always be in the following order: product number, variable number, both being separated by a comma. The value observed for product i on variable j will therefore be represented by $x_{i,j}$:

$$\text{observed value: } x_{product\_number,\ variable\_number}$$

Observed values are generally arranged in tables. Tables will play an important role in this book. They must therefore be fully understood.

Example:

A column is dedicated to a variable: it gives the values observed on all the products for this particular variable

A row is dedicated to a product: it gives the values of all variables observed for this particular product

$$\Downarrow$$

$$\Rightarrow \begin{Vmatrix} 23 & 45 & \dots & 52 \\ 47 & 15 & \dots & 37 \\ \vdots & \vdots & \vdots & \vdots \\ 67 & 17 & \dots & 75 \end{Vmatrix}$$

Such a table – the mathematical term of "matrix" will generally be used – will be represented by a letter inside two sets of two small bars, just to remind the reader it is a special entity, such as $||x||$ for the observed values.[8]

$||x||$ is the basic matrix or set of observed values. An element of this matrix is marked by two indexes giving its row number and its column number, both numbers starting at 1. The following basic rule will be always applied:

$$\text{element of matrix } ||x||: x_{row\_number,\ column\_number}$$

This means that a row is dedicated to an object, a column being dedicated to a variable.

$||x||$ contains the raw data directly given by the observations (the sample). However, several matrices, derived from this raw matrix, will have to be used in the computations. They will be referred to by a pre-index or exponent:

- $||^{+}x||$ is derived from $||x||$ by adding a first column filled with 1. This column represents a constant value which is "observed" on all objects.

[8] Symbol X (in capital letters) is generally used in most textbooks, sometimes in bold or italic characters. If you are not very familiar with matrices computations, it makes the reading confusing. To facilitate this reading for all cost analysts, the small vertical bars were added in this book.

- $||_c x||$ is derived from $||x||$ by "centering" all the quantitative data. This centering proceeds column per column (centering has no meaning for the rows), each column being dealt with independently from the other ones:
  - the mean or average value of column j is computed; it can be called $\bar{x}_{\cdot,j}$, the little hat reminding it is an average,
  - this mean value is subtracted from each value of the column.
- $||_s x||$ is derived from $||x||$ by "scaling" the data. This scaling proceeds column per column (scaling has no meaning at all for the rows), each column being dealt with independently from the other ones:
  - the standard deviation of column j is computed; it can be called $s_{\cdot,j}$,
  - each value of the column is divided by this standard deviation.

  Such a matrix is very rarely used, the data, before scaling being nearly always first centered.
- $||_{cs} x||$ is derived from $||x||$ by centering and then scaling all the quantitative data. In this process an element $x_{i,j}$ becomes:

$$x_{i,j} \rightarrow {}_{cs}x_{i,j} = \frac{x_{i,j} - \bar{x}_{\bullet,j}}{s_{\bullet,j}}$$

The major advantage of this process is that now all the variables have the same unit, whatever they represent (mass, energy, speed, etc.).

These pre-indexes can be used together. For instance matrix $||^{+}_{cs}x||$ represents the matrix derived from $||x||$ by adding a first column of 1, centering and scaling the quantitative data. *Note*: In the centering and scaling process, the column of 1 – as well as the qualitative variables – remains unchanged: it is not concerned by these two processes.

We may have to use matrices from which a row (a product) or a column (a variable) is deleted. The following symbols will be used:

- $||x_{[i,\cdot]}||$ represents the matrix $||x||$ when row i is deleted.
- $||x_{[\cdot,j]}||$ represents the matrix $||x||$ when column j is deleted.

Mathematical Symbols

log or $\log_{10}$, sometimes used for the sake of clarity, represents the logarithm in base 10.

ln represents the natural logarithm (base $e = 2.71828$).

What You Need to Know About Matrices Algebra

You do not need to know so much ...

If you want to study the subject in depth, one can recommend Pettofrezzo [45] as an introduction, Lichnerowicz [36] and Golub and Van Loan [31] as full – sometimes complex – developments.

Matrices Are First a Stenography

Matrices are first used because they are a very simple and powerful stenography: it is always easier to mention the set of values as $||x||$, instead of displaying the whole table of these values.

Matrices, in this book, always contain real – or ordinary – numbers. The set of all these real numbers is represented by the symbol $\Re$ (another stenography), the set of all positive numbers being noted $\Re^+$. A matrix has a size given by its number of rows, let us call it I, and the number of columns, let us call it J (or $J + 1$ if a column of 1 is added); it therefore contains $I \times J$ elements. All matrices containing this number of elements (which are real numbers) are said belonging to the set $\Re^{I \times J}$. A particular matrix of this size is therefore said to belong to $\Re^{I \times J}$, or simply to be a matrix $\Re^{I \times J}$, I being the number of lines, J the number of columns.

General Properties About Matrices

The row "rank" of a matrix is the largest number of linearly independent rows; the column "rank" is the largest number of linearly independent columns. It can be demonstrated that for any matrix both ranks are equal: so one can speak only about the **rank** of the matrix.

A square matrix is said to be "full" rank if its rank is equal to the smallest of I or J (as $J = I$). A square matrix (the notion of singularity applies to square matrices only) is said to be **singular** if it is not full rank; its determinant is then equal to 0.

A matrix of which determinant is close or equal to 0 is said to be "*ill conditioned*".

Particular Matrices

A matrix such as $I = 1$ and $J = 1$ (it has just one element) is a scalar. We will consider it as an ordinary number.

A matrix such as $J = 1$ (one column then) is said to be a **column-vector**, or simply a vector; an example was given by the set of the y_i (the list of the costs for a set of products). It is represented either by $||y_i||$ or more commonly by $\vec{y}$.

A matrix such as $I = 1$ (just one row then) is said to be a **row-vector**. An example is given by the values of all the variables observed for a particular product. If we call z these values, such a row-vector is represented by:

$$|| z_0, z_1, \ldots, z_j, \ldots, z_J || \text{ or } \vec{\vec{z}}$$

Neither confuse a column-vector and a row-vector. They are different entities, as the examples given illustrate: the column-vector groups homogeneous values (costs for instance), whereas the row-vector groups inhomogeneous values (for instance: mass, power, material, etc.). The product of $\vec{y}$ by $\vec{\vec{z}}$ has a meaning (if the number of elements is the same) whereas the products of $\vec{y}$ by $\vec{u}$ or $\vec{\vec{v}}$ by $\vec{\vec{w}}$ have no meaning at all, even if they have the same number of elements.

A matrix having the same number of rows and columns is said "**square**", or $\Re^{I \times I}$.

A **diagonal** matrix is a matrix of which all the elements are 0 except the elements in the first diagonal. The following example illustrates what the first diagonal is:

$$\begin{Vmatrix} d_1 & 0 & \ldots & 0 & \ldots & 0 \\ 0 & d_2 & \ldots & 0 & \ldots & 0 \\ \vdots & \vdots & & \vdots & & \vdots \\ 0 & 0 & \ldots & d_i & \ldots & 0 \\ \vdots & \vdots & & \vdots & & \vdots \\ 0 & 0 & \ldots & 0 & \ldots & d_I \end{Vmatrix}$$

One can also define bidiagonal or tridiagonal matrices, but you do not need them, except if you want to compute by yourself the SVD (which stands for "singular values decomposition" of a matrix).

The "*trace*" of a square matrix is the sum of the elements of its first diagonal. For the matrix just defined, the trace is equal to $\sum_i d_i$.

A triangular superior matrix is a matrix of which only all the elements of the superior triangle are different from 0, as for instance the following matrix:

$$\begin{Vmatrix} 0 & 7 & 2 & 8 \\ 0 & 0 & 5 & 3 \\ 0 & 0 & 0 & 1 \\ 0 & 0 & 0 & 0 \\ 0 & 0 & 0 & 0 \end{Vmatrix}$$

A **symmetric** matrix is a matrix of which values are symmetrical in respect to the first diagonal.

An **orthogonal** matrix is a matrix of which the inverse is equal to its transpose (both terms are defined later on): $||M||^{-1} = ||M||^t$.

An **idempotent** matrix is a matrix of which square (the product – the term is defined below – of the matrix by itself) is equal to it: $||M||^2 = ||M|| \otimes ||M|| = ||M||$. An idempotent matrix has special properties:

- The trace of an idempotent matrix is equal to its rank.
- Its eigenvalues are only 0 and 1.

Algebra of Matrices

We are still in the domain of stenography. Here we just need rules.

The **inverse** of $||f|| \in \Re^{I \times J}$ is the matrix noted $||f||^{-1} \in \Re^{J \times I}$ (notice the dimensions) which is such that:

$$||f|| \otimes ||f||^{-1} = ||1|| \quad \text{with here } ||1|| \in \Re^{I \times I}$$

An important theorem about inverse is that the inverse of a product is given by the inverse of each matrix, the product being taken in the reverse order (this is obvious to save the rule about the matrices dimensions):

$$(||f|| \otimes ||g||)^{-1} = ||g||^{-1} \otimes ||f||^{-1}$$

The **transpose** of a matrix $||f|| \in \Re^{I \times J}$ is a matrix noted $||f||^t \in \Re^{J \times I}$ obtained by interchanging the rows and columns: column 1 of $||f||$ becomes row 1 of $||f||^t$, etc.

An important theorem about transposition is that the transpose of a product is given by the transpose of each matrix, the product being in the reverse order (this is obvious to save the rule about the matrices dimensions):

$$(||f|| \otimes ||g||)^t = ||g||^t \otimes ||f||^t$$

With the rule given for transposition, the transpose a row-vector is a column-vector, and the reciprocal. This is an important application of the transposition.

Operations on Two Matrices

Two operations can be defined on matrices: addition and multiplication. Addition will use the symbol $\oplus$, multiplication the symbol $\otimes$; these symbols are just there to recall the reader that these operations are not "ordinary" operations.

You can **add** two matrices *ONLY IF* they have the same $\Re^{I \times J}$ type (the same size). The sum of two matrices is a matrix of the same type, of which element *i,j* is the sum of the corresponding elements of the original matrices: the operation $||u|| \oplus ||v||$ gives a matrix $||w||$ with the simple rule $w_{i,j} = u_{i,j} + v_{i,j}$. The "neutral" matrix for the addition is the matrix, noted $||0||$, of which all elements are equal to 0:

$$||u|| \oplus ||0|| = ||u||$$

if, of course $||u||$ and $||0||$ have the same type.

You can **multiply** two matrices *ONLY IF* the number of lines of the second matrix is equal to the number of columns of the first one: the product $||f|| \otimes ||g||$ *in this order* where $||f|| \in \Re^{I \times K}$ and $||g|| \in \Re^{K \times J}$ gives a matrix $||h|| \in \Re^{I \times J}$: the mnemonic rule is that, when you write the matrices in the order you want to multiply them, the indexes which are "in the middle" (here *K*) must be equal and disappear in the operation. Note that the multiplication is not commutative ($||g|| \otimes ||f||$ if it is possible, is different from $||f|| \otimes ||g||$): both operations are of course possible if both matrices are square and of the same size. The element $h_{i,j}$ (row *i*, column *j*)

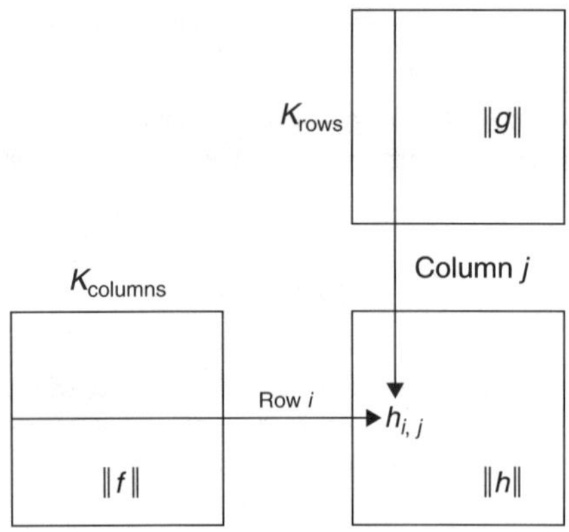

Figure 1 Multiplying two matrices.

of the matrices product is given by the sum of the products, term to term, of the elements of row i of matrix $||f||$ by the elements of column j of matrix $||g||$:

$$h_{i,j} = f_{i,1} \cdot g_{1,j} + f_{i,2} \cdot g_{2,j} + \cdots + f_{i,k} \cdot g_{k,j} + \cdots + f_{i,K} \cdot g_{K,j} = \sum_{k=1}^{K} f_{i,k} \cdot g_{k,j}$$

Maybe it is easier to remember the rule with a graph (Figure 5).

The "neutral" element for the product of two matrices is, for the matrix $||f|| \in \Re^{I \times J}$, the diagonal matrix $||1|| \in \Re^{J \times J}$. One can write, if the dimensions are respected, $||f|| \otimes ||1|| = ||f||$.

An important theorem about multiplication of matrices is that this operation is associative:

$$||a|| \otimes (||b|| \otimes ||c||) = (||a|| \otimes ||b||) \otimes ||c|| = ||a|| \otimes ||b|| \otimes ||c||$$

So, the parenthesis may be deleted.

An Exception

A scalar was defined as a matrix belonging to $\Re^{1 \times 1}$. With this definition it is impossible to make the product of a matrix by a scalar (the dimensions rule is not satisfied). Therefore the product of a matrix $||f|| \in \Re^{I \times J}$ by a scalar a is specially defined by a matrix belonging to $\Re^{I \times J}$ where all elements of the first one are multiplied by a. This is a small defect in the stenography!

Decompositions of a Matrix

A matrix can take different forms and mathematicians worked a lot about transforming forms. The advantages of these other forms are to present, in a simple way, interesting characteristics of a matrix. The most useful decompositions are the QR and the SVD.

Assume the matrix we are interested in belongs to $\Re^{I \times J}$.

The **QR decomposition** is the fact that any matrix can be written as the product of two matrices, the first one, noted $||Q||$, being orthogonal and the second one, noted $||R||$, being triangular superior.

The **SVD decomposition** is the fact that any matrix can be written as the product of three matrices:

1. The first one is orthogonal and belongs to $\Re^{I\times J}$. It is generally called $||U||$.
2. The second one is diagonal and belongs to $\Re^{J\times J}$. It is generally called $||D||$. It is a square matrix which has as many lines and columns as the number of parameters. The values which are in the diagonal matrix are called the "singular values" of the matrix; they are represented by the symbol d_j.
3. The third one is also orthogonal and belongs to $\Re^{J\times J}$. It is generally called $||V||^T$.

Norm of a Matrix

There are several definitions of a matrix norm. The most frequently used is given by:

$$\left|M\right|_q = \sup_{\vec{a}\neq 0}\frac{\left|M\otimes\vec{a}\right|_q}{\left|\vec{a}\right|_q}$$

where $q = 2$ is the most common. This norm is then based on the usual norm of the vectors, defined as:

$$|\vec{a}|_2 = \left(\sum |a_i|^2\right)^{\frac{1}{2}}$$

A quick definition of the matrix norm is the norm of the largest vector which can be obtained by applying the matrix to a *unit* vector.

Matrices, as Mathematical "Objects", Also Have Special Properties

These properties come from looking at matrices as operators. This simply means that a matrix can be seen as an operator which transforms a vector into another vector; we write:

$$\vec{f} = || M || \otimes \vec{g}$$

According to the multiplication rule, if $||M|| \in \Re^{I\times J}$, then $\vec{g}$ must have J rows (type $\Re^{J\times 1}$) and $\vec{f}$ will be a vector with I rows (type $\Re^{I\times 1}$).

Eigenvalues and Eigen Vectors of a Square Matrix $||M|| \in \Re^{J\times J}$

Eigen vectors (also called "characteristic vectors") are vectors which are transformed by matrix $||M||$ into vectors parallel to themselves: if $\vec{E}$ is an eigen vector, then:

$$|| M || \otimes \vec{E} = \lambda\cdot\vec{E}$$

where λ is a scalar, called an eigenvalue (or a "latent root", or "characteristic root or value").

The Case of Full Rank, Symmetric, Matrices

These are the only matrices for which we are going to search the eigen vectors and the eigenvalues.

For these matrices:

- there are J different eigen vectors;
- these eigen vectors are orthogonal;
- the eigenvalues are real;
- the sum of the eigenvalues is equal to the trace of the matrix, and their product is equal to its determinant (consequently if an eigenvalue is very small, the determinant of the matrix may be small as well and its inverse may be quite large).

If $||M||$ is non singular, the eigenvalues of $||M||^{-1}$ are the inverse $(1/\lambda_j)$ of its eigenvalues.

Relationships with the Singular Values

The eigenvalues are the squares of the singular values:

$$\lambda_j = d_j^2$$

One advantage of the singular values on the eigenvalues is that the search for the singular values does not require the inversion of the matrix, whereas it is necessary for the search of the eigenvalues: if the matrix $||M||$ is ill conditioned, the latter may not exist, when the first ones always do.

Part I

Population and Sample

Part Contents

Our domain of interest is all the products, or objects, or "things", whatever the name you want to use, made by man. These things can be anything, such as roads, or bridges, or houses, or cars, or cameras, or software, etc.

It can also be the activities carried out by man, as activities can be dealt with exactly the same way as products.

The set of all these products or activities can be seen, from our point of view, as potentially infinite. In this set, we follow, for cost estimating them, the common practice of grouping them in specific subsets and we will investigate each subset independently of all the other ones.

A subset we are interested in is called a "population". This term will be defined in Chapter 1. Presently consider a "population" as the set of all the objects we would intuitively consider together for comparisons purposes. As you will certainly not compare together bridges and cameras, a population can be here defined as either as a set objects fulfilling the same function(s), or a set of similar activities. In the first case, a population will therefore also be called a "product family", in the second case an "activity family", the first term being used as a generic term.

As we are not concerned by human population, we will generally use the words "**product family**": it clearly express what we have in mind, but the term "population" will still be used from time to time. Potentially this product family is an infinite set of "similar" (we will insist on the definition of this word in Chapter 1) products.

The "distribution" of the cost inside this product family is what we are interested in for decision-making purposes. It is of course unknown.

In order to get an idea about it, we gathered in our database (statisticians would say: "we draw a sample from the population"), a, sometimes small, set of products we consider belonging to this product family.

This whole volume is dedicated to answering the question:

What can we infer about the product family from the knowledge of the sample?

This first chapter briefly explain the path we will follow to answer this question.

The basic idea is the following one: the distribution of the cost can be a function of a lot of variables. Such a function will be very difficult to work with. Therefore our objective will be to replace it by a distribution of just one variable, distribution which will be much more practical for our objective.

The basic idea for attaining this end will be made in three steps:

1. We will look for the "center" of this distribution in the sample.
2. Once we know it we will study the distribution of the costs, always in the sample, around this center. This distribution is much easier to study because it does not depend anymore on many variable.

3. We will then extrapolate these results from the sample to the "population" (or product family).

To implement this idea it is important to start with the study of populations which do not depend on any variable. This study will introduce most if the concepts we will extensively use afterward.

This is the purpose of this Part I.

1 From the Sample to the Population

Summary

An "object" in this book is something which has to be manufactured (for products) or more generally realized by using a specific process. So the word "object" must be understood in a very large way: it can be a part of an equipment (such a mechanical part, or an electronic board, or even an electronic chip), the equipment itself (such as a printer, or a reactor), the software, the system it belongs to (such as an airplane), a building or part of it, a trench, a tunnel, etc. or even an activity (such as boring a hole in a plate, painting a room or performing a surgical operation).

The term "population" is generally used for naming the whole set of objects the statistician works on. The same name is kept in this book; however our populations are very special: a population is the set of objects which constitute what will be called a "product family". Such a family must be as homogeneous as possible, the degree of homogeneity being let to the cost analyst.

In a product family, objects may more or less differ, depending on the level of homogeneity. Their differences are quantified, or more generally described, by variables. A rule of the art is "the less homogeneous the product family, the more variables you need". At the minimum, the size of the objects – by their physical size, or by their functional size – must be described. This chapter first presents a few definitions.

The purpose of this book is to forecast something about any object of a product family (our population): it can be the cost of manufacturing it, or the time to do it, or the tooling which is needed, or anything else. In order to be able to make this forecast, we get a sample (it is the data we start from) from which we are going to extract the information we need. The logic for doing it is exposed in Section 1.4 which is an important section of this chapter: it shows how the study of a complex distribution of several variables can be solved by studying the distribution of one variable only.

The concept of distribution of one variable will therefore be present in any part of this book: it must be fully understood by the user and, for this reason, Chapter 2 presents different ways for describing such a distribution: the purpose of this description is that it would be extremely difficult to continuously work with the full distribution: it is much more easier to use a limited set of descriptors.

Chapter 3 is devoted to the description of several "standard" distributions, which are very well known: if any of our distributions looks similar to ones of them, solutions are immediately available.

1.1 The Population

A Few Basic Definitions

The term **"product"** is a generic term used to call any item we are interested in: it can be an equipment, a spare part, a software, a building, a tunnel, etc.

A **population**[1] is defined in this part as the set of products we are interested in. The number of products which constitute a population can be finite or infinite; most often it will be regarded as infinite for a reason which will appear later on.

A **variable** is one of the characteristics of the products belonging to this population. To be really interesting this variable must be defined for all the products. The number of variables attached to a product is finite: it can be the cost, the mass, the material, the time to carry out an activity on the products, etc.

1.1.1 The Concept of Product Family

The concept of product family is one of the most important concepts when dealing with data for preparing a tool for cost, or any other thing, forecasting.

A **product family** is a population constituted of homogeneous products. Products are homogeneous if:

1. they fulfill the same function (homogeneous functionality),
2. they fulfill it about the same way (homogeneous design),
3. they are prepared about the same way (homogeneous preparation).

Consequently and ideally (but it will be shown that one can slightly deviate from this ideal with the use of variables, quantitative, qualitative or subjective) a product family is a set of products:

1. fulfilling the same function(s),
2. designed the same way,
3. manufactured from the same materials (for hardware products),
4. using the same manufacturing process,
5. and which, consequently, differ only by their size.

A **formula** (or CER, which stands for "cost-estimating relationship") is, nowadays, an algorithm which can be used for cost estimating of all the products belonging to the same product family. The formula is part of a "**specific model**": a model because its purpose is to mathematically modelize the behavior of a variable (the cost for us), specific because it addresses one and only one product family.

The first characteristic of such a model is therefore its *specificity*, which means that it deals with homogeneous products; the first consequence of this statement is that a model can only be used for the product family it was built for.

Experience shows that many models give poor results because they were created from heterogeneous data. The first responsibility of the cost analyst is then to check

[1] The word comes from the fact that statistics were, and still are, developed for the study of human populations. For the same reason, the objects of the population are generally called "individuals".

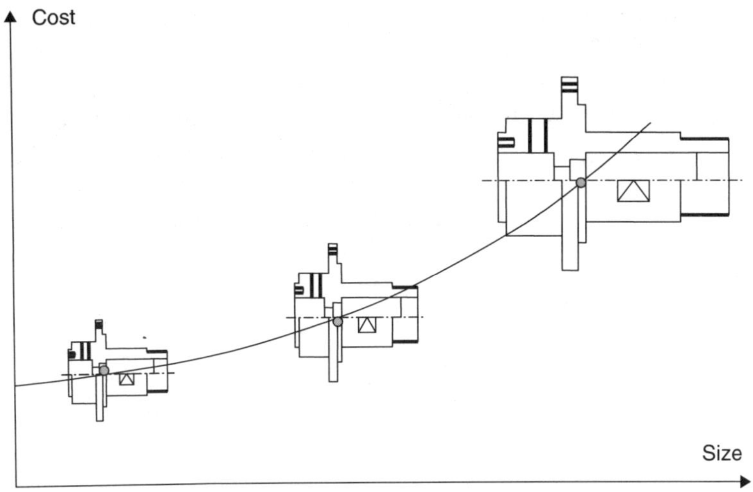

Figure 1.1 How the cost changes in an ideal product family.

the homogeneity of the products he/she puts in a given product family. It is not always as simple as it may look.

Figure 1.1 illustrates an ideal product family.

Using the same preparation (or manufacturing for hardware products) technique(s) – unless a dedicated variable is used for this purpose – is an important condition: *in a specific model one does not have to describe how the product is manufactured.* This simplifies a lot the process of preparing the model: the idea is to compare together products costs; this can only be done if costs are homogeneous; one of the conditions of this homogeneity is that the manufacturing methods be about the same.

The first, and important, consequence of working inside a product family is that it is not possible to cross the product family border: one does not know how to go from a product family to another one. The fact to have a specific model for bicycles, for example, says nothing about washing machines. It is an important restriction to the use of specific models: when you want to use such a model in order to estimate the cost of another product, the first thing you must do is to check that this new product really belongs to the product family the model was built for. It is not always so easy.

> The first thing that the cost analyst must do is to make sure about the homogeneity of the products he/she wants to compare.

1.1.2 The Variables

The **variables** are the second important concept which has to be investigated in order to build high-quality cost-estimating models.

The strict definition of the product family we gave in the previous section seriously limits the potential use of a model. It is rare in the industrial world that we can dispose of a set of completely homogeneous products. Variables are used to overcome this problem.

The purpose of the variables (quantitative or qualitative), beyond the size effect, is to palliate some inhomogeneities inside the product family.

This must be emphasized: the purpose of the variables is not to describe the products in "absolute terms" as we did it when dealing with general models, but in "relative terms". This means that these variables are there to mention only the *differences* between the products.

Three different types of variables can be considered: quantitative, qualitative and subjective.

Quantitative Variables

Quantitative variables are information which express the result of a **measurement**; the mass of product, its volume, its power, its load, ... can be measured: they are all quantitative variables.

The result of a measurement is given by a number which can vary continuously: the mass of a product can be 2 or 3.6 or 5.78 kg, etc. Consequently it will be considered in the computations that it is a continuous variable, which means that:

1. arithmetically the number may have as many decimal as the user wants,
2. mathematically it is possible to compute the derivative of the function giving the relationship between the cost and these variables.

The idea of considering that a product family includes a potentially infinite number of products comes from this assumption.

One must add that considering that the quantitative variables are continuous greatly simplifies the computations ...

Qualitative Variables

Experience shows that qualitative variables are not generally well understood by cost estimators and even cost analysts.

A qualitative variable expresses a **fact** which cannot be measured. An equipment can be made in steel, in aluminum, or in titanium: the material is a qualitative variable. It can be produced by company A, or B, or C, ...: the manufacturer is a qualitative variable. The technology used can be mechanical machining, or electro-erosion, or chemical erosion, or anything else: the manufacturing technology is a qualitative variable.

Since a qualitative variable cannot be measured:

1. Its value is given by a name or an adjective. In order to avoid the word "value" which generally refers to a number, the name "modality" will be used in this book. If for instance a manufacturer may work from three different types of materials, we will say that the qualitative variable "material" has three modalities: steel, aluminum, titanium for instance.
2. The set of modalities is finite and discrete. This is one of the major difference between a quantitative variable and a qualitative one, difference which can be used for discriminate these types of variables. The number of modalities is the important input when defining qualitative variables.

Two points must be added at this stage in order to avoid any confusion between the variables types:

1. It is not because modalities of a variable can be expressed by numbers, as they sometimes are, that such a variable is defined as quantitative.
2. Not all qualitative variables are qualitative by nature.

These points are commented on the two following examples.

Suppose you work with electrical engines and that you have in your database engines with 4, 6 or 8 poles. The number of poles is obviously given by a number; but, as it is not a continuous variable (you cannot have an engine with 4.7 poles), it should be defined as a qualitative variable. You could of course define it as a quantitative variable; what we mean by that is the algorithms used for computing a formula can very well work with integers (even if they do consider them as continuous) and will therefore deliver a result. But we will see, in Chapter 5 dealing with qualitative variables, that it is not the best solution from a cost-estimating point of view.

Let us take another example: if you work with two materials, for instance steel and aluminum, you can only consider the material as a qualitative variable (named "material") if the products are entirely made out of steel or aluminum. But you may very well have a product which is made 30% out of steel and 70% made out of aluminum, another one made 60% out of aluminum and 40% out of steel, etc. In such case you better consider other, quantitative, variables now named "amount of steel" and "amount of aluminum". Such a solution is theoretically possible, but may not give good results if products are a mix of different technologies: it could be difficult to compare two products – even if the percentages of steel and aluminum are the same – if for one product the steel is used for low-technology parts and for the other one for high-technology parts. The conclusion is of course that the cost analyst should have some information about the products and decide accordingly: defining a variable as quantitative or qualitative is sometimes a question of judgment.

Let us conclude this section by saying that a variable is often not qualitative by nature, but by decision of the cost analyst; his/her decision rests on the fact that he/she considers that the variable for all the products – existing or potentially existing – in the studied product family can only take a few vales; the fact that these values can be associated with a number (such as the example with the poles) or an adjective (such as high quality or not) is not relevant to this decision.

Subjective Variables

This is a special type of quantitative or qualitative variables.

They are different by the fact that they do not express "facts" but "**judgment**" or "**opinion**". A good example is given by the word "complexity" which tends to be widely used by some cost analysts. It can be defined as a continuous variable and quantified as such (the complexity of this product is 7.3 for instance) or as a discrete variable (with a few modalities such as from "very complex" to "very simple"). The important thing about this complexity – unless it is derived from an algorithm, which is another question – is that it does not describe a fact such as "steel", or "aluminum" but tries to quantify a judgment.

Using such a variable has two important consequences.

1. First of all it can be considered as an attempt to mix, in the same product family, heterogeneous, sometimes very heterogeneous, products. To readdress the example given earlier on bicycles and washing machines, one may try to aggregate both in the same family, saying that bicycles have a "complexity" of 2.3 and washing machines a "complexity" of 3.5. This could be possible if we were sure that the way the cost changes the same way with the size, let us say the mass. But the purpose of using the data we have is precisely to discover that. Handling these data the conventional way will force this change to be the same; we are not discovering something: we are just trying to force nature to enter in a narrow corridor.
2. The difficulty about the judgments is that we are never sure, even when they are made by the same person, about their consistency. When they are made by different persons, we are never sure if these persons apply the same set of, generally informal, rules for going from the product description to this complexity; the main problem comes from the fact that complexity takes into account a lot of hidden variables, and that we are not sure if everybody considers the same variables, if they understand them the same way and if they aggregate them the same way.

There are some procedures which may alleviate this difficulty, such as the Delphi method. One is discussed in Chapter 6 of Volume 1. But the problem must not be overlooked.

This type of variables should be avoided as much as possible. If you cannot avoid them, prepare a list of "sub-parameters" to be used (it does not solve the problem if these sub-parameters also are subjective but allows to discuss between experts on more homogeneous territories), and the way the notations on each will be aggregated (What is the "weight" of such or such sub-parameter?). The minimum is to get a scale of complexity with several well-known examples positioned on this scale.

When using subjective variables, try to avoid to use quantitative values on them: judgment or opinion cannot be easily quantified ...

1.1.3 Formula or Analogy?

What are the advantages of using a formula instead of analogy?

Conceptually the major advantage comes from the fact that cost estimating using a formula is built on all the data present in the family, whereas analogy is built on the two or three data which are in the vicinity of the product to be estimated: the foundations of the estimate based on a formula are much more stronger.

Practically:

- A formula is easy to prepare, once the analysis of the data has been seriously made.
- It is, and this may be the most important point, an extremely powerful tool for communication. The person who receives the information immediately understands its validity if the way the cost changes with the parameter(s) is displayed to him/her. Other examples of communication are for instance:
  - looking for products of which cost seems abnormal,
  - comparing our products to the competition,
  - choosing between several possible solutions, etc.

- It is also an excellent tool to visualize the quality of an estimate. When one parameter only is involved, displaying the curve of the nominal cost – what will be called the "dynamic center" – with the data makes immediately appear the level of confidence we may have in an estimate. If several parameters are involved, computing the standard deviation of the residuals may be more informative.

1.1.4 Breaking the Symmetry

Data analysis deals with variables: it makes comparisons, studies dependencies, etc.

Mathematically speaking all quantitative variables play the same role. Practically, we break the symmetry by deciding which is the "dependent" variable, which are the "causal" variables. Once again such a decision has nothing to do with the theory, it is only made for practical reasons: very briefly we are interested in forecasting the cost when the mass (of a new product inside the product family) is known, not to estimate the mass when the cost is known!

It must be added that, when the design to cost started to be implemented in various companies, some people, starting with the cost objective, tried to estimate the size of the product. If this size is defined by the mass, this effort has little interest; but if the size is defined by the technical specifications of the product, it might be a good process. Nevertheless in such a case, the symmetry was not really restored: the basic relationship was prepared to relate the cost to some functions of the other variables, and this relationship was simply "inverted" in order to compute a mass from the cost.

Breaking the symmetry has importance consequences.

It allows us to define the cost (or anything else) as the "dependent" variable, and to speak about the other variables as "causal" – commonly called "parameters", or "cost drivers" (all these terms are synonymous) by the cost analysts. These terms are not accepted by the mathematicians who say that nothing allows us to decree what is dependent and what is not. Let us hope that they will forgive us "this language abuse".

In order to make this break very clear, the dependent variable will always be called[2] Y, and the values it takes noted in the population $Y_1, Y_2, \ldots, Y_i, \ldots$ whereas the causal variables are called $V_1, V_2, \ldots, V_j, \ldots$ and the values they take for a specific product belonging to the product family $X_{i,j}$ (value of variable V_j for product A_i). The name V_0 will represent a constant, if it is needed.

Causal variables – quantitative, qualitative or subjective – will be called "**parameters**", or "**cost drivers**". The dependent variable will be generally called the cost in this book, but it may be anything else such as the duration, or even the mass if you are trying to estimate the mass from technical specifications.

Note About the Word "Parameter"

Be careful about the word "parameter": it is used in different disciplines with different meanings. This is particularly the case in statistics, which is a domain very

[2] The capital letter has two meanings: it is the name of the dependent variable on the one hand (such as the cost or the duration of an activity, etc.), the value of this variable for an object in the population on the other hand. As there is only one such variable in a study, no confusion will arise, and the notations are kept simpler. The value of this variable in the sample will always be noted by a small y.

close to ours. In statistics a parameter is a characteristic of a distribution[3], such as the arithmetic mean, the standard deviation, etc. A distribution of which the shape is known – for instance to be normal, or log-normal – is said to be "parametric". In mathematics, a parameter is just an auxiliary variable.

In this book, a "parameter" means a variable the cost analyst considers as relevant to his/her study. But the name will be avoided as much as possible.

1.1.5 What Could, Should, be the Causal Variables?

Consequently we have three types of variables:

1. One, and only one, dependent variable (the cost in general for us).
2. One or several quantitative causal variables.
3. Zero or several qualitative causal variables.

First of all, and this is obvious – even if it may be overlooked by cost analysts – variables are choosen in order to be known when an estimate is requested: it does not serve any purpose to work on a variable that has a very limited chance to be known when it will be requested.

Among the quantitative variables, there is one which has nearly always to be there: it is the object size; the size answers the question: Is it small or large? Except in extremely rare circumstances a product family contains different products of different sizes and the size is nearly always one of the most important cost drivers. Some cost analysts try to use a quantitative "complexity" – which, as explained upward, is a subjective parameter – without taking into account the size, the logic being that this complexity includes the size. Experience shows that, even in this case, using the size always improve the quality of the model.

An interesting question is often: Should we use functional or physical descriptors? A functional descriptor is a variable which is interesting for the object user; examples are given by the power, the capacity, the number of seats, the range, the tolerances, etc. A physical descriptor is a variable seen by the object designer, such as the mass, the volume, the number of parts, etc. Our recommendation will be to capture both:

- *Functional variables* will be really helpful at the early stages of a project on one hand (at these stages these variables are the only one to be known), for the managers on the other hand: managers are not especially interested in the object mass, but they have to understand the relationship between functional characteristics and cost in order to carry out trade-offs analysis and so are able to offer the best compromise between these characteristics and the cost.
- *Physical variables* may be helpful later on when the design is more advanced: at this time these variables may be known and they will help the design engineers to reduce the manufacturing cost, and sometimes the development cost.

This discussion means that the cost analyst should always think, when developing a cost model, to develop two cost models at the same time, one using functional variables, the other one physical variables. Both will be helpful during the project life.

---

[3] In this book we keep the word "characteristic" of a distribution for this usage.

1.2 The Distribution Φ of Y for the Population

Being able to tell the cost of any product belonging to the population is equivalent to knowing the distribution of the cost variable Y for the entire population. This distribution is called Φ.

This distribution is of course unknown to the cost analyst.

To each product are associated the values of quantitative and maybe modalities of qualitative variables. The challenge is then to use this known information to get some knowledge – not a complete one but a sufficient one – about this distribution Φ.

If the parameters are properly selected, they will be known at the time we want to make an estimate for a new product.

The way we can do that is based on a belief about cost.

What Do We Believe in?

We believe in two things:

1. Cost does not arrive by chance, even if there is a small randomness in human activity.
2. Cost depends on a lot of variables.

The consequence of these beliefs is the following one: if we were able to know all the variables which influence the cost, its value could be forecasted with a great accuracy.

Any scientific research is based on the assumption that there is a relationship between the dependent variable Y and something else (the variables or cost drivers). In other words, the value of this variable Y is not a random value, arriving just by chance. To be completely exact the assumption is that the order of magnitude (for the time being, just consider the current sense of the terms) of this variable is not random, as some uncontrollable events may add some randomness around this order of magnitude, this randomness being small compared to the order of magnitude.

This belief sounds reasonable when we deal with cost. *It is the foundation of our studies.*

But we have to go one step further: we believe that there is a relationship between this variable Y and some parameter(s) belonging to the product. As strange as it may seem once we are used to study these relationships, this belief is not shared by everybody: many people still think that the only way to make a cost estimate of something is to go in many details. As it is impossible to convince anyone by arguing, we may only recommend to these persons to try it.

But for the time being, it is only a **belief.** A very strong belief, based on all the results which have been accumulated for years, but nevertheless a belief.

This belief is nowadays expressed by a mathematical relationship: we believe there is a relationship (capital italic letters are used for the population):

$$Y = F(X_1, X_2, \ldots; B_0, B_1, \ldots)$$

where $X_1, X_2, \ldots$ are the values of what we call the parameters or the "cost drivers" we are supposed to know, and $B_0, B_1, \ldots$, some coefficients which, for the present time, we do not know. These coefficients are specific to the product family (our population).

Do not be afraid by the apparent complexity of the relationship: we will need very rarely so many variables and coefficients: it was just written to be as general as possible.

1.3 Drawing a Sample from the Population

The way to find out:

1. the shape of function $F()$,
2. the values of the coefficients $B_0, B_1, \ldots$

is to make some observations and to "extract" them from these observations.

The observations – the ones we already got from the database for our product family – is, in statistics, called the "**sample**".

A sample is characterized:

1. By its size (the number of individuals extracted from the population), which is called I;
2. By the distribution of the cost values, distribution noted φ. This distribution is as complex as the distribution Φ of the population and cannot therefore be used without some work on it.

Many things have already been written on sampling in order to test a relationship (as it is the case when a new medicine is discovered and tested for example) or to find out some characteristics of a population (as it is the case before an election). These things can be grouped into two sets:

1. How to choose a sample which will not give biased results?
2. How to go from the values observed in the sample to the values we are interested in for the whole population (the coefficients $B_0, B_1, B_2, \ldots$)?

The first set is of no interest to us. It would be nice to, randomly, sample the cost of a few products (the theory could help us to select the "right" sample), but this is clearly impossible. We have to accept the data as they are (once they are normalized): they are our sample. We generally have no way to select another sample – from the same population or product family – or, most often, even to add another data to our sample. Either it is impossible, or it would cost too much: we have to be happy with what we have and infer from it.

However the second set is extremely important to us: it is our objective in saving our data. This volume will explore some of the most important concepts derived by mathematicians and statisticians in order to achieve this result. How it is done practically will be developed in the following chapters; these chapters will use all the concepts introduced here.

A Preliminary Warning

You will learn in these chapters that most of these concepts – and of course the results we arrive at when applying them to real situations – are only true if the sample is really randomly selected from a population. For this reason, we want to warn you that care should always be taken before accepting some statistical results from computations made on your sample: maybe they are based on hypotheses which are not satisfied in this sample. As everybody knows (but it has to be recalled):

> Before accepting the results of a computation check if the hypotheses on which it is based are satisfied.

This is the basis of science, but it is often overlooked when dealing with cost. It often happens that some computations are used because they are the only ones available, on the first look, without checking if they are relevant. We will insist often in this volume on this subject; for the present time, do not consider that the result given by an algorithm found in a book is true without checking when it can be accepted as such.

Independent Observations

Independence is a very useful concept in the theory of probability: two events are said to be independent (Ref. [49], p. 32) if they do not mutually interact or are not jointly influenced by other events. Independence greatly simplifies the computations when dealing with probabilities.

Observations put in a sample are always assumed to be independent. This is generally the case, in the cost domain, when a sample includes products manufactured by different companies, or made by different departments of the same company, or, at different times, inside the same department.

However similar products manufactured at the same time by the same company are probably not independent: if a problem occurred during the manufacturing of one of them, we can reasonably expect that the same problem occurred for all of them.

Dependent observations may introduce a bias in the relationship we want to establish, or produce an autocorrelation between the residuals.

1.4 Using the Sample Values

1.4.1 The Three Possibilities

The only reason for collecting values (which constitute our sample) is to use them for estimating the costs of new products belonging to the same product family.

There are three ways[4] and probably only three ways, very different, to use the sample data for our purpose. They are called: the "CBR", the "frequentist" approach, and the "Bayesian" approach. The present section reminds the foundations on which they are built and describes the basis of their application.

The "CBR" or "Case Based Reasoning"

The axiom of the CBR is the following one: there is no difference between the sample and the population: the population is the sample. There is no need to build a

[4] All of them are built in EstimLab™.

specific model for this population, because we already know the cost of each individual of the population.

Any new product – what is to be estimated – is a "foreigner": it does not belong to the population and, consequently, we cannot apply to it any model which could be built on the population value. Nevertheless this foreigner is described by the same variables or parameters as any member of the population.

What can then be done for this "foreigner"? We can do something about it if it is not too different from someone belonging to the population. In order to see if it is not too different, we have to quantify a "distance" between the individuals, distance which will be applied to the foreigner.

A distance is not too difficult to define if only quantitative variables are used: there are several possibilities which will be described in Chapter 3.1. But what can we do if we have a mix of quantitative and qualitative variables?

The "distance" will be used for finding the individual of the population which can be considered as the most similar to the foreigner: starting from the variables which describe the foreigner, we will search the individual of which the set of variables – considered as a whole – are as the most similar to the ones of the foreigner. If this set of variables include qualitative ones, we will limit the search to the subset of the same qualitative variables.

How should this search be organized: if the qualitative variables include five modalities, called A, B, C, D and E, should we start the search with A or with B, etc. ... The answer to this question can only be found by a reorganisation of the data base according to the degree of significance of each modality. This is the major difficulty of the CBR. Generally this degree is found by using the concept of entropy.

The "Frequentist" Approach

The frequentist approach is based on the following axiom: the sample was randomly from a population in which the dependent variable Y (the cost for us) is linked to the causal variables by a relationship:

$$Y = F(X_1, X_2, \ldots; B_0, B_1, B_2, \ldots)$$

in which the coefficients $B_0, B_1, B_2, \ldots$ are **fix**, well defined, **values.**

Of course we do not know these values. In order to find at least an approximate value of them (we will call them "estimates" of the true values) we will look at the frequencies the cost values in the sample change with them: the name of the approach comes from this strategy.

This is the classical solution.

The "Bayesian" Approach

The Bayesian is based on the same axiom: the sample was randomly from a population in which the dependent variable Y (the cost for us) is linked to the causal variables by a relationship:

$$Y = F(X_1, X_2, \ldots; B_0, B_1, B_2, \ldots)$$

but now the coefficients $B_0, B_1, B_2, \ldots$ are defined as **random variables.** This means that they may – slightly – change from one individual to another one.

This approach, still rather unconventional, does make sense: how can we be sure that the coefficients have fix values inside a product family? We saw in Volume 1 that the cost of a product, due to the way, we, as humans, work may fluctuate (we call it "chance" in Chapter 18 of this Volume 1): one way of expressing this "chance" is to consider that the coefficients may slightly change from one product to another one. The Bayesian approach is therefore more closer to the reality than the frequentist approach.

But this is not all: we said in Chapter 17 that any experienced cost estimator has a preconceived idea about what a product should cost. Pushing the analogy a little bit farther, we can say that any experienced cost analyst has a preconceived idea about the value of (maybe only some of them) the coefficients $B_0, B_1, B_2, \ldots$. How can we express this preconceived idea: very simply – because this cost estimator has only a limited knowledge about the product family – by saying that this person has an "a priori" knowledge of the distribution of the these random coefficients; this shape of this distribution is generally "normal".

How does the Bayesian approach works then? The cost analyst starts from this a priori distribution of the coefficients and use the sample to see if it confirms it or not. More precisely the sample values are used, taking into account the *a priori* distribution, to compute a "*a posteriori*" distribution. This approach uses Bayes' theorem, well known in the theory of probabilities.

Experience shows that the confidence interval of a cost estimate will be generally smaller with a Bayesian approach than with the frequentist approach, the logic being that information is added to the information contained in the sample.

This book is dedicated to the conventional "frequentist" approach: we believe it has to be known by the cost analyst, before other approaches can be investigated. It is therefore a good start. Other approaches will be dealt with in another volume.

1.4.2 The Logic of the "Frequentist" Approach

Working with the distribution φ of the cost in the sample would be difficult.

The logic is to replace for both the sample and the population the costs distributions[5] by a set of two things:

1. A relationship (φ for the sample, Φ for the population) related to the *center* of these distributions. This relationship quantifies how the center changes with the variables: for this reason it will be called a "dynamic" center.
2. The scattering of the costs *around this center*: the relevant distributions are called ψ for the sample, Ψ for the population.

The work will be done into two distinct phases: the first phase works with the sample, the second "extrapolates", the results observed in the sample to the whole population.

We therefore work in two steps:

[5]Small letters [φ] are used for the sample, capital letters [Φ] for the population.

1. **We substitute to φ two things which are easier to handle:**
   - The *center* of this distribution. This center will be a fix number, called $\hat{y}$, if no parameter is considered: it will then be called a "static center". It will be a "dynamic center", called $\hat{y}_i$ (the symbol y reminding it is a cost, or more generally, the value of a dependent variable, the index reminding it is a value computed for a particular product and the little "hat" it is the center we are looking for).

     The computation of $\hat{y}_i$ may be rather complex. It will be made, at its turn, in two steps:
     - A *decision* on the form of the assumed relationship $F()$. No algorithm is available for automatically finding this form: the cost analyst will have to make a decision, based on his/her experience, on the shape of the data (for this reason being able to visualize the data is an important part of the data analysis) and maybe on the theoretical relationship between the cost and the parameters (if a theory is available for the studied product family).

       More exactly, as the reader will see, the decision is made for function $f()$, based for instance on the shape of the data, and then the same function will be used for the population.

       This assumed relationship may include several coefficients, called $B_0, B_1, B_2, \ldots$. The value for product A_i is of course $\hat{y}_i$.
     - The *values* of these coefficients for the sample – they will be represented by small letters, such as $b_0, b_1, b_2, \ldots$ – will be computed by minimizing the "**distances**" between y_i and $\hat{y}_i$.

       Computing these distances will require the definition of a metric – which is only the way distances are computed, as there are several ways to do it.
   - The *distribution* ψ of what is left when this value $\hat{y}_i$ is "removed" from the cost y_i of product A_i. This "what is left" will be called the **residual** corresponding to this product A_i. Residuals can be expressed in several forms:
     - An additive form:

$$e_{i+} = y_i - \hat{y}_i$$

     - A multiplicative form:

$$e_{i*} = \frac{\hat{y}_i}{y_i}$$

     - Or any other form such as $(\hat{y}_i/y_i) - 1$ or anything else.

   The reader must pay attention to the fact that **residuals and distances are two different concepts.** It may happen that the same values are used, but it is not the general case and certainly not a reason to confound them.
2. When this is done, **the results ($\hat{y}_i$ and ψ) will be applied to the whole population,** resulting in two things:
   - A *formula* $F()$, based of course on the $f()$ relationship.
   - The *distribution* Ψ of the deviations of the true values of the costs in the population from the relationship $F()$.

   Note that the model is the sum of these two things.

The reader may note that the word "residual" is used for the sample, the word "deviation" being preferred for the population. This preference has three reasons: the first one is not to confuse what is done for the sample and what is done for the population; the second one is that we try to develop *f*() from the sample data, but, as this operation cannot take all the information existing in the sample, some "residuals" do exist; the third one comes from the fact that we believe that the relationship *F*() is, on the average, the right one, but that individual products may deviate from this average relationship.

Figure 1.2 synthesizes the approach.

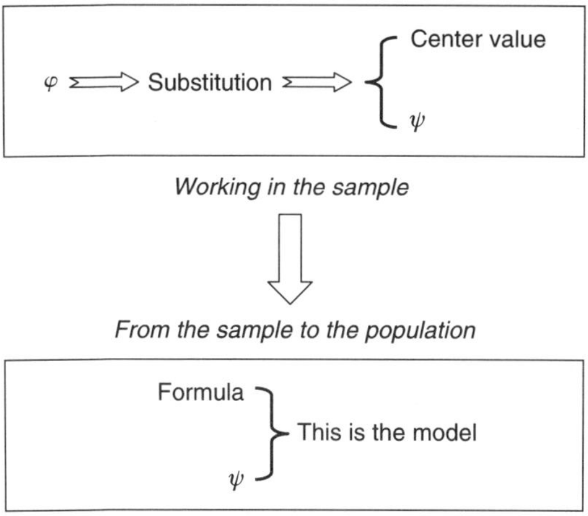

Figure 1.2 The logic of the work.

1.5 How Do Probabilities Creep into Our Business?

Basic probabilities will be often used in this volume. What do they have to do in our business which is so far away from randomness and probabilities?

We are so used at using probabilities that we rarely ask ourselves the question: How do probabilities come into the picture?

The answer to this question is here rather simple: you have certainly noticed that the word "random" or "randomly" has already been used several times. This is how probabilities penetrate our domain: we consider that the data we have (our sample) were randomly selected among an infinite population (we know this is not exactly true, but there is at least a part of truth: let us say that we happened to know the cost of several products, and these products were not selected with an objective in our mind[6] from the set of products we have inside the product family belonging to our database).

If the sample was randomly selected, we could have observed (always for the time being in the same population) other products and probably different costs. From these observations we would have drawn different conclusions for our coefficients $B_0, B_1, B_2, \ldots$

[6]This is the reason why we are always very skeptical about models built on a sample when we know that several products were removed from the sample: by removing data which are considered as trouble makers, you can prove anything! If a data has to be removed, the reason for its removal must be completely explained…

Different yes! But *how much different*? All the mathematics developed for manipulating random numbers could probably be useful to answer this question and that will be shown.

And, from now on, probabilities are here to stay…

1.6 Conclusion

The beautiful thing about this approach is to be able to replace something which might be very difficult to handle by something much easier to handle:

- A formula.
- The distribution of one quantity only (the residuals for the sample or the deviations for the population) around this formula. Even if several variables or parameters are attached to the products, this distribution plays with *one variable only*.

This makes the approach extremely useful for practical purposes.

2 Describing the Population

Summary

An "object" is in this book something which has to be manufactured (for products) or more generally realized by using a specific process. So the word "object" must be understood in a very large sense: it can be a part of an equipment (such a mechanical part, or an electronic board, or even an electronic chip), the equipment itself (such as a printer, or a reactor), the software, the system it belongs to (such as an airplane), a building or part of it, a trench, a tunnel, etc. or even an activity (such as boring a hole in a plate, painting a room or performing a surgical operation).

The term "population" is generally used for naming the whole set of objects the statistician works on. The same name is kept in this book; however, our populations are very special: a population is the set of objects which constitute what will be called a "product family". Such a family must be as homogeneous as possible, the degree of homogeneity being let to the cost analyst.

In a product family, objects may more or less differ, depending on the level of homogeneity. Their differences are quantified, or more generally described, by variables. A rule of the art is "the less homogeneous the product family, the more variables you need". At the minimum, the size of the objects – by their physical size, or by their functional size – must be described.

This chapter first presents a few definitions.

The purpose of this book is to forecast something about any object of a product family (our population): it can be the cost of manufacturing it, or the time to do it, or the tooling which is needed, or anything else. In order to be able to make this forecast, we get a sample (it is the data we start from) from which we are going to extract the information we need. The logic for doing it is exposed in Section 4 of Chapter 1 which is an important section of this book: it shows how the study of a complex distribution of several variables can be solved by studying the distribution of one variable only.

The concept of distribution of one variable will therefore be present in any part of this book: it must be fully understood by the user and, for this reason, this chapter presents different ways for describing such a distribution: the purpose of this description is that it would be extremely difficult to continuously work with the full distribution: it is much more easier to use a limited set of descriptors.

The next chapter is devoted to the description of several "standard" distributions, which are very well known: if any of our distributions looks similar to one of them, solutions are immediately available.

Distributions of values of one variable only play an important role in this book: as explained in the previous chapter, what we are trying to do is to replace complex distributions involving several variables by a mono-variable distribution.

Several characteristics of these distributions will be frequently used. We found it convenient to group them in one chapter.

A distribution is defined here as a set of I values that takes a variable here called z. These values will be called $z_1, z_2, \ldots, z_i, \ldots, z_I$.

Such a set is difficult to handle. To simplify computations based on the sample, mathematicians found a few concepts that can be substituted to it. The purpose of these concepts is to describe the distribution by a few characteristics; for most of the applications, these few characteristics describe this distribution with a sufficient accuracy for practical purposes. Therefore, most of the computations can be made on them instead of the distribution itself.

There are two ways to build these concepts. The first one can be defined as "analytic", the second one as "global". There are obviously relationships between these two ways.

One Example

For illustrating the discussion of the different approaches, we will use the following example:

3.4
4.2
6.3
8.2
9.9
16.3
21.2
35.9
46.5
64.5
84.5

The approach works step by step. It analyzes the distribution into two components:

1. Its *center* on one hand.
2. The scattering of the values *around* this center, on the other hand. This scattering can be, in its turn, studied in two complementary ways:
   - The general spread of the values (How much are, generally speaking, scattered the values around the center?).
   - The shape of this spread (Is it symmetrical around the center for instance?). Generally speaking, two characteristics only are considered as sufficient to describe this shape.

A distribution can therefore be characterized either by the information it contains or by four characteristics (quite often the first two are only used).

2.1 The Center of a Distribution

Note: 1. This chapter is just an introduction to the subject. A detailed analysis of the concept is postponed to Part III, which describes, from a mathematical point of view, this concept.

Note: 2. When dealing with one variable only, the center should be called the "static center" in order not to confuse it with the "dynamic center" developed in Part III.

Note: 3. In order to simplify the notations, the center will be generally written as $\hat{z}$, would it be static or dynamic.

The first important characteristic of any distribution is its center $\hat{z}$; it has therefore to be determined with care. You may have an intuitive perception of what the center of a distribution is. This is nice, but in order to work with it (calculate it and use it), we need a formal definition of what the center is. You will then discover that the concept of center is far from obvious and that it is impossible to define the center in an unambiguous way: *there are as many centers as you may think of*, and each center has a special purpose.

The general **definition** of the center of a distribution can be the following one: the center is "**a value which is as close as possible to all values present in the database**".

2.1.1 A First Approach

What do we mean by "close"? Two definitions of the word can be given at this stage (more comments will be done on this concept in Part III) $\hat{z}$ can be said to be close to z_i if the difference of their values is close to 0, or if the ratio of the values is close to 1.

Using the Differences: The Arithmetic Mean or "Expected Value"

If the differences are used, the center is the value for which the sum of all the differences between this center and all the data is equal to 0:

$$\sum_i (\hat{z} - z_i) = 0$$

The value can be then immediately computed as:

$$\hat{z} = \frac{1}{I}\sum_i z_i$$

which is called the "arithmetic mean" or the "arithmetic average". As it is very often used, it is generally called just the "mean" and receives a particular symbol: $\bar{z}$. For the example given at the beginning of this chapter, $\bar{z} = 27.355$.

The **expected value** of a random variable, defined by:

- for a discrete variable $E(z) = \sum_i z_i P(z_i)$ where $P(z_i)$ is the probability to get the value z_i (if all the values are different, as it is generally the case in a sample, then $P(z_i) = 1/I$);

- for a continuous variable (which is generally the case for the populations) $E(z) = \int_{-\infty}^{+\infty} z f(z) dz$ where $f(z)$ is the probability density function;

is equivalent to the arithmetic mean. This explains why this mean is so often used for defining the center of a distribution.

What Happens to the Arithmetic Mean When the Variable Is Transformed by a Linear Relationship?

Suppose we make a linear transformation of the variable. Let us assume that z_i is replaced by:

$$z_i' = \frac{z_i - k}{r}$$

if for instance we translate the scale and change the unit (for instance replacing € by k€). If $k = \bar{z}$ and $r = 1$, the variable z' is said to be "centered"; it will be noted ${}_c z_i$. If $k = \bar{z}$ and $r = s_z$ (s_z is the standard deviation of the distribution, defined below) the variable z' is said to be "centered and scaled"; it will be noted ${}_{cs}z_i$. These variables are often used in the computations.

It is easy to show that

$$\overline{z'} = \frac{\bar{z}}{r} - \frac{k}{r}$$

which means that the value of the center is divided by the change in the unit (this is normal) and translated by the translation expressed in the new unit (this is also understandable).

Using the Ratio: The Geometric Mean

If the ratios are used, the center is the value for which the *product* of all the ratios between the center value ${}_g\hat{z}$ and all the data is equal to 1:

$$\prod_i \left(\frac{z_i}{{}_g\hat{z}} \right) = 1$$

Here also the value is immediately computed as:

$${}_g\hat{z} = \sqrt[I]{z_1 \times z_2 \times \cdots \times z_i \times \cdots \times z_I}$$

which is called the "geometric mean"; as this value is little used, there is no need to give it a special symbol. In the given example ${}_g\hat{z} = 16.39$.

These values are rather different! But both can legitimately be called the "center" of the distribution. It is a first perception of the fact that the center is a value which results from conventions, or from a deliberate choice.

For our purpose this procedure for finding the center is a bit "rough", as – due to the sign effect in the first case, to the quotient effect in the second case – values may compensate each other. We would like to define the center with more flexibility in order to adjust it to the particular problems we have to solve.

2.1.2 Other Approaches

Several other measures for the center of a distribution have been defined and are widely used.

The Median

The median – often symbolized, due to its importance, by the symbol $\tilde{z}$ – is a very important characteristic when working with scattered data. One of reasons is that this characteristic is much more "robust" than the arithmetic mean (the concept of robustness is defined and largely used in Part III).

The median is defined as the value which has as many values in the sample greater than it as it has values smaller that it. For a odd number of data, the median receives a value belonging to the sample (in our example $\tilde{z} = 16.3$); if this number is even, we choose the half sum of the values close to it for instance deleting the value 84.5 in the example will give a median equal to 13.1.

When the data are scattered – and it is not rare in the domain of cost – the median gives a more intuitive idea of the center of a distribution: just look at the data and decide where you would like the center of the distribution to be! The problem with the mean is that it is very sensitive to values which are away from the center of the distribution. For instance, if the data point 84.5 is deleted in the example, the mean goes from 27.4 to 21.6: a change of about 6, whereas the median sees a change of about 3.

We will therefore have to return to the median when dealing with cost distributions.

The Harmonic Mean

The harmonic mean is defined as:

$$\frac{1}{\hat{z}} = \frac{1}{I} \sum_i \frac{1}{z_i}$$

This value is rarely used (the reader will find in Ref. [49], p. 88 examples of usages) in the domain of cost. For the example, the harmonic mean is equal to 10.01.

The Mode

In a continuous distribution, the mode is the value which appears the most frequently. In a discrete distribution, especially when each value appears only once, another definition must be found. In order to find one, the range of the values is divided in a few intervals of equal size and the mode is defined as the middle of the interval which has the maximum number of data points.

What can be the width of an interval? A rule due (quoted in Ref. [49], p. 97) to Sturges suggests to take

$$\frac{R}{1 + 3.32 \log I}$$

where R is the range (the difference between the maximum and the minimum values). This gives for the example 18.19. We therefore have six intervals given below. The mode of this distribution would be the middle of the first interval, or 14.

Interval	Quantity
3.4–24.6	7
24.6–39.8	1
39.8–57.9	1
57.9–76.2	1
76.2–94.4	1

2.1.3 Conclusion

One can define many centers for a distribution; the following figure illustrates the different values computed. As said earlier, all of these values may pretend to be the center of the distribution. The choice could therefore be a matter of personal choice; but it is mainly dictated by the easiness of computations and by the mathematical properties of the selected value.

This explains why the mean is so often used, despite its defects in our domain, where, as it will be illustrated below, the median would certainly be more appropriate.

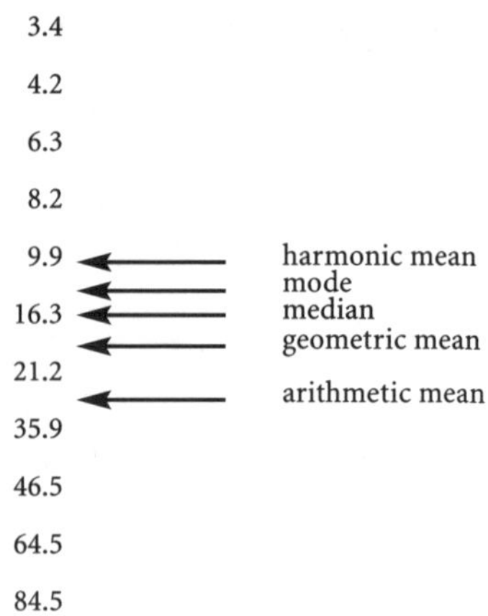

You can see in this figure how scattered are the various centers of the distribution, and especially that the arithmetic mean is "far away" from the intuitive center.

2.2 The Spread of a Distribution Around the Center

The center is an interesting piece of information, but it does not reveal a lot of things about a distribution. What can be done now is to study the scattering of the data around this center: the word "around" is a very important word that we will often meet in this book. The spread is not defined in absolute terms but in respect to the center, which could now take a value of 0 as it does not play a role anymore (data translated for having a center equal to 0 are said to be "centered").

We start by quantifying the spread in the most frequent way, through the standard deviation and will then mention a few other measures of the spread.

2.2.1 The Standard Deviation

The standard deviation is defined by the formula, which clearly indicates that the spread is measured around the mean:

$$s_z = \sqrt{\frac{\sum_i (z_i - \overline{z})^2}{I}}$$

The standard deviation is then equal to the square root of the average of the square of the distances to the mean; it clearly uses the Minkowski metric with $\alpha = 2$ (this metric will be mentioned in Part III) and is therefore *consistent with the usage of the mean*; its unit is the same as the one of the data and therefore its interpretation is easy.

The standard deviation uses the symbol s, always with an index reminding the variable from which it is computed: the usage of the standard deviation is so largely used that an index is a must. It is certainly the most used measure of the spread around the mean.

Pay attention to the fact that some authors define the standard deviation as:

$$s'_z = \sqrt{\frac{\sum_i (z_i - \overline{z})^2}{I - 1}}$$

This comes from the fact that this s' is used as an estimate of the standard deviation of all the products that could be incorporated in the same product family (the population), standard deviation we will name S, with a capital letter. It is possible to demonstrate that s is a biased estimator for this S, whereas s' is not. But this is true when only one parameter is used and we do not see any reason to get away from the definition of a distance average for such a particular case.

The square of the standard deviation is called the **variance**:

$$s_z^2 = \frac{1}{I}\sum_i (z_i - \overline{z})^2$$

Its unit is the square of the unit of the data.

Variance is often defined as the second "moment" around the mean. The moments are defined in more details in Chapter 18 of Volume 1; we just need here the definition of the moments around the mean, also called "centered moments". The centered moment of order k is given, for a discrete variable (the definition can easily be extended to continuous variables), by

$$\mu_k = \frac{1}{I}\sum_i (z_i - \overline{z})^k = \mathrm{E}((z - \overline{z})^k)$$

From this definition, it is clear that $s_z^2 = \mu_k = \mathrm{E}((z - \overline{z})^2)$.

An important theorem about variances is related to the variance of a sum of two random variables z and w. This theorem requires the definition of the covariance of random variables. The covariance expresses the fact that two variables do vary in the same direction together (more is given in Chapter 5 of this volume, paragraph 3.1.):

$$\mathrm{cov}(z, w) = \mathrm{E}[(z - \overline{z})(w - \overline{w})]$$

Then the variance of a sum of two random variables is given by:

$$\mathrm{var}(z + w) = \mathrm{var}(z) + \mathrm{var}(w) + 2\,\mathrm{cov}(z, w)$$

What Happens to the Standard Deviation When the Variable Is Transformed by a Linear Relationship?
Suppose we make a linear transformation of the variable:

$$z_i' = \frac{z_i - k}{r}$$

It is then easy to show that:

$$\mathrm{var}(z') = \frac{1}{r^2}\,\mathrm{var}(z)$$

The variance is divided by the square of the scaling factor.

2.2.2 Other Measures of the Spread

The **range**, equal to the difference between the highest and the lowest value, is very sensitive to "outliers".

The **interquartile range** is more stable than the range. A 25% quartile is a value which has 25% of the data smaller than it; a 50% quartile corresponds to the median; a 75% quartile is a value which has 25% of the data higher than it. The interquartile range is the difference between the 75% and the 25% quartile.

The **average absolute range** is defined by:

$$\text{average absolute range} = \frac{1}{I}\sum_i |z_i - \overline{z}|$$

It is the average distance of the data to the mean.

The **average deviation around the median** is defined by:

$$\tilde{D} = \frac{1}{I}\sum_i |z_i - \tilde{z}|$$

This measure of the spread uses the Minkowski distance (defined in Chapter 8) with $\alpha = 1$. It is therefore *consistent with the use of the median*: if you use the median as the value of the center of a distribution, you should use $\tilde{D}$ as the measure of the spread.

2.3 The Shape of the Distribution

The standard deviation quantifies the spread of a distribution, but does not say if it is symmetrical around the mean, or more or less flat. The skewness and kurtosis fill this gap.

Both characteristics are easily defined from the centered moments defined upwards:

$$\mu_k = \frac{1}{I}\sum_i (z_i - \overline{z})^k$$

2.3.1 The Level of Asymmetry (Skewness)

Both distributions represented below have the same mean and the same standard deviation, but they are not identical.

The skewness of a distribution is given by (the notations are standard):

$$\gamma_1 = \frac{\mu_3}{s_z^3}$$

This skewness is equal to 0 for a symmetrical distribution; if the left part of the distribution is more extended than the right part (as it is for the dotted line in Figure 2.1), the distribution is said to have a positive skewness: the probability of finding a value higher than the mean is greater than the probability of finding a value below the mean. A good example is given by the log normal distribution (see Chapter 3).

2.3.2 The Level of Flatness (Kurtosis)

It is defined as:

$$\gamma_2 = \frac{\mu_4}{s^4}$$

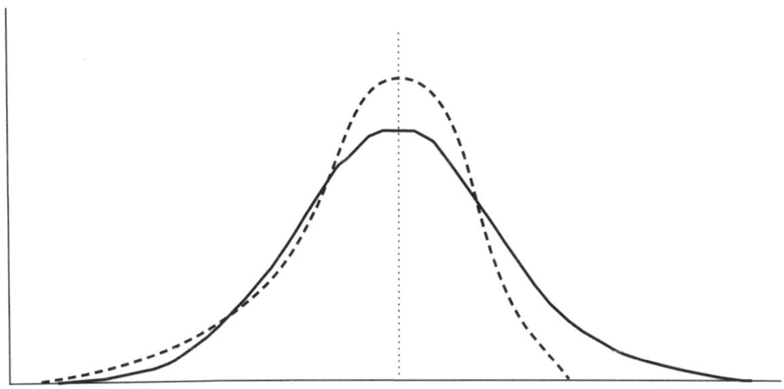

Figure 2.1 Symmetrical and non-symmetrical distributions.

The "reference" of kurtosis is given by the normal distribution for which $\gamma_2 = 3$. A distribution of which $\gamma_2 > 3$ is less peaked than the Normal distribution: a good example is given by the Student's distribution (see Chapter 3); this distribution is always flatter than the Normal distribution and therefore has a kurtosis greater than 3. Note that some authors define the kurtosis as $\gamma_2 - 3$ in order to have a 0 kurtosis for the normal distribution (the reference): with such a definition, they say that a more peaked distribution has a negative kurtosis.

2.3.3 Using Higher Moments

It would be possible to use moments of higher order than 4 in order to get a more precise representation of a distribution, but it is never done in practice, as the amount of information they would provide is negligible.

2.4 The Concept of Degrees of Freedom

In the three-dimensional space, a point has 3 degrees of freedom: its three coordinates may take independently all the values. If it has to remain on a surface, one degree disappears because the three coordinates are linked by the equation of the surface: if two coordinates are randomly selected, the third one is automatically computed; it is said to have 2 degrees of freedom. If it has to remain on a curve, then 2 degrees of freedom disappear because its coordinates are linked by two equations.

A similar idea is used in statistics. A sample of size I has I degrees of freedom, and so has any value, such as the mean, computed only from the data themselves. Now in the formula:

$$s_z^2 = \frac{1}{I}\sum_i (z_i - \bar{z})^2$$

$\bar{z}$ is supposed to remain constant, as we are computing the spread *around the mean*. Due to that, the I values z_i are not now totally free, because their sum must keep the same value. The variance therefore has $I - 1$ degrees of freedom: it has less "liberty" to move than the mean.

Generally speaking, the degrees of freedom of the result of the computation based on a sample of size I is equal to I minus the degrees of freedom "absorbed" by the values which are included in its formula. A formula which includes the mean and the variance has $I - 2$ degrees of freedom.

3 Typical Distributions

Summary

This chapter presents a few distributions which are often used in statistical analysis. More details are given in specialized manuals and the purpose of this presentation is only to facilitate the reading of this book.

All the distributions reminded here deal with one variable only.

The reasons for these reminders are the following:

- Several of these distributions are useful for discovering some properties of the extrapolations of the values computed for the sample to the whole population.
- Several characteristics of these distribution are very useful: they will be immediately available to the reader.
- If the distribution you use is similar to one of these typical distributions, you will find easily the response to some of the questions you ask.
- Several mathematical packages allow to write a formula and compute from it. Using the formulae as described in this chapter will allow you to easily calculate.

The reader will find more information about these distributions in most books on statistics.

3.1 The "Normal", or Laplace–Gauss, Distribution

The Laplace–Gauss distribution, often called the "normal" law, is certainly the most frequently used distribution (this does not mean that other distributions are abnormal ...). Its principal interest is that if a stable phenomenon which gives one output value at each run (for instance the cost of same part machined on a repetitive process) is subject to many small independent perturbations, each one being unable to modify in a sensible way the phenomenon, then the distribution of the output values follows a normal distribution.

3.1.1 Mathematical Expression

The general expression is given by:

$$N(x,m,s) = \frac{1}{s(2\pi)^{\frac{1}{2}}} e^{-\frac{(x-m)^2}{2s^2}}$$

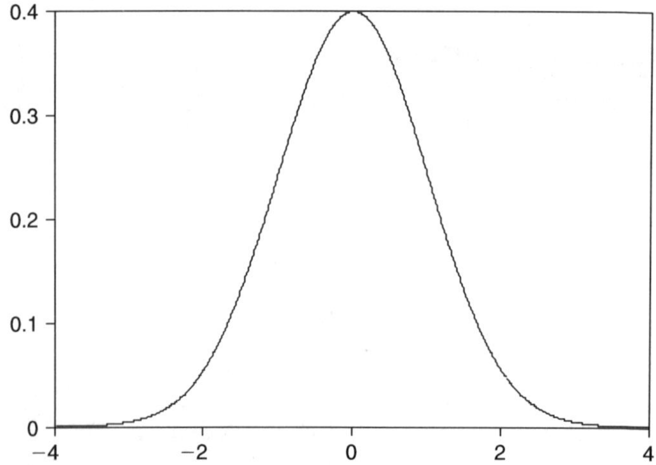

Figure 3.1 Normal distribution.

The main characteristics of this distribution are its mean, m, its variance, s^2, its skewness, 0, and its kurtosis equal to 3.

The "normalized" expression (with a mean equal to 0 and a variance equal to 1) is given by:

$$N(x,0,1) = \frac{1}{(2\pi)^{\frac{1}{2}}} e^{-\frac{(x)^2}{2}}$$

3.1.2 Geometrical Perspective

The distribution is represented in Figure 3.1.

3.1.3 Cumulative Distribution *CN*(*x*, 0, 1)

This is the most useful distribution; it gives the area under the curve from $-\infty$ to x (Figure 3.2).

Important Values for the Cumulative Distribution

$$\begin{aligned}
CN(-1.96) &= 0.025\\
CN(-1.6449) &= 0.05\\
CN(-1.2816) &= 0.10\\
CN(0) &= 0.50\\
CN(1.2816) &= 0.90\\
CN(1.6449) &= 0.95\\
CN(1.96) &= 0.975
\end{aligned}$$

Consequently 95% of the area is included between -1.96 and $+1.96$, and 90% of the area between -1.645 and $+1.645$. These values are the most frequently used when dealing with cost.

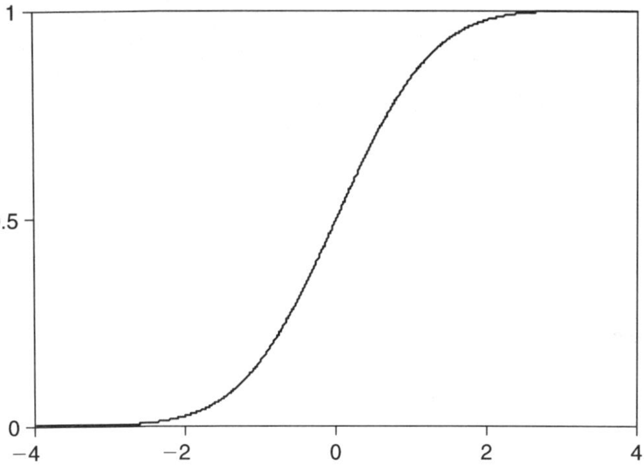

Figure 3.2 The cumulative normal distribution.

3.1.4 Other Moments

Centered moments of order k are given by:

- if k is odd, $\mu_k = 0$
- if k is even, $\mu_k = (k-1) \times (k-3) \times \cdots \times 3 \times 1 \times s^r$.

3.2 The Log-Normal Distribution

This is the distribution of a positive x variable such as its LN – let us call it ξ – follows a normal distribution $N(\mu, \sigma)$. The distribution of ξ is then given by:

$$\frac{1}{\sigma\sqrt{2\pi}}\exp\left[-\frac{(\xi-\mu)^2}{2\sigma^2}\right]$$

3.2.1 Mathematical Expression

The distribution of x depends on the two parameters μ and σ; it is given by (pay attention to the x in the denominator):

$$LN(x,\mu,\sigma) = \frac{1}{x \times \sigma\sqrt{2\pi}}\exp\left[-\frac{(\ln x-\mu)^2}{2\sigma^2}\right]$$

Its mean equals $\bar{x} = e^{\mu+(\sigma^2/2)}$ and its variance $s^2 = e^{2\mu+\sigma^2} \times (e^{\sigma^2} - 1)$. Its mode is given by $e^{\mu-\sigma^2}$ and its median by $\tilde{x} = e^{\mu}$. The median is here an important characteristic of this distribution, because it is not symmetrical.

3.2.2 Geometrical Perspective

The distribution is represented[1] in Figure 3.3, for $\mu = 0$ and $\sigma = 1$ (full line) and, $\sigma = 0.5$ (dotted line).

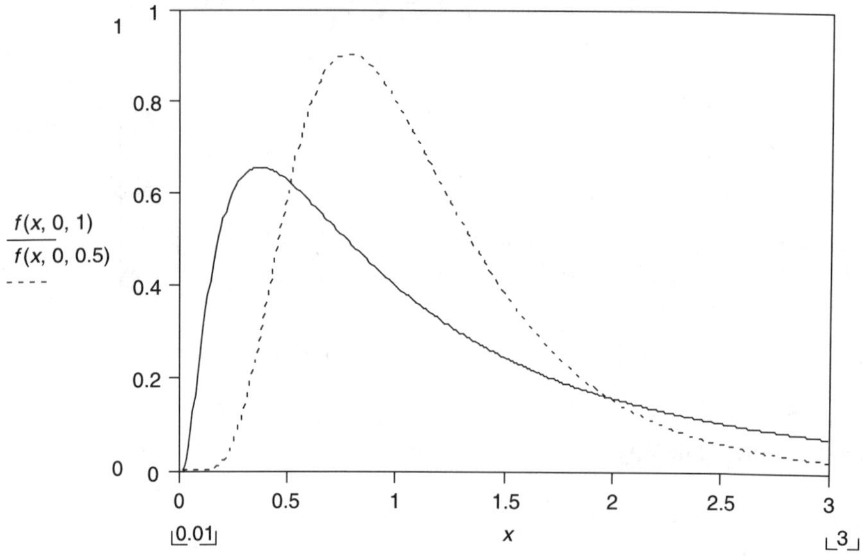

Figure 3.3 The log-normal distribution.

3.2.3 About the Moments

The multiplicative moment of order k is given by:

$$M_k = e^{k\mu + \frac{1}{2}k^2\sigma^2}$$

From this formula the skewness and the kurtosis can be computed:

- Skewness $= (e^{\sigma^2} + 2) \times (e^{\sigma^2} - 2)$: this distribution is always skewed (and the skewness is always positive), as the skewness can only be $\sigma = 0$.
- Kurtosis $= e^{4\sigma^2} + 2e^{3\sigma^2} + 3e^{2\sigma^2} - 3$.

3.3 The χ^2 Distribution

3.3.1 Definition

There are two definitions of the χ^2 distribution:

1. Suppose we have a population with a variance S^2. From this population is extracted a sample of size I of which variance is noted s^2. Then the random variable $((I - 1)s^2)/S^2$ follows an χ^2 distribution with $\nu = I - 1$ degrees of freedom.

[1] This distribution is especially important for us with $\mu = 0$. For the full line, the mean is equal to $e^{0.5}$ or 1.65, the mode to 0.37 and the median to 1.

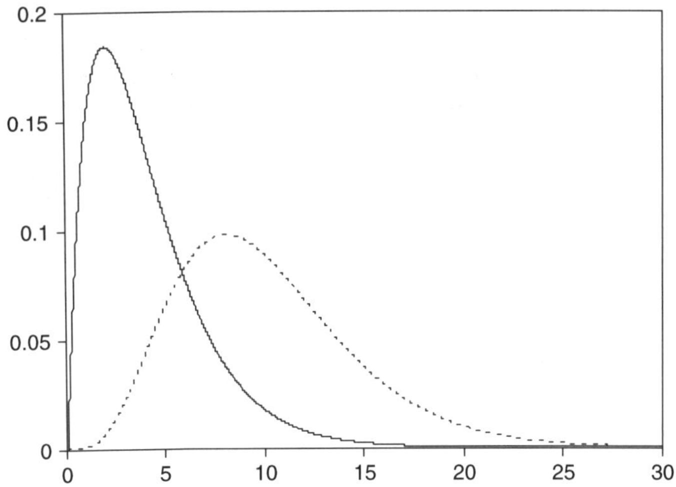

Figure 3.4 The χ^2 distribution.

2. Suppose you have now ν independent "standard" normal variables $N(0, 1)$, named $z_1, z_2, \ldots, z_\nu$. Then the sum $\Sigma_{i=1}^{\nu} z_i^2$ follows a $\chi^2_{(\nu)}$ distribution with ν degrees of freedom.

3.3.2 Mathematical Expression

The χ^2 distribution has one parameter, ν, called the number of degrees of freedom. It is defined for $x \geqslant 0$. The mathematical expression is given by:

$$\chi^2_{(\nu)} = \frac{1}{2^{\frac{\nu}{2}}\Gamma\left(\frac{\nu}{2}\right)} e^{-\frac{x}{2}} x^{\frac{\nu}{2}-1}$$

The coefficient of skewness is given by $2^{2/3}\nu^{-1/2}$ and the kurtosis by $3 + (12/\nu)$.

The multiplicative moment of order k is equal to $M_k = 2^k \times [\Gamma(k + (\nu/2))]/[\Gamma(\nu/2)]$.

3.3.3 Geometrical Perspective

The distribution is given by Figure 3.4 for $\nu = 4$ (full line) and for $\nu = 10$ (dotted line).

3.3.4 Important Properties

Characteristics

The arithmetic mean is equal to ν and the variance to 2ν.

Additivity

If two independent random variables have χ^2 distributions with ν_1 and ν_2 degrees of freedom, then their sum also follows a χ^2 distribution with $\nu = \nu_1 + \nu_2$ degrees of freedom.

3.4 The *F*-Distribution

3.4.1 Definition

There are two definitions of the *F*-distribution:

1. Suppose you have two populations normally distributed. Two samples, of sizes n_1 and n_2, are extracted from these populations; these samples show variances s_1^2 and s_2^1. Then the random variable $x = s_2^1/s_2^2$ follows an *F*-distribution with $\nu_1 = n_1 - 1$ and $\nu_2 = n_2 - 1$ degrees of freedom.
2. If you have two χ^2 distribution noted $\chi^2_{[\nu_1]}$ and $\chi^2_{[\nu_2]}$, the ratio $x = \chi^2_{[\nu_1]}/\nu_1/\chi^2_{[\nu_2]}/\nu_2$ follows a *F*-distribution.

3.4.2 Mathematical Expression

The mathematical expression is a bit complex:

$$F(x,\nu_1,\nu_2) = \frac{\Gamma\left(\frac{\nu_1}{2}+\frac{\nu_1}{2}\right)}{\Gamma\left(\frac{\nu_1}{2}\right)\Gamma\left(\frac{\nu_2}{2}\right)} \times \frac{\left(\frac{\nu_1}{\nu_2}\right)^{\frac{\nu_1}{2}} x^{\frac{\nu_1}{2}} - 1}{\left(1+\frac{\nu_1}{\nu_2}x\right)^{\frac{\nu_1+\nu_2}{2}}}$$

The mean value equals $\nu_2/(\nu_2 - 2)$ and the variance is $2 \times \frac{\nu_2^2}{\nu_1} \times \frac{\nu+\nu_2-2}{(\nu_2-2)^2(\nu_1-4)}$

3.5 The Student Distribution

This is an important distribution for the practical use of specific models.

3.5.1 Definition

Suppose you have two variables:

1. x_1 following a normal distribution $N(x_1, 0, 1)$.
2. x_2 following a χ^2_ν distribution.

Then the variable $t = x_1/(\sqrt{x_2}/\nu)$ follows a Student distribution. The mean equals 0, the variance is $\nu/(\nu - 2)$ (which tends toward 1 if n is large), the skewness is 0 and the kurtosis 3 $((\nu - 2)/(\nu - 4))$. ν is called the number of degrees of freedom. These values show that the Student distribution tends toward the normal one when ν becomes large.

3.5.2 Mathematical Expression

The Student distribution is given by:

$$St(x,\nu) = \frac{\Gamma\left(\frac{\nu+1}{2}\right)}{(\pi\nu)^{\frac{1}{2}}\Gamma\left(\frac{\nu}{2}\right)\left[1+\frac{x^2}{\nu}\right]^{\frac{\nu+1}{2}}}$$

The moments are given by complex formulae; as they are never used, they are not given for this distribution.

3.5.3 Geometrical Perspective

The shape of the distribution is given in Figure 3.5 for $\nu = 3$ (full line) and $\nu = 20$ (dotted line).

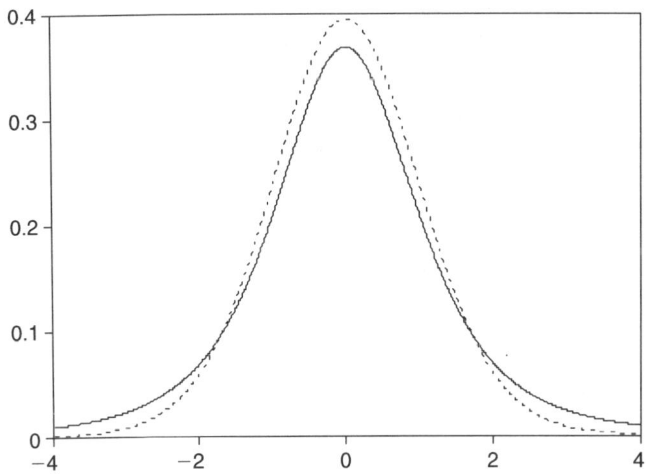

Figure 3.5 The Student distribution.

3.5.4 Cumulative Distribution

The following table gives the values of x corresponding to the cumulative distribution starting from $-\infty$:

Cumulative value	Number of degrees of freedom 5	10	20	∞
0.025	−2.571	−2.228	−2.086	−1.960
0.05	−2.015	−1.812	−1.725	−1.645
0.10	−1.476	−1.372	−1.325	−1.282
0.50	0	0	0	0
0.90	1.476	1.372	1.325	1.282
0.95	2.015	1.812	1.725	1.645
0.975	2.571	2.228	2.086	1.960

This table gives the value of x for a given cumulative value: for instance if the number of degrees of freedom is equal to 10, and $x = -1.812$, then the area under the curve on the left of this value is equal to 0.05, which means that the area on the right is equal to 0.95 (the total area being normalized to 1). If the number of degrees of freedom is very large (column "∞"), the Student distribution is equivalent to the Normal distribution; from the values of the table, one can clearly see that as soon as the degree of freedom exceeds about 10, the Student distribution can, for practical purpose in our domain, be replaced by the Normal distribution.

The values of the table are said to be "one tail". If you want the range – symmetrical around 0 – in which x must be in order to get 90% of the surface, it means that each tail must have a surface of 5%, which means that x must be inside the range $[-1.725, +1.725]$ or $|x| \leqslant 1.725$.

3.6 The Beta Distribution

3.6.1 Definition

The beta distribution is a unimodal distribution defined between two values l and h. It depends on two parameters called α and β, thanks to which the distribution can take a lot of different forms. The easiness with which these forms can be written makes this distribution often used when the range of the variable is limited.

3.6.2 Mathematical Expression

It is given by:

$$B(x;\alpha,\beta) = \frac{1}{h-l}\frac{\Gamma(\alpha+\beta)}{\Gamma(\alpha)\Gamma(\beta)} \times \left(\frac{x-l}{h-l}\right)^{\alpha-1} \times \left(\frac{h-x}{h-l}\right)^{\beta-1}$$

where $\Gamma(u) = \int_0^{+\infty} t^{u-1} e^{-t}\,dt$ generalizes the factorial function: when u is an integer, then $\Gamma(u+1) = u!$

If we normalize the interval of variation to $[0,1]$, the function is easily written as:

$$B(x;\alpha,\beta) = \frac{\Gamma(\alpha+\beta)}{\Gamma(\alpha)\Gamma(\beta)} \times (x)^{\alpha-1} \times (1-x)^{\beta-1}$$

where the constant $\Gamma(\alpha+\beta)/\Gamma(\alpha)\Gamma(\beta)$ is there just to insure that the area under the curve is equal to 1.

The main characteristics of the normalized distribution are given by:

$$\text{mean} = \frac{\alpha}{\alpha+\beta}$$

$$\text{variance} = \frac{\alpha\beta}{(\alpha+\beta)^2(\alpha+\beta+1)}$$

$$\text{mode} = \frac{\alpha-1}{\alpha+\beta-2}$$

The multiplicative moment of order k is given by $\mathrm{M}_k = \prod_{i=0}^{k-1} \frac{\alpha+i}{\alpha+\beta+i}$.

3.6.3 Geometrical Perspective

The following graph shows the distribution for different values of α and β (Figure 3.6).

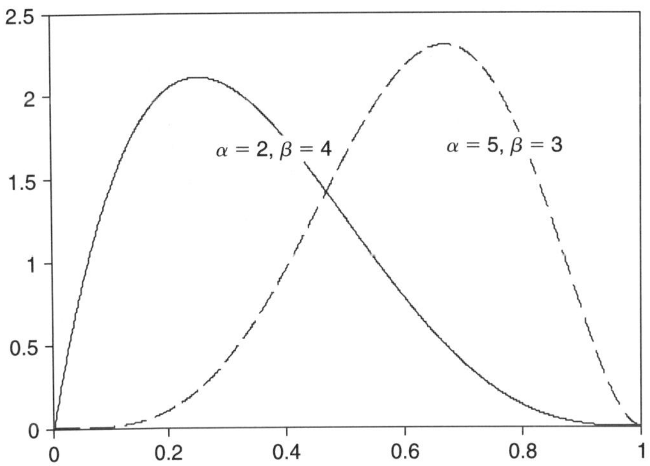

Figure 3.6 The beta distribution.

Part II

Data Analysis Precedes the Search for a Specific Model

Part Contents

Introduction
The main steps for carrying out a data analysis.

Chapter 4 **Data Analysis on One Variable Only**
Only one variable is considered; this variable can only be the cost (or the specific cost, or any other thing) of several products belonging to the same product family.

This is a simple case, of which purpose is mainly to introduce several concepts.

As there is only one variable to consider, only steps 1 and 3 are relevant.

Chapter 5 **Data Analysis on Two Variables**
Each variable can here be investigated independently of the other one.

But it is also important to consider both variables simultaneously. In this domain three steps are now relevant: steps 1, 3 and 4.

Chapter 6 **Simultaneous Data Analysis on $J + 1$ Quantitative Variables**
The variables are now the cost, or any other thing, plus $J + 1$ quantitative parameters.

This is the general case: all the steps are now relevant.

Chapter 7 **Working with Qualitative Variables**
Qualitative variables are important in cost estimating, because it is extremely rare that a product family is homogeneous enough to avoid them.

Therefore, they have to be studied as well.

Summary

A sample is a collection of normalized data, the normalization process being there in order to make the data comparable. But this process does not guarantee that no mistake was made during the data collection or that any problem will not occur during the following steps of which purpose will be to "extract" from the data the information which can be used for cost-estimating new products belonging to the same product family.

Never start building a specific model without making first an analysis of your data. This step is especially important when dealing with several quantitative variables.

The logic for the data analysis is triple:

1. checking for any mistake or "abnormal" data;
2. detecting, before any other computation is made, any potential problem;
3. obtaining an understanding of the data structure in order to be able to decide on the work that can be done on the data and the results which can be expected.

Generally speaking an analysis of a set of data drawn from a product family must always follow the following steps:

1. Search for outliers (the definition of an outlier is given in p. 43).
2. Search for possible multi-collinearities between the causal variables (collinearity between the causal variables and the cost is what we are looking for and therefore is not dealt with here).

3. Visualization of the data, in order to understand their structure, which means answering some questions such as:
   - Is the set of data homogeneous enough, or should we split it into sub-families? If, for instance, we find that the distribution has two modes (it is then said "bimodal"), then the product family should probably be broken into two "sub-families".
   - Are there any outlier that step 1 could not detect? Some outliers are for instance detected in step 1 with an algorithm based on the distance; but distant points are not necessarily outliers.
   - How are the data related (if at least two variables are simultaneously checked)?
   - Finding a standard distribution curves (the most frequent distribution curves are listed in Chapter 3) that may usefully be used instead of the real distribution. The advantage of doing so is that all the characteristics of these standard distributions have already been computed: this simplifies the work. For instance a distribution curve may look very similar to a normal distribution. As this normal distribution is very important, you will find in Chapter 15 some tests allowing to check the "normality" of a given distribution.
4. Quantification of the perceived relationships between the variables. If we perceive relationships between the variables, it is convenient to quantify these relationships in order to decide on the future use of these data.

This chapter describes how these steps may be carried out, depending on the number of variables which the cost analyst considers relevant. The description is here limited to quantitative variables only; other information for other cases will be given in the dedicated chapters.

Introduction

You have collected data; these data have been normalized and are organized in a database; now you want to use the data for preparing your future cost estimates.

The starting point of this part is a product family, as it is described in Chapter 1. You believe that the products inside this product family are homogeneous enough to be dealt with **simultaneously**; simultaneously being the key word: you are not going to consider – inside the product family – each "data point" independently of the other ones, but consider all the data points together.

Data analysis must precede any attempt to build a specific model. It is extremely important; unfortunately many cost estimators forget about it.

The purpose of the analysis is to "discover" your data in order to prepare, in the most efficient way, their treatment.

This implies to distinguish the dependent and the causal variable(s): any relationship between the first one and any other one is a good thing (it is what we are looking for!). Any relationship between the causal variables can be dangerous.

Analyzing the data generally involves four major steps:

1. *Discovering if there is some "abnormal" data*: Abnormal data – frequently called "outliers" – are data, of which quantity is very limited (otherwise they would not be "abnormal"), which by their sole presence may seriously modify any relationship between the "normal" data. The seriousness will of course have to be

quantified: what do we mean by serious? And the potential damage will have to be quantified.

2. *Dealing with a possible multi-collinearity between the data*: Of course, this concerns only data involving several independent variables. It is, in such a case, one of the major problem of data analysis, because it may produce instabilities in the relationship between the dependent variable and the causal ones.
3. *Visualizing the data*: This is something important. An algorithm will always produce a result – except in rare circumstances (for instance if a matrix is ill conditioned) – but this result must be interpreted. We do possess a wonderful "sensor" for interpretation: the eye! As often as we can it is always extremely useful to present the data in a visual form: the human brain is rather poor in interpreting a table of figures; it is extremely powerful at interpreting a figure. The purpose is to find out the relationships between the data.
4. *Quantifying these relationships*: The human eye is a very interesting sensor for possible relationship, but our mind is rather poor for quantifying them. The purpose of this quantification is to compare them. This quantification will produce a few figures which will:
   - confirm the impression given by the visualization;
   - make the data set as a whole easy to handle: it is always easier to handle a few "summary figures" than the whole set of data.

Once this is done, you will be well prepared to establish a quantified relationship that will be used for cost predictions.

Once you have finished this part, you have not yet any tool for making a cost estimate. You are only confident that the data you have for the product family you are working with are homogeneous, reliable, sufficient and proper to extract something that will become your cost-estimating tool for this family. The process by which you will extract this something will be fully discussed later on.

4 Data Analysis on One Variable Only

Summary

This chapter is just an introduction to the subject: when dealing with a distribution involving just one variable, data analysis is rather straightforward!

Nevertheless some algorithms may be useful when the number of data becomes large.

These algorithms deal with two subjects:

1. Search for outliers.
2. Visualization of the data, in order to understand their structure, which means answering some questions such as:
   - Is the set of data homogeneous enough, or should we split it into sub-families? If, for instance, we find that the distribution has two modes (it is then said "bimodal"), then the product family should probably be broken into two "sub-families".
   - Are there any outlier that step 1 could not detect? Some outliers are for instance detected in step 1 with an algorithm based on the distance; but distant points are not necessarily outliers.
   - Finding a standard distribution curves (the most frequent distribution curves are listed in Chapter 3) that may usefully be used instead of the real distribution. The advantage of doing so is that all the characteristics of these standard distributions have already been computed: this simplifies the work. For instance a distribution curve may look very similar to a normal distribution.

When a population is defined by one variable only, this variable is the one we are interested in, the one which was defined as the "dependent" variable, even if here it is considered that it does not depend on anything. Most often it is for us the cost, but it can be the specific cost or anything else.

It is labelled Y and the observed values are called y_i, the index i varying from 1 to I (I being the number of products). This "empirical" distribution, which is the distribution of the values inside the sample, is named φ.

Working with just one variable rests on the assumption that this variable does not depend, in the studied domain, on another variable. A sample of data is collected in order to answer these questions:

1. Is this assumption valid?
2. What is the distribution of the values of this variable, distribution which will be used to forecast the value of a new product belonging to the same product family?

The answer to these questions will be given in Chapter 8. Before we answer them from the data collected in the sample, we must analyze these data in order to prepare these answers.

Let us give a few technical examples of situations where a population can be defined by just one variable:

- the duration to accomplish several times the same task (as for going from your home to work);
- cost (more exactly price) proposals made by different companies for the same work;
- cost of the same product made several times, etc.;
- plus a frequent situation in cost estimating: the wish, or the will, to use the specific cost (a discussion about the use of the specific cost is postponed to Part V) for making a forecast. The specific cost is the cost per unit of size, such as the cost per kilogram for mechanical or even electronic equipment, the cost per square meter for apartments, the cost per instruction for software (or its opposite which is called the "productivity" which is the number of instructions performed per unit of time), etc. Many people still believe it is constant.

This chapter deals with the analysis of the empirical distribution φ (the sample). This analysis will be used, in Chapter 15, for establishing results valid for the whole population from which the sample is drawn, which means for establishing some properties of the distribution Φ of the whole product family.

The theory will be here illustrated with the set of data of the Table 4.1 or Example A; it is representative of values which can be observed when measuring a relatively stable phenomenon.

Table 4.1 The set of data used for illustration (Example A).

Data number	Value
1	11.60
2	12.10
3	10.20
4	12.60
5	11.10
6	12.15
7	11.65
8	12.90
9	11.20
10	11.65
11	12.40
12	13.30
13	11.25
14	11.70
15	12.45
16	11.80
17	12.50
18	11.40
19	11.84
20	11.90

4.1 Looking for Outliers

Definition of an Outlier

An outlier is a value which, by its sole presence, profoundly changes the characteristics of the distribution. Such a concept is easy to illustrate when two variables are considered: refer to Figure 5.5 for a clear example. Here it designates a data point (this group of words will be used as a synonymous of "product" or "observed value") which is far away from the other data. The only question is to define what we mean by "far away".

An Algorithm

Sprent and Smeeton ([53], p. 409) after writing that many tests for outliers lack robustness, recommend to compute the "median absolute deviation" (MAD) of the distribution:

$$\text{MAD} = \text{median}\left| y_i - \text{median}(y_i) \right|$$

The word "median" is defined in Chapter 2 of this volume: if you sort the data in ascending order, it is the data which has as many values under it than over it. In the example given in Table 4.2, the median is equal to 11.82 (when the number of data is even, the median is the arithmetic average of the values which framed the median).

The absolute deviations around the median are here given by the second column in Table 4.2; they are sorted in the third column of the same table. The median of this new set equals 0.495, which is the MAD.

Table 4.2 Potential outliers.

Values sorted	Values less median	Values less median sorted
10.20	1.62	0.02
11.10	0.72	0.02
11.20	0.62	0.08
11.25	0.57	0.12
11.40	0.42	0.17
11.60	0.22	0.17
11.65	0.17	0.22
11.65	0.17	0.28
11.70	0.12	0.33
11.80	0.02	0.42
----------	----------	---------- Median line
11.84	0.02	0.57
11.90	0.08	0.58
12.10	0.28	0.62
12.15	0.33	0.63
12.40	0.58	0.68
12.45	0.63	0.72
12.50	0.68	0.78
12.60	0.78	1.08
12.90	1.08	1.48
13.30	1.48	1.62

They then propose to consider as potential outliers the data which are at more than 5 MAD from the median. In the example, this means that the data outside the interval [9.345, 14.295] should be viewed as outliers. According to this procedure, there is no outlier here.

This algorithm for detecting potential outliers looks conservative enough for our purpose: the risk to consider as an outlier a reasonable value is very limited.

Note

Instead of the value proposed by Sprent and Smeeton [53], we sometimes use the MAD defined from the mean:

$$\text{MAD} = \text{median}\left|y_i - \text{mean}(y_i)\right|$$

which is easier to compute. The result is very similar.

4.2 Visualizing the Distribution

The purpose of the visualization is to use our eyes, which are – with our brain – a fantastic device for capturing information from a picture, in order to discover potential problem in our data.

The visualization of a distribution is different if it is discrete or continuous. As both distributions are present in our analysis (the distribution of the sample values is discrete, whereas the distribution of the population, supposed to be infinite in size, is continuous), both visualizations must be considered.

4.2.1 Visualizing a Discrete Distribution

The data can be presented in values or in cumulative form.

In Values

Instead of the values themselves, it is usual to display their relative frequency.

If the set of possible values is limited, the bar graph is the obvious representation of the data. For instance if you play with a dice (six values only are possible), you may observe the number of occurrences in a set of 300 trials in Table 4.3.

In order to normalize the values (the purpose being to compare this distribution with others), it is usual to display the relative frequencies, as in Table 4.4.

As the number of possible values is very limited, the eye is here able to get a general picture; but it becomes difficult to get this general picture if this number is larger than 10. A bar graph, as the one displayed in Figure 4.1 (pay attention to the scale selected for the ordinates axis: it enlarges the visual differences between the relative frequencies), and helps to grasp the major features of this distribution.

Table 4.3 Values and number of occurrences.

Values	Number of occurrences
1	47
2	49
3	55
4	52
5	49
6	48

Table 4.4 Relative frequencies.

Values	Number of occurrences
1	0.1566
2	0.1633
3	0.1833
4	0.1733
5	0.1633
6	0.1600

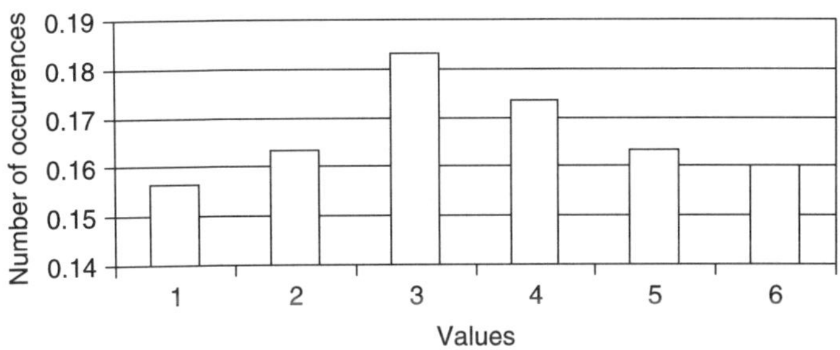

Figure 4.1 Bar graph: showing relative frequencies.

If the set of the values depends on the sample, and if, as it is generally the case in cost analysis, each value occurs only once, such a graph cannot be used. The idea is then to group the data in a limited number of intervals and to compute the number of data in each interval. The choice of the interval side depends on the desired precision and of the number I of data points. Sturges (quoted by Sachs [49], p. 53) suggested the following number of intervals:

$$\text{number of intervals} \approx 1 + 3.32 \times \log_{10} I$$

For example given in Table 4.1, one can use the following intervals: (10.0–10.5), (10.5–11.0) (11.0–11.5), (11.5–12.0), (12.0–12.5), (12.5–13.0) (13.0–13.5), and (13.5–14.0), with the rule that the upper side of the interval belongs to it, but not the lower side. For instance 11.5 belongs to the interval (11.0–11.5).

Our table takes now the form given in Table 4.5.

The distribution can now be easily displayed, as in the previous section (Figure 4.2).

Table 4.5 Percentage of data (in values) in each interval.

Interval	Number of data	%
(a) 10.0–10.5	1	5
(b) 10.5–11.0	0	0
(c) 11.0–11.5	4	20
(d) 11.5–12.0	7	35
(e) 12.0–12.5	5	25
(f) 12.5–13.0	2	10
(g) 13.0–13.5	1	5
Total	20	100

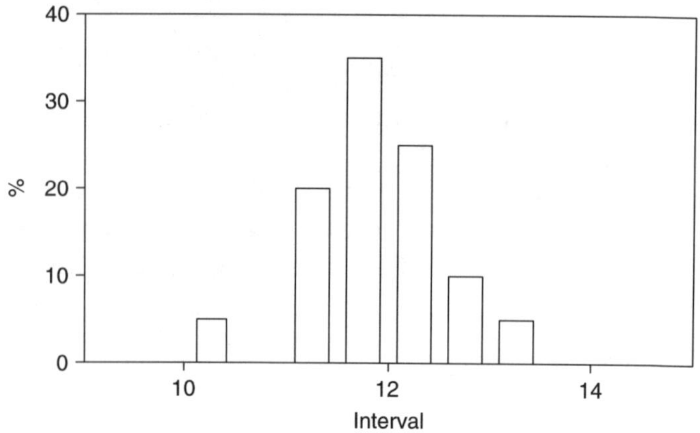

Figure 4.2 Bar graph: percentage of data (in values) in each interval.

In Totals

A first way to get an overview of the cumulative distribution is straightforward. It is illustrated on Figure 4.3 established from the values computed in Table 4.6.

A second way is to compute the "percentiles". Percentiles are based on ranks: a value which has p data lower then it is said to be at percentile $100(p/I)$ (I being always the number of data). For instance in the data displayed in Table 4.1 and sorted in Table 4.2, value 11.2 is at percentile 10, value 11.40 at percentile 20, etc.

Percentiles, because our number of data is generally limited, are often too detailed and we prefer to use quartiles. The first quartile corresponds to percentile 25[1] (11.50 in the example), the second to percentile 50 (it is also called the median), etc.

An information which sometimes useful is the "**interquartile range**": it is defined as the interval from percentile 25 to percentile 75 and therefore includes 50% of the data centered around the median. In our example, the interquartile range extends from 11.50 to 12.425; its value is 0.975.

[1] This percentile is sometimes called a "hinge", for instance by Mosteller and Tukey [43], as well as percentile 75.

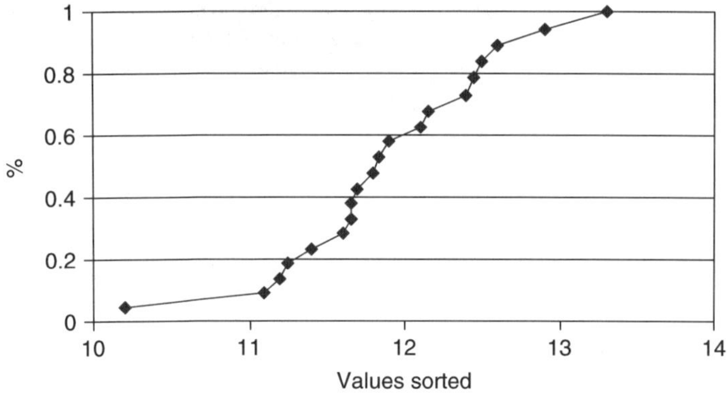

Figure 4.3 Bar graph: percentage of data (in total) in each interval.

Table 4.6 Percentage of data (in total) in each interval.

Values sorted	Values added	%
10.20	10.20	0.0429
11.10	21.30	0.0896
11.20	32.50	0.1367
11.25	43.75	0.1841
11.40	55.15	0.2320
11.60	66.75	0.2808
11.65	78.40	0.3298
11.65	90.75	0.3789
11.70	101.75	0.4281
11.80	113.55	0.4777
11.84	125.39	0.5275
11.90	137.29	0.5776
12.10	149.39	0.6285
12.15	161.54	0.6796
12.40	173.94	0.7318
12.45	186.39	0.7842
12.50	198.89	0.8368
12.60	211.49	0.8898
12.90	224.39	0.9440
13.30	237.69	1.0000

4.2.2 Visualizing a Continuous Distribution

A continuous distribution is obviously represented by a curve, as illustrated on Figure 4.4.

The total area under the curve is normalized to a value of 1. Consequently the area between two values, such as a and b, gives the relative frequency of values that can be found between these two limits.

Obviously the cumulative distribution can also be drawn: the shape is similar to the one illustrated on Figure 4.3, but is "smooth".

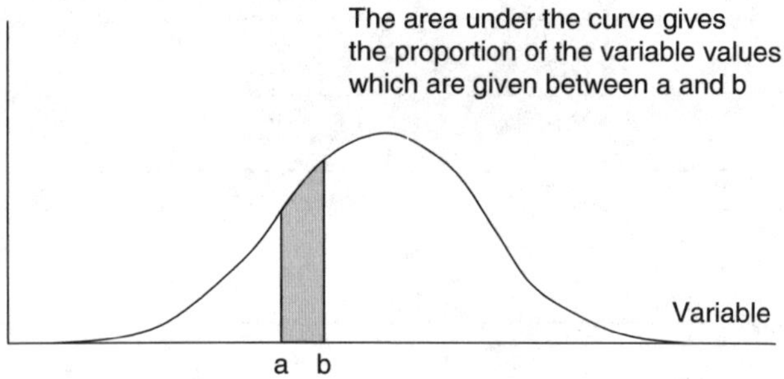

Figure 4.4 Distribution curve for a continuous variable.

5 Data Analysis on Two Variables[1]

Summary

This chapter deals with a set of data containing the values of two variables, one of which being the "dependent" or "explained" variable (for us, generally the cost) designated by the letter y. In this set, the analyst thinks there is a linear correlation between the variables, or at least wants to check the possibility of such a correlation.

What do we mean by "linear"? Let us remember what we are looking for: we are looking, in our sample, for a relationship which can be written:

$$y = f(\vec{b}, x)$$

where y represents the value of the dependent variable, x the value of the causal variable (unique in this chapter) and $\vec{b}$ a set of numeric constants, called the "coefficients", conveniently represented by a vector.

In most statistical books, "linear" means that the function f is a linear function of the elements of $\vec{b}$. In other words a relationship such as:

$$y = b_0 + b_1 x + b_2 x^2$$

is called "linear" and there is some logic in this appellation, as we will discover in the following chapters.

But it is not our definition of linearity at this stage: we are looking for a relationship which is linear in x. It is just a question of definition, because it will also be linear in terms of the coefficients; such a relationship can be called "bilinear":

$$y = b_0 + b_1 x$$

and such bilinear relationships are the subject of this chapter.

In order to be able to develop a bilinear relationship, we have to study the distribution – generally called the "empirical distribution" φ – of the dependent variable. This chapter is the first necessary step towards this direction.

[1] All the computations made in this chapter were performed with EstimLab™.

Analysis of a distribution of two variables must always follow the following steps:

1. search of outliers,
2. visualization of the data,
3. quantification of the perceived relationship between the variables.

All these steps will be discussed in detail one after the other in this chapter.

5.1 Looking for Outliers

Let us recall the definition of an outlier: it is a data point that, by its sole presence, seriously changes the relationship we try to establish between y and x.

The following example (unfortunately real) clarifies the idea:

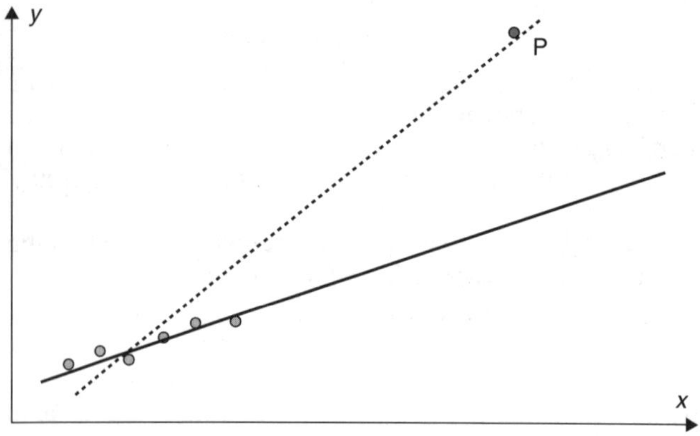

Figure 5.1 What is an outlier?

Suppose we are looking (because we have a preconceived idea about the relationship between x and y, as we would think differently if we were looking for an exponential or a quadratic relationship[2]) for a linear relationship between x and y:

- When data point P is present, the relationship may take the form of the dotted line.
- When it is absent, the relationship takes the form of the full line.

It is clear that point P, by its sole presence, profoundly modifies the considered (bilinear) relationship between x and y: it is a potential outlier.

Note that we call it a "potential outlier", first of all because we have a preconceived idea about the relationship, second because, if we are extremely confident about its value, maybe some of the other points are "true outliers".

Always remember that this visual approach and even the computations only give a symptom which must be corroborated by your judgment.

This Figure 5.1 allows distinguishing two kinds of outliers, that we call "outliers by position" and "outliers by cost".

---

[2] Such an exponential or quadratic relationship would certainly better fit with the data as they are. For selecting such a relationship we have to be sure that:

1. P is a realistic data,
2. it does make sense, technically speaking.

An "outlier by position" is a data point of which the causal variable is "far away" from the bulk of the data. For instance on Figure 5.1, data P is an outlier by position because the value of its causal variable is away from the other data points. Outliers by position only can be an excellent thing for the cost analyst, because it can confirm a general trend given by the bulk of the data. However, the attention of the cost analyst should be drawn to these data: Is it normal that in our database we have a data so far away, as far its causal variable is concerned, from the other data? Is it not due to a typing mistake?

An "outlier by cost" only is a data point of which the dependent variable is "far away" from the bulk of the data, although its causal variable does not differ so much from the other data. Such a data point is probably a true outlier and it should be carefully checked by the cost analyst. Outliers by cost are most often more dangerous than outliers by position: they can completely damage the searched relationship between cost and the causal variable.

Data point P in Figure 5.1 is an outlier both by position and by cost. Such data points, if the analysis reveals that the values are correct, are interesting because they suggest that the linear relationship we are looking for is probably not the right one. In the case of Figure 5.1, one would certainly prefer the "correction by constant" formula that will be explained later on in this book (Chapter 12 in Part III). Figure 5.1 clearly shows the interest of the graph for deciding on what to do with the outliers.

5.1.1 A First Approach: Looking at the Graph

The first way to search for outliers is consequently to look at the graph displaying the values of y as a function of x. It is simple but is more useful to detect errors (if most of the data are not too much scattered) than to find "true" outliers.

However, it may sometimes be difficult to use, as the following example illustrates.

Let us assume that our data have the following values (Example C) which are defined as matrices in Table 5.1.

The graph of these data is reproduced in Figure 5.2. The data points are rather scattered, but it so happens from time to time. Is it possible to decide, on this simple graph, what are the best candidates for being outliers? Probably not, although points E and F may appear as good candidates.

Clearly the eye, as powerful as it is, cannot sometimes decide between several potential outliers. Some computational aids are required; this is the purpose of the following paragraphs.

Table 5.1 The set of data used in Example C.

$$\|y\| = \left\| \begin{matrix} 799 \\ 865 \\ 1334 \\ 1270 \\ 1150 \\ 600 \\ 1420 \end{matrix} \right\|, \qquad \|x\| = \left\| \begin{matrix} 320 \\ 400 \\ 650 \\ 800 \\ 900 \\ 400 \\ 700 \end{matrix} \right\|$$

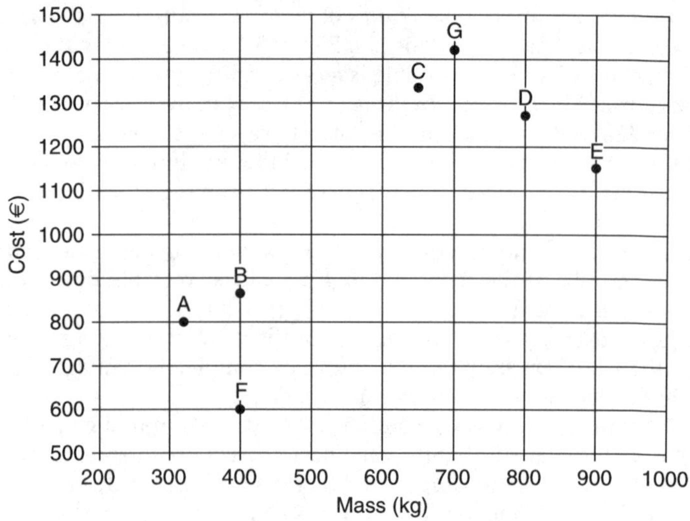

Figure 5.2 The data points (Example C).

Three different directions can be taken:

- The first one looks only at the causal variable, which means it attempts to detect "outliers by position". This idea is that if one value of the causal variable is "far away" from the other ones, we may have some doubt about the belongingness of the related product to the product family we put it in. Are we sure that the same technology was used for manufacturing this particular product? Quite often when the size (when one causal variable only is used, it generally quantifies the size of the product) of a product does increase too much, a different technology may have to be used.
- The second one looks at the dependent variable, which means it attempts to detect "outliers by cost". Looking at the values of this variable, we may ask the question: Is not one value, compared to the other ones and taking into account the difference in size, too large or too small? The difference may be due to a typing error, or to a change in the technology, or anything else, but it has to be noticed.
- The third one looks at both variables at the same time; it is based on the precision with which the coefficients b_0 and b_1 will be computed, precision which depends at the same time on the dependent and the causal variables; we can expect that, in the presence of outliers, this precision will be degraded.

The three directions are successively investigated.

5.1.2 Looking at the Causal Variable: Introduction to the "HAT" Matrix

The "HAT" matrix delivers a lot of useful information *about the causal variable and only about it*. This matrix is defined as

$$\|h\| = \|^{+}x\| \otimes \left(\|^{+}x\|^{t} \otimes \|^{+}x\|\right)^{-1} \otimes \|^{+}x\|^{t}$$

Table 5.2 (Example C).

$$\|^+x\| = \begin{Vmatrix} 1 & 320 \\ 1 & 400 \\ 1 & 650 \\ 1 & 800 \\ 1 & 900 \\ 1 & 400 \\ 1 & 700 \end{Vmatrix}$$

0.396	0.322	0.093	−0.044	−0.136	0.322	0.047
0.322	0.27	0.108	9.927 × 10⁻³	−0.055	0.27	0.075
0.093	0.108	0.153	0.18	0.198	0.108	0.162
−0.044	9.927 × 10⁻³	0.18	0.282	0.35	9.927 × 10⁻³	0.214
−0.136	−0.055	0.198	0.35	0.451	−0.055	0.248
0.322	0.27	0.108	9.927 × 10⁻³	−0.055	0.27	0.075
0.047	0.075	0.162	0.214	0.248	0.075	0.179

Figure 5.3 The "HAT" matrix for Example C (the values between the two dotted lines are the diagonal elements).

where $\|^+x\|$ represents the causal values matrix $\|x\|$ to which a column of 1 was added (unless we force the intercept to be 0, which is rather rare in the domain of cost), and $\|^+x\|^t$ its transpose (see the section called "What you need to know about matrices algebra?" located at the beginning of this volume). This "HAT" matrix is a square $\Re^{I \times I}$ matrix (I represents the number of data points) entirely defined by the causal variables.

For the example presently studied, the $\|^+x\|$ matrix is defined as Table 5.2 and its "HAT" matrix is computed as (note it is a square matrix with seven lines and seven columns) displayed on Figure 5.3.

For clarity purpose, the diagonal elements, which are the values we are interested in, of this matrix are placed between two straight lines.

How can we interpret these numbers?

The computation of one element of this matrix is relatively easy in the present case when there is only one causal variable:

$$h_{i,j} = \frac{1}{I} + \frac{(x_i - \overline{x})(x_j - \overline{x})}{\sum_{k=1}^{I}(x_k - \overline{x})^2}$$

and therefore the diagonal elements are given by:

$$h_{i,i} = \frac{1}{I} + \frac{(x_i - \overline{x})^2}{\sum_{k=1}^{I}(x_k - \overline{x})^2}$$

From this expression, it is clear that, when there is only one causal variable $\Sigma_i h_{i,i} = 2$, the sum of diagonal elements of this matrix is equal to 2.

At this stage these diagonal elements $h_{i,i}$ are particularly interesting. The denominator of its expression is equal to the variance of x multiplied by I, if we represent this variance by the symbol s_x^2 one can write:

$$Ih_{i,i} - 1 = \frac{(x_i - \overline{x})^2}{s_x^2}$$

Therefore $Ih_{i,i} - 1$ can be interpreted as a "normalized – Euclidian – distance" between product P_i and the "average product", average product which can conveniently be represented by the symbol $\overline{P}$.

This point is interesting for our search for outliers: when looking at the causal variable, we are interested in data points which are far away from the "average product" as Figure 5.1 illustrated.

How far can we consider that a product is an outlier? If the distribution of the x is normal (this is a strong assumption!), Belsley, Kuh and Welsh remind [14] that the expression:

$$\frac{I-2}{I}\left(\frac{Ih_{i,i} - 1}{1 - h_{i,i}}\right)$$

follows a $F_{1,I-2}$ distribution, with 1 and $I - 2$ degrees of freedom. This distribution is briefly presented in Chapter 3 of this volume. From that distribution a threshold can be computed.

Illustration

Returning to Example C, one should note first that the average value of the $h_{i,i}$ is given by $2/I$ or 0.286. We expect the values for all the products to be in the vicinity of this value: Figure 5.3 shows they are not far from it.

We can go one step further by computing, for each data point, the expressions defined upwards and checking their distribution. These expressions are displayed in Table 5.3 for each product, with $I = 7$.

Table 5.3 Values of $h_{i,i}$ and of the expression.

$h_{i,i}$	Expression
0.396	2.096
0.270	0.871
0.153	0.060
0.282	0.969
0.451	2.806
0.270	0.871
0.179	0.220

The value of $F_{1,5}$ which has a probability of 0.1 to be over passed is equal to 4.06. Consequently for this example no product could be considered "too far away" from the average product.

5.1.3 Looking at the Dependent Variable

We turn now to the search of data points for which the dependent variable takes a value which seems to be abnormally large or small. This search must take into account the fact that, as presented at the beginning of this section, the value of dependent variable is linearly related to the causal variable: its abnormality must then be checked taking into account the change in the size.

This is done, for each data point, by computing the expected change of the dependent variable with the size. The concept of the "dynamic center"[3] does that.

The search for potential outliers now proceeds in three steps: the results for Example C (data points are labelled in the order in which they appear in Table 5.1) are presented in Figure 5.4:

1. The "dynamic center" is computed with all data points being present. The residuals for the data points are computed (second column, called "Residual 1") as well as their standard deviation (not printed).
2. Then the "dynamic center" is recomputed without one of the data point. The new residual ("Residual 2") for the *same* data point is recomputed. This is done I times, successively for each data point.
3. For each data point, the difference between both residuals gives an information on how much the dynamic center is changed when this data point is present or not. For a normalization purpose, this information is divided by the standard deviation mentioned in step 1 (above), multiplied by a factor which can be chosen by the analyst (the default value of this factor is 1, which means that a change in the residuals of ± 1 standard deviation is not considered as abnormal). It is then multiplied by 100 to be read as a percentage. As an outlier is defined as a data which, by its sole presence, profoundly changes the relationship we are looking at, we expect that this percentage will be small for most of the data points, except for the outliers. A threshold can be set for detecting potential outliers; in the following

Name	Residual 1	Residual 2	Relative variation	R^2
A	32.389	53.588	11.746	0.542
B	12.514	17.148	2.567	0.569
C	213.157	251.558	21.278	0.643
D	−11.858	−16.506	−2.576	0.566
E	−239.201	−435.464	−108.745	0.781
F	−252.486	−345.970	−51.798	0.567
G	245.485	299.013	29.659	0.637

Figure 5.4 Looking at the changes in the residuals (Example C).

[3] The concept of the "dynamic center" is developed Part III. For the time being, assume it is the result of a linear regression.

Figure 5.4 the threshold has been set at 100% for the example: the potential outliers are indicated by values displayed in grey.

For Example C, only data point E is, according to this computation, a potential outlier.

4. Simultaneously, the value of the coefficient (this coefficient is defined in Chapter 16 (Part V)) of determination (R^2) is computed
   - first of all when all points are present: its value for Example C equals 0.603,
   - then when each data point is successively deleted in the computation.

For Example C, one immediately remarks that this R^2 dramatically changes when data point E is removed, as its value goes up to 0.78, whereas its average value when the other data points are deleted is around 0.6. This means that removing data point E improves considerably the quality of the formula we can compute: this data point is certainly an outlier.

This example shows that these rather simple computations are a very efficient tool for detecting the "true" outliers.

What happens when data point E is deleted in the computations?

Name	Residual 1	Residual 2	Relative variation	R^2
A	91.646	53.588	−26.490	0.542
B	36.527	17.148	−13.489	0.569
C	127.032	251.558	86.675	0.643
D	−164.065	−16.506	102.706	0.566
F	−228.473	−345.970	−81.783	0.567
G	137.333	299.013	112.535	0.637

Figure 5.5 Search for outliers when data point E is deleted.

Now data points D and A appear as potential outliers. This is quite understandable when looking at Figure 5.2. However note that there is no dramatic change in the R^2 any more.

Simultaneously, the same analysis can be done on the logarithms of the values, in order to see if results can be improved, as this will suggest to use a "multiplicative formula" for the "dynamic center". The results are given in Figure 5.6.

The residuals, based now on the logarithms of the values, are obviously much smaller. The interesting point is that now no outlier was detected by the procedure; you may notice at the same time, that the R^2 were improved. Both results are a strong incentive, when a specific model will have to be built, to look for a "multiplicative formula" instead of an "additive" one.

Name	Residual 1	Residual 2	Relative variation	R^2
A	0.113	0.21	55.733	0.65
B	0.049	0.066	9.708	0.637
C	0.17	0.204	19.529	0.661
D	−0.013	−0.017	−2.77	0.616
E	−0.188	−0.303	−65.898	0.736
F	−0.317	−0.427	−62.817	0.722
G	0.185	0.23	25.857	0.646

Figure 5.6 Search for outliers on log values.

5.1.4 Looking at the Variance of the Coefficients

We turn now to an analysis which takes into account both the dependent and the causal variables.

The purpose of this section is to compute a formula such as:

$$y = b_0 + b_1 x$$

in which b_0 and b_1 will be computed from the data contained in the sample we have.

An important question will be related to *the precision* with which these coefficients will be computed: if the data are scattered, it is obvious that this precision will be low. Similarly, one can expect that the presence of outlier(s) may degrade this precision. One way to express this precision is the variance of these coefficients; intuitively the variance expresses the "fuzziness" of a coefficient.

Anticipating on the results given in Chapter 15, the variances are given, always in the linear relationship hypothesis, plus other hypotheses, by:

$$\mathrm{var}(b_0) = \frac{S^2}{I} \times \frac{\sum x_i^2}{\sum (x_i - \bar{x})^2}$$

$$\mathrm{var}(b_1) = \frac{S^2}{\sum (x_i - \bar{x})^2}$$

where I is the number of products, S the standard deviation of the deviations inside the population (the meaning of these terms will be developed in Chapter 15) and $\bar{x}$ the arithmetic mean of the causal variable.

It is clear, at least for $\mathrm{var}(b_1)$, that their values depend:

- *On the causal variable*, via the standard deviation of its values: the larger this standard deviation, which is closely related to $\Sigma(x_i - \bar{x})^2$, the smaller will be the variances (this is logic: if you measure a slope – and b_1 quantifies a slope – between points distant by 10 cm and points distant by 1 m, the precision of your measurement will be better in the second case than in the first one, if the points are known with the same precision).
- *On the dependent variable* via the term S. Presently this term is unknown, but we will see in Chapter 15 that an estimated value can be made from the standard deviation of the residuals in the sample; both values are proportional. Consequently the more the values of the dependent variable are scattered around the dynamic center, the greater will be the variances.

These variances could therefore be used as an global indicator of the presence or the absence of outliers: one should study what happens to them when each data point is successively deleted from the computations. This could take some time.

A shortcut is possible: it uses the variance–covariance matrix (this matrix will be discussed in Chapter 6 of this part) of which terms are these variances, plus their covariances (the term is explained thereafter). The idea for comparing these (square) matrices when a data point is deleted is to look on the change of their determinants; this is not really a perfect comparison, but it is sufficient for our purpose. An easy way to quantify this change is to make the ratio of these determinants.

Consequently the ratio of the determinants of the variance–covariance matrices with and without a data point is an easy way to search for outliers.

Name	Change in the determinant of the variance–covariance matrix when each data point is successively removed
A	2.480
B	2.133
C	0.904
D	2.168
E	0.198
F	0.456
G	0.655

Figure 5.7 Change of the determinants of the variance–covariance matrix.

For Example C, these ratios are given by Figure 5.7.

We will see in Chapter 6 of this part that it is possible to compute, in the framework of some hypotheses of course, an interval outside which a ratio can be considered as abnormal. For this example, the interval is equal to [0.391–3.646]. According to this computation, only product E is considered as a possible outlier. Looking at Figure 5.2 reveals that it is a good candidate indeed.

5.1.5 A Synthesis

It is interesting at this stage to compare the results proposed by the three different ways for detecting the outliers. This is easily displayed on Figure 5.8.

Name	Search for outliers based on		
	The dependent variable	The causal variable	Both variables
A			
B			
C			
D			
E	X		X
F			
G			

Figure 5.8 Synthesis of the outliers detection.

For Example C, the same outlier was detected by two procedures, but not by the second one, using the "HAT" matrix. Why did this last procedure not detect this data point E: the reason for, that is, as we previously said, that the "HAT" matrix procedure is only concerned by the values of the causal variable. A look at Figure 5.2 clearly shows that the mass (the causal variable) for data point E is not really "far away" from the bulk of the other masses. Consequently this data point could not be seen as an outlier by this procedure.

This data point E could only be detected by a procedure which uses either the dependent variable only, or both variables. The result confirms that these procedures correctly found it.

5.1.6 Conclusion

When looking for outliers, pay attention to two points:

1. The procedure dealing with the "HAT" matrix only looks at the causal variable: the dependent variable is not considered at all.
2. The other algorithms look for outliers from a linear relationship point of view between the causal and the dependent variable; *linearity is a very important hypothesis*, not always met in cost behavior. A data, considered by the algorithms as an outlier, may very well reveal that the linearity hypothesis is not the right one: this was suggested in the comments for Figure 5.1. The consequence is: never consider that a data point detected as a possible outlier by the last two algorithms without checking if the hypothesis of linearity is correct.

The three approaches, the use of the "HAT" matrix, the change in the residuals when one data point is removed, and the change in the variances–covariances matrix, are complementary. The recommended approach is therefore the following one:

- Start with studying the "HAT" matrix. This study will reveal the data points of which the causal variables are far away from the bulk of the data. If there are some, check with the other algorithms if the related costs are really disturbing the linear relationship (if this linear relationship is the relevant one). The data which satisfy all criteria are very likely outliers; otherwise the data points which are far away from the bulk of the data are rather interesting and should be kept.
- Check then for all the other data points in order to see if their costs do not cause any trouble, in which case the data points which cause trouble are very likely potential outliers.

It is always a good practice to delete the potential outliers from the procedures and to redo the computation. New phenomena can be disclosed and the analyst must check if these phenomena cause serious problems or not. A modern software (such as EstimLab™) does these computations so quickly that there is no reason not to perform them.

What to Do About Outliers?

The algorithms mentioned here only signal "potential" outliers. It is up to the cost analyst to decide what to do with them. He/she may:

- Return to the source of information in order to get a confirmation of the value, or to get a corrected figure.
- Unselect the candidates for future computations.
- Keep them and maintain the linear hypothesis. This can be the case for Figure 5.1 for instance if the small values are not very reliable[4] whereas the high value is considered as really representative, and if future cost estimates will be in the vicinity of this data point P; this obviously assumes that you are very confident about its value.
- Keep them, disregard the linear hypothesis and chose another one. An easy way to do it is to make the same search on the log values instead on the values

[4] It is well known by most cost analysts that the costs for small products are generally more scattered than costs for large products. This is due to the cost measurement process.

themselves (but not on the HAT matrix for which it has no purpose). This can of course be automated; it has been in EstimLab™.

5.2 Visualization of the Data

With only two variables, the visualization of the data is very easy as Figure 5.1 demonstrates.

Visualization has here four aims:

1. Confirming or invalidating the outliers. Outliers have been found in Section 5.1.2 with an algorithm based on the distance; but distant points are not necessarily outliers.
2. Testing visually the linearity of the relationship between y and x, and investigating if a non-linear relationship would be more satisfactory (a quick check can be made, e.g. on the graph if the user can quickly choose between linear or logarithmic scales for the coordinates, but do not forget that true linear and linear based on log are not the only relationships to be tested).
3. Checking the homogeneity of the product family. If, for instance, we find that the distribution has two modes (it is then called "bimodal"), then the family should probably be broken into two "sub-families".
4. Getting an idea on the scattering of the data and then on the accuracy that can be expected from a model based on these data points.

Other graphs, which are presented in Chapter 6 in the case of several causal variables, can also be used for one causal variable, as illustrated on the Figure 5.9 with

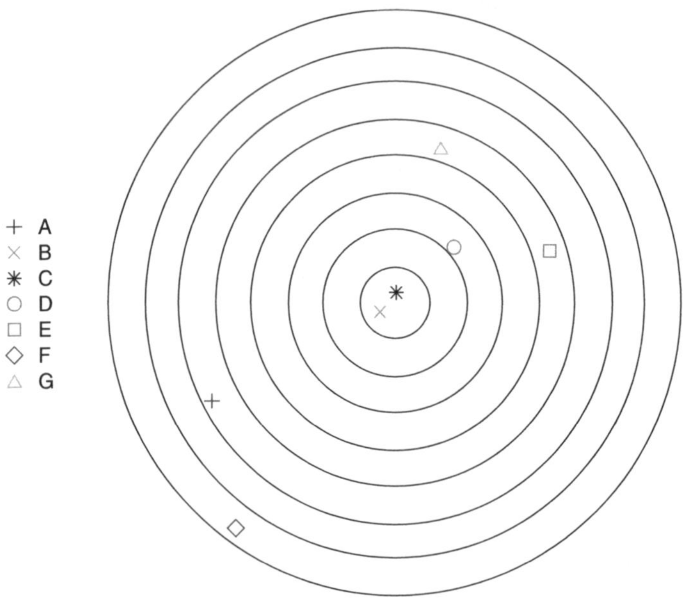

Figure 5.9 The star diagram.

the "star diagram". However, the simple graph, Figure 5.1 conveys all the information; nevertheless using other diagrams with just one causal variable is a good way to get used with them.

5.3 Quantification of the Perceived Relationship

Once the data have been looked at, it is interesting to quantify the visual impressions. The purpose is then to find a single characteristic value that will give a global idea on how well y and x vary together, how well the variations are correlated.

Generally speaking the inventors of such characteristics always try to get a number in the interval $[-1, +1]$, $+1$ meaning that y and x vary in complete relationship together, -1 they vary exactly in opposite direction, 0 that they are completely independent, any other value suggesting some correlation.

5.3.1 The Covariance and the Bravais–Pearson Correlation Coefficient

The covariance is a first step in this direction. It is defined by:

$$s_{xy} = \mathrm{cov}(x, y) = \frac{1}{I}\sum_i (x_i - \bar{x}) \times (y_i - \bar{y}) = \frac{1}{I}\sum_i {}_cx_i \cdot {}_cy_i$$

where ${}_cx_i$ and ${}_cy_i$ represent the "centered" coordinates (meaning the distance from the arithmetic mean) as illustrated (on another example) in Figure 5.10.

The logic of the covariance is very easy to understand: it is based on the sum of the products ${}_cx_i \cdot {}_cy_i$. Each time ${}_cx_i$ and ${}_cy_i$ are in quadrants 1 or 3 (see Figure 5.11), this product is necessarily positive and contributes to increase s_{xy}; it is the contrary for data points which are in quadrants 2 or 4. The covariance is all the more important as data points are all located in opposite quadrants. Let us consider now the average:

$$s_{xy} = \frac{1}{I}\sum (x_i - \bar{x})(y_i - \bar{y})$$

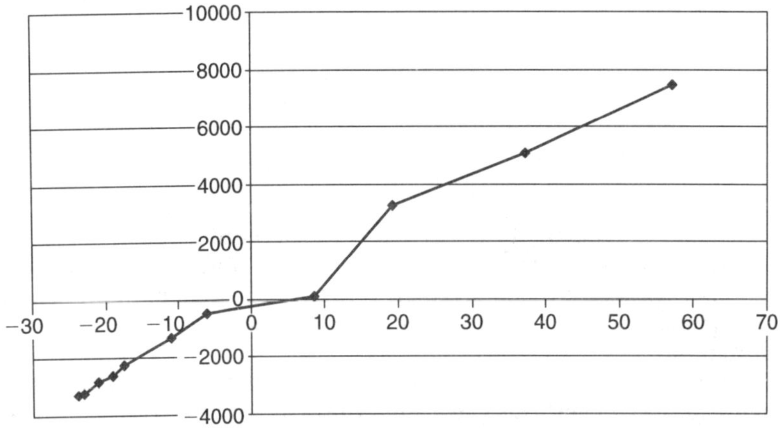

Figure 5.10 Simultaneous evolution of the centered coordinates ${}_cx_i$ and ${}_cy_i$.

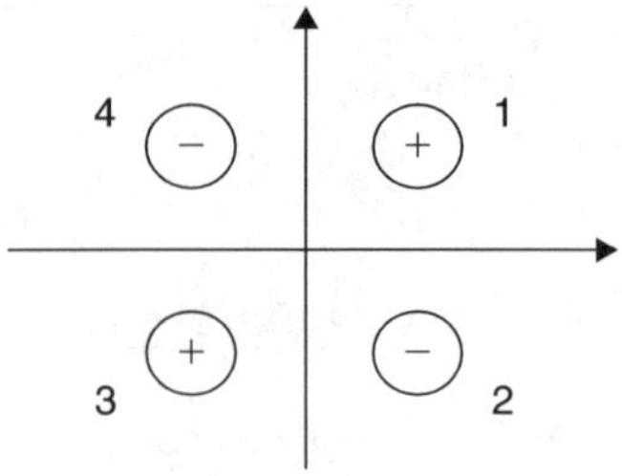

Figure 5.11 The four quadrants of the y–x plane.

of all these products. It is clear that if this average is positive, the majority of the data points are in quadrants 1 and 3, whereas, if it is negative, the majority is in quadrants 2 and 4. If it is equal to 0, this means that there are about as many data points in all quadrants: data appear as very scattered.

What Can be Said About the Size of the Covariance?
For data sets using the same units (e.g. the size being given by the mass in kilogram and the cost in euros), the value of the covariance will be higher if the sign of all the products is the same and if the value of the data is larger (in kilograms and in euros). In such a case, the larger the covariance, the more the variables change in the same direction: the word "covariance" is well chosen indeed.

In our example, the covariance amounts to 99 809.3€ × kg (a strange unit!).

The Bravais–Pearson Correlation Coefficient

The covariance is difficult to handle because its value depends on the units in which the values are given: consequently two covariances computed on different populations or samples cannot be compared.

The correlation coefficient was built by Bravais and Pearson to avoid this problem. It is defined by:

$$r_{BP} = \frac{s_{yx}}{s_y s_x} = \operatorname{cov}({}_{cs}x_i, {}_{cs}y_i) = \frac{\frac{1}{I}\sum (x_i - \bar{x})(y_i - \bar{y})}{\sqrt{\frac{1}{I}\sum (x_i - \bar{x})^2}\sqrt{\frac{1}{I}\sum (y_i - \bar{y})^2}}$$

where ${}_{cs}x_i$ and ${}_{cs}y_i$ represent the centered and scaled coordinates.

The covariance appears in the numerator, the denominator being used for normalization of the result: s_x and s_y are the standard deviation of the variables. For our example $s_x = 27.33$, $s_y = 3677.3$ and consequently $r = 0.99308$. There is of course no unit.

Note that the correlation coefficient is nothing else that the covariance of the centered and scaled data. This correlation coefficient varies between -1 (perfect anti-correlation) and $+1$ (perfect correlation), the value 0 corresponding to no correlation at all.

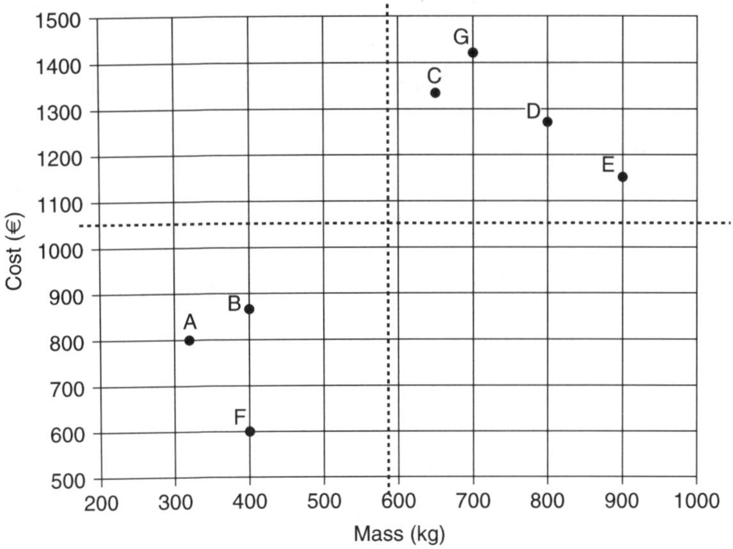

Figure 5.12 Computing the Bravais–Pearson correlation coefficient for Example C.

A Coefficient to be Used with Care

The Bravais–Pearson correlation coefficient is an interesting characteristic but should be used with care:

- It uses the same metric as the mean and suffers from the same inconveniences: it is very sensitive to data points which are far away from the means $\bar{x}$ and $\bar{y}$, the data close to these means having little or even very little influence. We will say that this characteristic is not "robust", in the sense that it is too sensitive to outliers.
- Consequently it may give a false feeling of comfort: for instance for Example C given in Section 5.1.1 of this chapter, a value of 0.777 is found: this "high" value, rather unexpected on a first look on the graph, is explained by the fact that the data are all in quadrants 1 and 3 and are away from the means (represented in Figure 5.12 by dotted lines).
- It is also well known that it measures not the correlation between the variables, but the *linear* correlation between them: variables may be highly correlated (for instance $y = 10 + x^2$ is a perfect correlation) and produce a low Bravais–Pearson correlation. The reason for that is explained in Chapter 4.

The message is always the same: always look at the graph in order to interpret this global characteristic.

5.3.2 More General Correlation Coefficients

The Bravais–Pearson correlation coefficient quantifies a linear correlation between our variables. This may be a drawback when costs are concerned, because non-linearities frequently appear in the cost domain. Other correlations, able to quantify non-linear relationships, have therefore been developed:

- The correlation coefficient of Spearman.
- The correlation coefficient of Kendall.
- The correlation around the medians.
- The coefficient of monotony.

The Spearman Correlation Coefficient

This coefficient is, as the previous one, computable for a set of data defined by two quantitative variables. The data are sorted, for example, according to the x values (the causal variable). The objective is then to examine if the y values grow simultaneously with the x values, *disregarding the rate of growth*; one can also say that we are looking if the ranks of y follows the same path that the ranks of x. This means that we are not interested anymore with the linearity concept. This allows to introduce a more general correlation coefficient: it is quite possible to have a low Bravais–Pearson coefficient associated with a high Spearman correlation coefficient.

In order to quantify it, one looks at the couples of variables (x_i, y_i) sorted in two different ways:

1. First with the increasing x_i: the rank of each couple is noted $\rho_{1,i}$, starting with the rank $i = 1$.
2. Then with the increasing y_i: the rank of each couple is noted $\rho_{2,i}$, starting with the rank $i = 1$.

Then the difference of the ranks $d_i = |\rho_{1,i} - \rho_{2,i}|$ is computed for the same data, according to both sorts. The Spearman correlation coefficient is then given by:

$$r_S = 1 - \frac{6}{I^3 - I} \sum d_i^2$$

This coefficient has been built by Spearman in order to vary between -1 and $+1$. It is equal to 0 if no correlation between the ranks does exist.

It is clear that this correlation coefficient is more general than the previous one. A strong difference between both should attract the attention of the cost analyst: it certainly means, if both x and y increase together, that their relationship is not linear.

The Kendall Coefficient of Concordance

This coefficient is only concerned by the "concordances" and "discordances" between two sets of data: it is not really a coefficient of correlation and therefore less interesting in the domain of cost than the previous ones. It is nevertheless mentioned here in order to show to the reader that it is possible to look at the data from different points of view.

Data are sorted according to x. Then the y values are looked at:

- One start from y_1 and the values of two new variables called n_{1c} and n_{1d} (number of concordances and of discordances) are initialized to 0. One examines then y_2. If $y_2 > y_1$ (which is logic if there is a correlation between x and y) one says there is a concordance between both variables and one notes $n_{1c} := n_{1c} + 1$ (the symbol ":=" is used to mean "takes the value of"); otherwise one says there is a discordance and one notes $n_{1d} := n_{1d} + 1$. One goes on with $y_3, y_4, \ldots$ to find out the total number of concordances and discordances: n_{1c} and n_{1d}.

- One starts then from y_2 and one computes n_{2c} and n_{2d}, etc.

At the end of the process, one computes two synthetic variables:

$$n_c = \sum n_{ic} \quad \text{and} \quad n_d = \sum n_{id}$$

The Kendall coefficient of concordance is then given by:

$$r_K = \frac{n_c - n_d}{\frac{1}{2} I(I-1)}$$

This coefficient varies between -1 and $+1$, the value 0 meaning there are as many concordances as there are discordances: there is no link between x and y.

The Correlation Around the Medians

The median plays a more and more important role in the analysis of data for cost-estimating purposes and the reader is advised to look at this type of correlation.

First of all the medians of both x and y are computed: M_x et M_y. These values are plotted on Figure 5.13.

The number of data in each quadrant is computed (data which are exactly equal to the medians are ignored): a, b, c and d. Then one computes:

$$r_m = \frac{ad - bc}{\sqrt{(a+b)(c+d)(a+c)(b+d)}}$$

which also varies between -1 and $+1$.

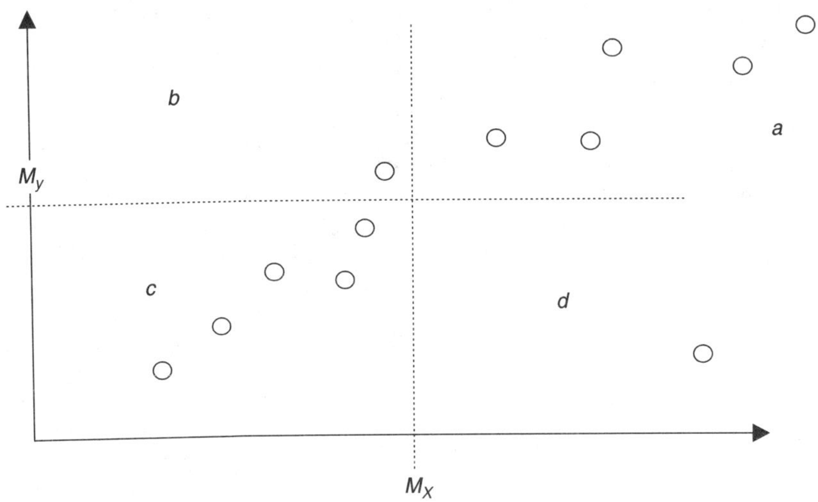

Figure 5.13 Displaying the data around the medians.

This coefficient does not provide a "rich" information. Its main advantage is that it is completely disconnected from any type of relationship between the x and the y.

The Coefficient of Monotony

The four coefficients described up to now are built for detecting if the variables grow in the same direction or not. Low values do not necessarily indicate a lack of correlation: a strong quadratic correlation will produce a Bravais–Pearson and, sometimes, a Spearman correlation coefficients rather low. Once again having a look at the graph is highly recommended.

In order to at least partly solve the question of non-linear correlations, we built a coefficient which only indicates if one variable changes in a monotonic way (not necessarily linear or even in the same direction) with another variable. Such a coefficient is adequate for all the major relationships in the domain of cost: log, exponential, correction by a constant or even parabolic (if it does make sense).

In order to build this coefficient, data (x_i, y_i) are sorted according to the x_i. Then for $i > 2$ one looks at the variation of the y_i:

- if $y_i > y_{i-1}$, one notes $+1$,
- otherwise one notes -1.

A table of notes is then created. In this table, the number of $+1$ (noted n_{+1}) and the number of -1 (n_{-1}) is computed. A coefficient m_f can then be created with the following constraints:

1. $m_f = -1$ if all the notes are equal to -1,
2. $m_f = +1$ if all the notes are equal to $+1$,
3. $m_f = 0$ if there is a continuous change from -1 to $+1$.

Between these key values, the number of changes is counted: let α be this number. Such an analysis assumes, of course, there are at least three data points. The maximum number of changes is equal to $\alpha_{\max} = I - 1$. The coefficient of monotony is then given by:

$$m_f = sign \times \left(1 - \frac{\alpha}{\alpha_{\max}}\right)$$

the sign being given by the majority of notes.

The major advantage of this coefficient is that it gives a high value to any type of correlation, linear or not.

6 Simultaneous Data Analysis on $J + 1$ Quantitative Variables

Summary

The purpose of this chapter is to analyze the data when the cost and several quantitative parameters are known. This is the most complete situation.

The four steps of analysis can now be completely investigated:

1. Search for outliers. In the case of several quantitative variables we must rely on algorithms, as looking to a table of figures does not generally reveal anything of that sort.
2. Search for possible collinearities between the causal variables (collinearity between the causal variables and the cost is what we are looking for and therefore is not dealt with here). This search is very important: when dealing with several quantitative variables the most serious problems are associated with such collinearities.
3. Visualization of the data, in order to understand their structure, which means answering some questions such as:
   - Is the set of data homogeneous enough, or should we split it into sub-families? If, for instance, we find that the distribution has two modes (it is then said "bimodal"), then the product family should probably be broken into two "sub-families".
   - Are there any outlier that step 1 could not detect? Some outliers are for instance detected in step 1 with an algorithm based on the distance; but distant points are not necessarily outliers.
   - How are the data related (if at least two variables are simultaneously checked)?
   - Does the visualization suggest a kind of relationship?
4. Quantification of the perceived relationships between the variables. If we perceive relationships between the variables, it is convenient to quantify these relationships in order to decide on the future use of these data.

Each step will be studied one after the other.

A New Example

In this section we will apply the techniques which will be introduced on the following example.

Cost	V_1	V_2	V_3	V_4
1278	6.83	1264	1274	10
724	2.18	1032	480	6
809	3.8	812	656	6
920	4.55	516	786	8
772	2.18	1032	480	6
877	2.11	1548	394	6
1064	4.67	2722	942	6
865	2.81	807	671	3
961	2.55	1598	872	6
856	1.68	737	450	5
1293	6.3	715	1400	19
717	1.98	186	430	7
648	1.25	228	257	6

V_1: represents the mass
V_2: the number of components
V_3: the number of connections
V_4: the number of boards

One product is described, inside the product family, by a row giving the value of its four variables, plus its cost

Figure 6.1 Example D.

This example, related to an electronic equipment, presents a cost column (which is a vector) and four quantitative variables[1] (grouped in a matrix, of which name is $||x||$). It includes 13 products. As illustrated on the Figure 6.1, one line describes one product.

From these values two matrices are established: the first one – which is also called a "vector" – is related to the cost (or the dependent variable), the second one to the causal variables.

For reasons which are explained in Chapter 15, the matrix of the causal variables must generally (unless we want to force the intercept to be 0) be written with the addition of a column of 1. This new matrix is noted $||^+x||$, the $+$ sign reminding the addition of one column (Figure 6.2).

Analysis dealing with one (the cost for instance) or two variables (the cost and the size for instance) can be made by using the graphs. It is of course impossible when dealing with 4 or 10 variables for instance. In such a case, the cost analyst must rely nearly exclusively on algorithms.[2]

We will however see that the eye can still be used, the question being to find a presentation which can easily be interpreted by our brain.

6.1 Looking for Outliers

Looking for outliers in the linear case, for quantitative variables, has been largely studied by different authors. This section gives the major results which can interest the cost analyst.

[1] Mass in kilogram, number of components, number of connections and number of boards.
[2] All the computations and figures were generated by EstimLab™.

$$\|{}^{+}x\| = \begin{pmatrix} 1 & 6.83 & 1264 & 1274 & 10 \\ 1 & 2.18 & 1032 & 480 & 6 \\ 1 & 3.8 & 812 & 656 & 6 \\ 1 & 4.55 & 516 & 786 & 8 \\ 1 & 2.18 & 1032 & 480 & 6 \\ 1 & 2.11 & 1548 & 394 & 6 \\ 1 & 4.67 & 2722 & 942 & 6 \\ 1 & 2.81 & 807 & 671 & 3 \\ 1 & 2.55 & 1598 & 872 & 6 \\ 1 & 1.68 & 737 & 450 & 5 \\ 1 & 6.3 & 715 & 1400 & 19 \\ 1 & 1.98 & 186 & 430 & 7 \\ 1 & 1.25 & 228 & 257 & 6 \end{pmatrix}$$

Figure 6.2 The matrix generally used in the computations.

We limit the search here to individual outliers. One could also search for couples of data which, as a group, are outliers. The logic is the same.

As we saw it in the previous chapter, which was an introduction to this one, there are several ways to look for potential outliers:

- The first procedure looks only to the causal variables, the purpose being to see if there are one or several products of which definition is far away from the bulk of the other products definition.
- The second one is not concerned by the causal variables: it looks only at the dependent value of each product in order to see if one such value does not change a lot the relationship between the causal variables and the dependent variable.
- The third one uses the values of both sets of variables.

6.1.1 Looking at the Causal Variables

The HAT matrix was defined in the previous chapter; its definition is recalled here:

$$\|h\| = \|{}^{+}x\| \otimes (\|{}^{+}x\|^{t} \otimes \|{}^{+}x\|)^{-1} \otimes \|{}^{+}x\|^{t}$$

As it was said in this previous chapter, the HAT matrix is very good at detecting outliers purely at looking at the causal variables: it completely disregards the cost values. The diagonal elements of this matrix, emphasized by a dotted line in Figure 6.3, are somehow related to the "distances" of each data points from the bulk of the data. The HAT matrix for the example is displayed on Figure 6.3: this is a square, symmetrical, matrix with 13 lines and columns (the number of lines of the $\|{}^{+}x\|$ matrix).

$$\begin{pmatrix}
0.496 & -0.051 & 0.214 & 0.283 & -0.051 & -0.103 & 0.186 & 0.165 & -0.047 & -0.069 & 0.148 & -0.046 & -0.125 \\
-0.051 & 0.128 & 0.054 & 0.023 & 0.128 & 0.185 & 0.091 & 0.014 & 0.056 & 0.107 & 0.016 & 0.104 & 0.145 \\
0.214 & 0.054 & 0.21 & 0.229 & 0.054 & 0.053 & 0.06 & 0.123 & -0.133 & 0.038 & -0.077 & 0.086 & 0.089 \\
0.283 & 0.023 & 0.229 & 0.291 & 0.023 & -0.02 & -0.032 & 0.122 & -0.182 & 0.017 & 0.043 & 0.113 & 0.091 \\
-0.051 & 0.128 & 0.054 & 0.023 & 0.128 & 0.185 & 0.091 & 0.014 & 0.056 & 0.107 & 0.016 & 0.104 & 0.145 \\
-0.103 & 0.185 & 0.053 & -0.02 & 0.185 & 0.413 & 0.302 & -0.184 & -0.081 & 0.055 & -7.357\times 10^{-3} & 0.049 & 0.152 \\
0.186 & 0.091 & 0.06 & -0.032 & 0.091 & 0.302 & 0.648 & -0.044 & 0.129 & -0.041 & -0.051 & -0.186 & -0.153 \\
0.165 & 0.014 & 0.123 & 0.122 & 0.014 & -0.184 & 0.044 & 0.437 & 0.289 & 0.157 & -0.194 & 0.083 & 0.018 \\
-0.047 & 0.056 & -0.133 & -0.182 & 0.056 & -0.081 & 0.129 & 0.289 & 0.673 & 0.185 & 0.124 & -1.822\times 10^{-3} & -0.066 \\
-0.069 & 0.107 & 0.038 & 0.017 & 0.107 & 0.055 & 0.041 & 0.157 & 0.185 & 0.166 & -0.02 & 0.145 & 0.154 \\
0.148 & 0.016 & -0.077 & 0.043 & 0.016 & -7.357\times 10^{-3} & 0.051 & -0.194 & 0.124 & -0.02 & 0.895 & 0.087 & 0.022 \\
-0.046 & 0.104 & 0.086 & 0.113 & 0.104 & 0.049 & -0.186 & 0.083 & -1.822\times 10^{-3} & 0.145 & 0.087 & 0.223 & 0.238 \\
-0.125 & 0.145 & 0.089 & 0.091 & 0.145 & 0.152 & -0.153 & 0.018 & -0.066 & 0.154 & 0.022 & 0.238 & 0.29
\end{pmatrix}$$

Figure 6.3 The HAT matrix of the example.

It is not difficult to show that this matrix is idempotent (the definition of this term is given in the introduction to this volume), because:

$$\|h\|^2 = \|^+x\| \otimes (\|^+x\|^t \otimes \|^+x\|)^{-1} \otimes \underbrace{\|^+x\|^t \otimes \|^+x\| \otimes (\|^+x\|^t \otimes \|^+x\|)^{-1}} \otimes \|^+x\|^t$$

By definition of the inverse of a matrix:

$$\|^+x\|^t \otimes \|^+x\| \otimes (\|^+x\|^t \otimes \|^+x\|)^{-1} = \|1\|$$

and therefore $\|h\|^2 = \|h\|$. Consequently the trace of the matrix (the sum of its diagonal element) is equal to its rank. Assuming this matrix is full rank, its trace is equal to the number of columns: $J + 1$ (J, if the intercept is forced to be 0).

Then we have $\Sigma h_{i,i} = J + 1$, or 5 in our example.

One would expect, if the causal variables of the data points are not too far away from the other ones, that each data point should received an equal or "fair" share of this total. This share should be equal to:

$$\frac{J+1}{I}$$

where J is the number of causal variables and I the number of products. In the example the "fair share" amounts to 0.3846 (or 5/13).

But we however can admit some deviations around this value, the problem being to decide how large could be the deviations, or, in other words, from which value an $h_{i,i}$ could signal a possible outlier. Belsley *et al.* [14], based on theoretical computations (which assume[3] that the causal variables are independent and distributed as multivariate normal variables), recommend to pay attention to all the values for which:

$$h_{i,i} > 2\frac{J+1}{I}$$

The value of this threshold is, for our example, 0.769.

[3] These assumptions are standard but rather strong in the domain of cost.

A	0.496
B	0.128
C	0.210
D	0.291
E	0.128
F	0.413
G	0.648
H	0.437
I	0.673
J	0.166
K	0.895
L	0.223
M	0.290

Figure 6.4 The diagonal elements of the HAT matrix.

All the objects which "consume" an unfair share of the total $\Sigma_i h_{i,i}$ are considered as potential outliers. In this example product K distinguishes itself (Figure 6.4).

Note: The reader will pay attention to the fact that this search based on the "HAT" matrix deals only with the causal variables, whereas, even if the causal variables of a product may help suspect an outlier, a true outlier is also concerned by the dependent variable.

6.1.2 Looking at the Dependent Variable

It is useful to introduce here a frequently used notation:[4] $||^+x_{(-i,\bullet)}||$ represents the matrix $||^+x||$ where row i (corresponding to object i) is deleted, the dot reminding the reader that the number of columns does not change. This matrix has the same number of columns than $||^+x||$, but $I - 1$ lines or products compared to this matrix.

Looking for residuals aims at computing what the residual of each product becomes when the dynamic center[5] of the cost distribution is computed from $||^+x||$ and $||^+x_{(-i,\bullet)}||$.

The result of the computation is given in Figure 6.5.

The first column gives the name of the products, the second one the residuals when all products are used for computing the dynamic center and the third one the residuals when each product is successively discarded. Figure 6.6 gives a visual presentation of the changes in the residuals.

[4]The reason for this notation is the following one: we deal with rectangular matrices; the first index refers to the row and the second one to the column. When an index is placed between parentheses, it means that the whole row is involved. The minus sign means that the whole row is deleted; a dot for the column index means nothing is changed about the columns.

[5]The term is defined in Chapter 8.

Residuals computed from:

$$||^{+}x|| \quad ||^{+}x_{(-i,\bullet)}||$$

Name	Residual 1	Residual 2	Relative variation	R^2
A	63.642	126.328	119.978	0.918
B	−72.522	−83.208	−20.452	0.934
C	−67.422	−85.375	−34.360	0.937
D	−17.514	−24.700	−13.753	0.928
E	−24.522	−28.136	−6.916	0.925
F	88.487	150.849	119.356	0.954
G	−35.759	−101.708	−126.224	0.930
H	19.099	33.930	28.386	0.928
I	−24.019	−73.446	−94.601	0.930
J	100.940	121.009	38.411	0.952
K	−3.114	−29.668	−50.823	0.890
L	−20.820	−26.787	−11.420	0.922
M	−6.476	−9.123	−5.067	0.914

Figure 6.5 The change in the residuals once each product is removed.

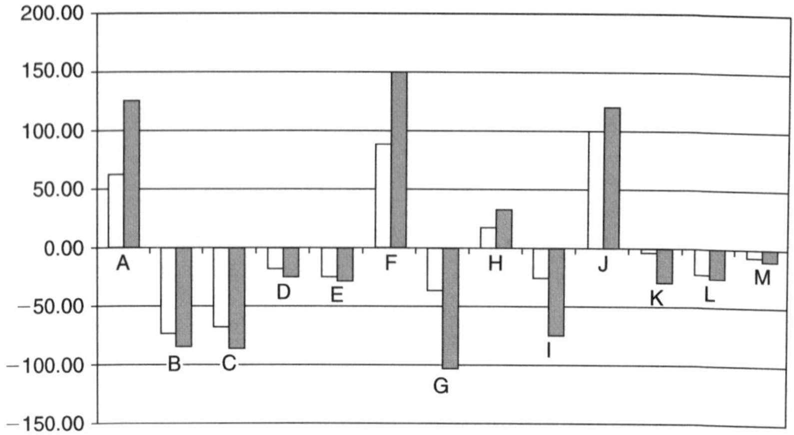

Figure 6.6 The change in the residuals related to each product when this product is removed (gray bar) or not (white bar) from the regression algorithm.

From this set of figures, two values can be computed:

1. The conventional approach just looks at the difference, given in percentage, of both residuals. But this mere difference, once divided by one value, cannot be a good indicator if this value is close to 0.
2. The proposed approach therefore looks at the differences of both residuals but considers that some differences are quite acceptable and consequently displays the differences divided by k times the standard deviation of the first

residuals: this standard deviation value is, in this example, 52.25. The logic is that the change in the residuals should not be more that one time the standard deviation, the value 1 being used for an experiment.

The results are given in Figure 6.5.

A simple look at this table would suggest, using this last approach, that data points A, F and G are potential, but not "serious" outliers. Why not serious? It is true that their deviations become large but we do not see a large change in the values of the R^2 when the considered data points are there (then $R^2 = 0.926$) or not.

The use of the conventional approach would suggest that data point K could be an outlier, but this comes from the fact that its first deviation is very small: dividing any value by a small quantity will always produce a great value.

Looking at Figure 6.1 which gives the data, it is difficult to see why these data points A, F and G are potential outliers: it is true that some values are a bit high or a bit low, but it is very difficult to get a definite conclusion. This is the advantage of using an algorithm.

It might be interesting to look at the results when data points A, F and G are discarded from the computations. The result appears on Figure 6.7 with the same value of the coefficient *k*.

Name	Residuals 1	Residuals 2	Relative variation	R^2
B	−29.400	−83.208	−143.555	0.934
C	−18.355	−85.375	−178.804	0.937
D	13.221	−24.700	−101.169	0.928
E	18.600	−28.136	−124.686	0.925
H	−1.334	33.930	94.083	0.928
I	−28.207	−73.446	−120.693	0.930
J	96.345	121.009	65.801	0.952
K	8.461	−29.668	−101.723	0.890
L	−44.086	−26.787	46.152	0.922
M	−15.244	−9.123	16.330	0.914

Figure 6.7 The change in the residuals when products A, F and G are discarded.

The result, looking at the variation of the residuals columns, does not appear fantastic. This is due to two facts:

1. The deviations in the first column are much smaller than in the previous computation – with the exception of product J (also note that discarding this object improves the R^2). As we start from lower values – and therefore a smaller standard deviation – we can expect larger values in column 4.
2. The number of data is seriously reduced: this means that the regression has, when one data point is removed from the computation, more possibilities to change.

The cost analyst, when analyzing the data should keep in mind all these comments, without forgetting that the results obtained here take only into account the dependent variable.

6.1.3 Looking at All Variables

We saw in the previous chapter that the variances–covariances matrix can help to get a general picture of the data as far as the outliers are concerned. This matrix is defined as:

$$\text{COV} = S^2 \times (||^+x||^t \otimes ||^+x||)^{-1}$$

where S^2 is the variance of the deviations inside the whole population. As this value is not known, it is replaced by an estimated value noted $\hat{S}^2$ as described in Part IV.

Its name comes from the fact that its elements give, on the diagonal, the variances of the coefficients $B_0, B_1, \ldots, B_j, \ldots$ which appear in the model, plus the covariances of these coefficients. As an outlier, according to its definition, may seriously change the model, we may expect that it will change both the formula itself (this was the subject of Section 6.1.2) and the variance – which means "how well they are defined" – of its coefficients. It therefore deserves our attention.

The idea is the same as the one which was developed in Section 6.1.2. We compute this matrix when all products participate to the construction of the formula, and look to what happens to this matrix when each product is successively removed from this computation. When all data points are present we have:

$$\text{COV} = \hat{S}^2 \times (||^+x||^t \otimes ||^+x||)^{-1}$$

and when data point i is discarded we have:

$$\text{COV}_{(-i,\bullet)} = \hat{S}(-i)^2 \times (||^+x_{(-i,\bullet)}||^t \otimes ||^+x_{(-i,\bullet)}||)^{-1}$$

How can be compared these two square matrices? The easiest way is to compare their determinants by computing their ratio:

$$\text{determinants ratio}_{(i,\bullet)} = \frac{\hat{S}(-i)^2 \times \left|(||^+x_{(-i,\bullet)}||^t \otimes ||^+x_{(-i,\bullet)}||)^{-1}\right|}{\hat{S}^2 \times \left|(||^+x||^t \otimes ||^+x||)^{-1}\right|}$$

There are of course as many such ratios as there are products.

If there is no change in the variances–covariances matrices, the ratio will be equal to 1 of course. Nevertheless a change generally occurs, small or large; the larger the change, the more we can consider the related product (the product which has been removed from the computation) as a potential outlier. What could be the interval in which we consider the change is too small for considering the product as a potential outlier? Belsley *et al.* [14] recommend a formula for this interval.

In Example D we get the following results displayed on Figure 6.8; the computed interval is (0.203, 8.499). Values outside this interval signal data points which are potential outliers; in this example: A, F and J.

Name of discarded data point	Variation of the determinant
A	0.083
B	0.678
C	0.639
D	2.436
E	1.967
F	0.021
G	0.210
H	2.604
I	0.814
J	0.141
K	6.416
L	2.202
M	2.702

Figure 6.8 The change of the variances–covariances matrix of the example.

6.1.4 Conclusion

As we have several ways to detect potential outliers, it is useful to prepare a synthesis of the detections (Figure 6.9).

Object	Looking for outliers		
	Based on the dependent variable	Based on the HAT matrix	Based on the variance–covariance matrix
A	X		X
B			
C			
D			
E			
F	X		X
G	X		
H			
I			
J			X
K		X	
L			
M			

Figure 6.9 Synthesis of the detection of potential outliers.

Looking for outliers requires now some judgment:

- according to the change in the residuals, products A, F and G are potential outliers;
- according to the "HAT" matrix, product K is a good candidate;
- according to the change in the variances–covariances matrix, products A, F and J are also good candidates.

The reason for K being selected by the "HAT" matrix is due to the fact that its causal variables are a bit far away from the bulk of the data, mainly on account of its high value of the number of boards. This is the purpose of the "HAT" matrix to detect such data; it does not mean that its presence changes a lot the formula which may be computed – as both other procedures do not show that.

Data points A and F are probably "true outliers": their causal variables are inside the bulk of the data, but their costs are probably too high, as a detailed examination of Figure 9.1 may reveal.

The conclusion of the search for outliers is that the cost analyst should not rely on one algorithm only before deleting one data.

The reader must also not forget that the last two algorithms are based on an assumed linear relationship between cost values and the causal variables and that, consequently, the search for outliers may also be considered as a check of this assumed linearity.

6.2 Dealing with Multi-Collinearities

6.2.1 What Is the Problem?

The fact that two or more parameters are correlated is the most serious problem in preparing a specific model with several quantitative variables. Saying that two parameters are correlated means that they quantify about the same characteristic of the products: a trivial example of correlated parameters is given by – generally but not always – the mass and the volume.

Figure 6.10 gives a geometrical illustration of two correlated parameters: V_1 and V_2, which are the causal variables, are correlated as their values on the plane (V_1, V_2) shows it.

In this example, but it does not appear every time like that, the correlation implies that line "ab" is well defined, but the intercept – represented by small squares – is

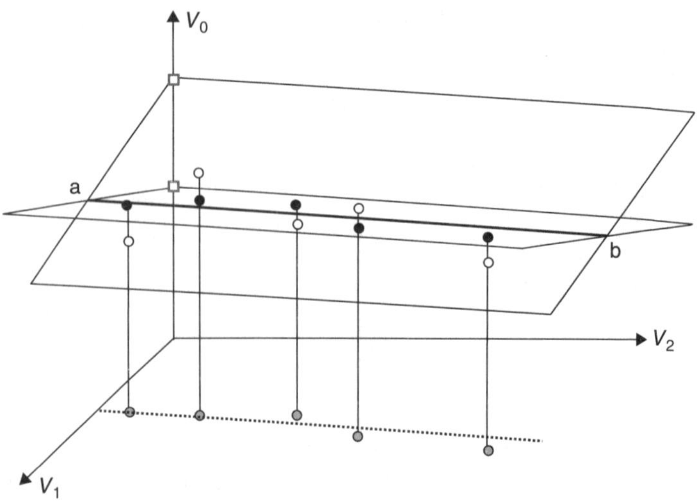

Figure 6.10 The geometric perspective when two parameters are correlated.

not: this means that a small change either of the value of one causal variable or even of the cost can completely change the plane we are looking for. This means also that the formula, which will be given by the equation of the plane, could be used if the correlation between the cost drivers is satisfied in the inputs, because, in such a case, the plane is in fact reduced to the straight line "ab".

However the plane itself, defined by line "ab" and its intercept, is not well defined at all. This means that the estimated value for a new product – if variables V_1 and V_2 are defined independently – will be extremely unstable.

Multi-collinearity can therefore be a very serious problem when several parameters are used.

What Are the Symptoms?

The first symptom which alerts the cost analyst is that the relationship he/she establishes does not really make sense. If, for instance, a multiplicative relationship has been selected, the data given in the example lead to the following formula:

$$\text{cost} = 69.7 \times \text{mass}^{-0.004} \times \text{boards}^{0.123} \times \text{connections}^{0.295} \times \text{components}^{0.062}$$

Nobody will believe that, if the mass is increased keeping constant the value of the other parameters, the cost will go down!

But the situation is not always so obvious. The same data, if an additive relationship is selected, give the formula:

$$\text{cost} = 479 + 13.6\text{mass} + 7.5\text{boards} + 0.39\text{connections} + 0.052\text{components}$$

where nothing particularly attracts the attention, unless you are familiar with this type of equipment and expect, from your experience, something else.

The mathematical reason for this situation is that, in the presence of multicollinearities, it may be very difficult for the algorithms to distinguish the influence of some variables, or more exactly to discriminate between them: the algorithms do what they can but if two parameters are strongly correlated it does not "know" exactly what is due to one and what is due to the second one. Consequently a coefficient may be computed too low and another one too large: the error on both may be quite large.

The second symptom is the fact that the standard errors (for the time being, consider that this term quantifies the accuracy with which the coefficients are computed) associated with the cost drivers are poor. If, for instance the standard errors of the coefficients are computed for the additive formula just presented, the following values are found (see Table 6.1).

Let us examine the values related to the mass: the coefficient is 13.6 and the standard error 30.4! Working with a number when we know that its standard error is more than the double of its value is not a comfortable situation!

Table 6.1 The standard errors of coefficients.

Parameter	Coefficient	Standard error
Constant	479	52.6
Mass	13.6	30.4
Components	0.052	0.036
Connections	0.39	0.180
Boards	7.5	8.521

The third symptom is the extreme sensitivity of some coefficients to the values of costs which are observed for the products entered in the database: a small change in one value – it does happen that we get a better information about one product and want to update the relationship – may completely change the formula. This is unfortunate if you presented to a large audience the result of your work a few days ago! One generally expects that all data contribute equally to the formula and that, consequently, a change in one data will only produce a small change of the formula.

As these symptoms are not always obvious, a detection procedure is recommended for checking the capacity of the data to generate a satisfactory relationship. We then need first such a procedure; this is the purpose of Section 6.2.2.

Once multi-collinearities have been found, it is necessary to decide on what to do about them; this will be dealt with in Section 6.2.3.

6.2.2 Detection of the Multi-Collinearities

Correlations between parameters here means there exists a *linear* relationship between two or more parameters; a non-linear correlation is not really damaging when we look for a linear formula (nevertheless it should be taken into account when using the formula); this correlation may be more or less strong, but it does exist.

This section assumes that the standard linear least squares algorithm is used for establishing the formula giving the dynamic center; this standard algorithm is defined in Part III. The residuals between the formula and the data points are defined to be additive and therefore noted as e_+.

Introducing the Subject

The coefficients of the linear formula we are interested in can be represented by a vector which is written in the general case of J parameters (each coefficient is related to a parameter, b_0 being a constant which would disappear if one decides to force the intercept to be 0):

$$\vec{b} = \left\| \begin{matrix} b_0 \\ b_1 \\ \cdots \\ b_j \\ \cdots \\ b_J \end{matrix} \right\|$$

Anticipating on the results developed in Chapter 15, let us admit that the variance of the coefficients can be – in the framework of several hypotheses – found by the following formula:

$$\text{var}(\vec{b}) = \hat{S}^2 \times (\|^+x\|^t \otimes \|^+x\|)^{-1}$$

where

$$\hat{S}^2 = \frac{\sum_i e_{i+}^2}{I - J - 1}$$

(unless we force the intercept to be 0, in which case the "-1" disappears) is an estimated value of the variance of the population deviations. It must be noted first that Johnston [34] established that this term $\hat{S}^2$ should not be seriously affected by the multi-collinearities effect. Consequently we must concentrate on the matrix:

$$(||^{+}x||^{t} \otimes ||^{+}x||)^{-1}$$

The variance of each coefficient is then given by the relevant diagonal element of this matrix (underlined by a straight line on Figure 6.11), the standard error being given by the square root of each diagonal element; the other terms of the matrix give the covariances between the coefficients. This is the way the standard errors given in Table 6.1 were computed, the var($\vec{b}$) having for our example the following form (the reader will check that the square roots of the diagonal elements are equal to the standard error written in this table).

$$\begin{pmatrix} 2.768 \times 10^{3} & 6.834 & -1.009 & 0.137 & -210.379 \\ 6.834 & 926.214 & 0.109 & -4.554 & 1.592 \\ -1.009 & 0.109 & 1.3 \times 10^{-3} & -2.666 \times 10^{-3} & 0.165 \\ 0.137 & -4.554 & -2.666 \times 10^{-3} & 0.032 & -0.697 \\ -210.379 & 1.592 & 0.165 & -0.697 & 72.604 \end{pmatrix}$$

Figure 6.11 The variance–covariance matrix of our example.

We mentioned, at the beginning of this section, the fact that multi-collinearities have two major consequences:

1. damaging the variances of the coefficients,
2. destabilizing the formula.

Both phenomena must be studied:

1. Section "Examining the coefficients variances" will examine the coefficients variances and answer to the following questions: are the variances really high? Are the multi-collinearities really the cause of their high values?
2. Section "Studying the stability of the formula" will investigate the stability of the formula.

Examining the Coefficients Variances

The variances were computed in the previous section. Let us display on a table the values we found for the various coefficients for our example (Table 6.2).

Table 6.2 Values of the variances.

Coefficient of		Variance
Intercept	b_0	2738
Mass	b_1	926.2
Number of components	b_2	0.0013
Number of connections	b_3	0.032
Number of boards	b_4	72.6

The reader will recognize that the variances are the values which appear on main diagonal of the variances–covariances matrix.

Are the Variances Really High? Table 6.2 is difficult to interpret: a large variance can correspond to a very large coefficient and being then quite reasonable. Consequently, it seems more useful to display the ratio between the coefficients values and their standard error; such a ratio is generally called "*t*", which was the symbol used by Gosset when he studied it (this will be explained in more details in Chapter 15). Such a presentation is illustrated on Figure 6.12:

Name	Coefficient value	*t*
Cost	479.275	9.109
Mass	13.654	0.449
Components	0.052	1.438
Connections	0.393	2.185
Boards	7.547	0.886

	Mass	Components	Connections	Boards
Mass	1.000	0.260	0.935	0.689
Components	0.260	1.000	0.319	−0.123
Connections	0.935	0.319	1.000	0.723
Boards	0.689	−0.123	0.723	1.000

Figure 6.12 The "*t*" values and the Bravais–Pearson correlations.

- The upper part of the figure gives the value of the coefficients (computed for an additive relationship in this example, but it could also be computed for a multiplicative relationship) and their "*t*" values, which is simply the ratio between the value of each coefficient and its standard error.
- Simultaneously the lower part of the figure displays all the Bravais–Pearson correlation coefficients between the couples of variables. These correlation coefficients are defined in Chapter 5.

The first thing which appears on this table is the large differences between the "*t*" values: one would expect that these values, given the data, would be rather similar. This is an important symptom that something did happen.

Now about the values themselves: the "*t*" value should be – as it will be explained in Chapter 15 – larger than 2 in order to get usable coefficients. One can immediately see on the figure that the "*t*" values of the mass and the number of boards are not satisfactory at all. The "*t*" value for the components is also not very good.

In order to decide if the low "*t*" values may come from multi-collinearities problems and therefore if we must go on with this study, it is useful to have an immediate look at the correlation coefficients between all couples of variables. This information appears on the symmetric table visible at the bottom of Figure 6.12 (it is symmetric because the correlation between mass and connections for instance is of course equal to the correlation between connections and mass). One immediately notices that both mass and number of boards are strongly correlated with the number of connections; this may help explain the poor "*t*" of these coefficients and encourages to go on with the study.

Breaking Down the Variances Let us recall the way the variances are computed:

$$\operatorname{var}(\vec{b}) = \hat{S}^2 \times (||^+x||^t \otimes ||^+x||)^{-1}$$

In order to investigate why some variances are quite high, it is necessary to have a look at both parts of this formula.

There is nothing that we can do about the term $\hat{S}^2$: it comes from the values of the residuals around the dynamic center of the distribution φ of the costs in the sample. Maybe changing the formula would slightly reduce it, but we decided in this chapter to use the standard linear regression.

We have then to turn to the second term, which is the matrix:

$$(||^+x||^t \otimes ||^+x||)^{-1}$$

Computations being easier if the data are centered and scaled, it may be interesting to do so.

What happens on these variances if the data are centered? Centering the data is geometrically equivalent to moving the coordinate's axis parallel to themselves until their center coincides with the center of the data. Vector $\vec{b}'$ which can be computed from the centered data is exactly the same as vector $\vec{b}$ except that[6] centering the data forces the intercept to be 0. On data centered and scaled, the variances of the coefficients are given by:

$$\operatorname{var}(\vec{b}) = \frac{\hat{S}^2}{I} ||R||^{-1}$$

where

$$||R|| = \frac{1}{I} ||_{cs}x||^t \otimes ||_{cs}x||$$

This analysis bears on a matrix $||R||^{-1}$ which is the inverted of matrix $||R||$ built on the data centered and scaled.

[6] We will see in Part III that one of property of the linear regression is that the dynamic center necessarily passes through the center of the data.

$$\|R\| = \frac{1}{I} \times \|_{cs}x\|^t \otimes \|_{cs}x\| \quad \text{and} \quad \|R\|^{-1} = I \times (\|_{cs}x\|^t \otimes \|_{cs}x\|)^{-1}$$

Returning to the formula:

$$\text{var}(\vec{b}) = \frac{\hat{S}^2}{I} \|R\|^{-1}$$

the variance of each b_j is given by $(\hat{S}^2/I)R_{jj}$, where R_{jj} is the diagonal element corresponding to b_j.

If all parameters were orthogonal (which is synonymous to "uncorrelated"), then the variances of all elements would be equal to $\hat{S}^2/I$; this is the reason why R_{jj} is called the "variance inflation factor" (VIF) associated with coefficient j; it is often noted as VIF_j. The link with the correlations comes from the fact that:

$$R_{jj} = \frac{1}{1 - R_j^2} = \text{VIF}_j$$

where R_j^2 is the square of the multiple correlation factor between variable j and the $J - 1$ other variable. This proves that correlation "damages" the variance of the coefficients: the closer R_j will be to 1, the more damaging will the effect be.

Illustration

Let us return to our example.

The VIF of each coefficient is displayed on Figure 6.13.

Variable	VIF
Mass	8.028
Components	1.585
Connections	10.504
Boards	2.951

Figure 6.13 Computations of the VIF for all coefficients.

One can see that the variance of the coefficient related to the mass is multiplied by 8 due to the effect of multi-collinearities: this explains its low "t".

Just as an experience, let's delete the mass and look to what happens to the VIFs: the results are given on Figure 6.14. On this figure it appears that the VIFs are much

Variable	VIF
Components	1.569
Connections	3.236
Boards	2.951

Figure 6.14 Computations of the VIF when mass is deleted.

smaller, but nevertheless that the VIF on the number of connections is still a bit high. This is due to the fact that the correlation between the connections and the boards is not negligible (0.723): there is still an important correlation between the number of connections and the number of boards.

What does happen if the number of boards is, at its turn, deleted (we prefer not to delete the number of connections because this parameter is highly correlated with the mass and therefore carries an important information about the products size). The results appear on Figure 6.15.

Variable	VIF
Components	1.113
Connections	1.113

Figure 6.15 Computations of the VIF if the mass and the number of boards are deleted.

Now the VIF, close to 1, are nearly perfect (the slight difference with 1 comes from the slight correlation between these two variables: 0.319): it means that the variances of the coefficients are minimized.

Studying the Stability of the Formula

We turn now to the second effect of the multi-collinearities: the lack of stability of the coefficients when a slight change is made to one data point. As previously indicated, we expect the formula to be equally built on all the data points: a small change of one data point should only slightly change the coefficients. Is it true in the presence of multi-collinearities?

When using the standard linear regression, it will be established in Chapter 9 that the coefficients of the formula are given by (note that we find again the variance–covariance matrix in this formula):

$$\vec{b} = (||^{+}x||^{t} \otimes ||^{+}x||)^{-1} \otimes ||^{+}x||^{t} \otimes \vec{y}$$

The question therefore is: how does a slight change of matrix $||^{+}x||$ affects $\vec{b}$?

About the Matrix Norm The response to this question supposes an understanding of the concept of matrix norm. The matrix norm is related to square matrices being used as operators, which means for transforming a vector into another vector; the norm of a matrix gives an information on how different will be the transformed vector from the original vector. Suppose a vector $\vec{A}$ is transformed into a vector $\vec{B}$ by a square matrix $||M||$; we write:

$$\vec{B} = ||M|| \otimes \vec{A}$$

In order to quantify the norm of matrix $||M||$ we look at the size (their Euclidian norm noted $|\vec{B}|$) of all the vectors $\vec{B}$ which can be obtained when vector $\vec{A}$ takes all possible values but keeping a norm equal to 1. The norm of matrix $||M||$ is therefore given:[7]

$$\left|||M||\right|_2 = \sup_{|\vec{A}|_2 = 1} \left|||M|| \otimes \vec{A}\right|$$

[7] The index 2 reminds, as several norms can be defined, that it is the Euclidian norm (see Ref. [31], p. 54).

The norm of a square matrix can therefore be defined as its "expansion rate" of transforming a vector into another vector.

The norm obeys to two relations:

- $\left|\|M\| \otimes \vec{A}\right| \leq \left|\|M\|\right| \times |\vec{A}|$
- $\left|\|M\| \otimes \|Q\|\right| \leq \left|\|M\|\right| \otimes \left|\|Q\|\right|$

Is it possible to easily quantify this norm? Fortunately yes, using the so-called singular value decomposition (SVD) of matrix $\|M\|$. The SVD of a matrix means that any matrix can be decomposed[8] into a set of three matrices, called $\|U\|$, $\|S\|$ and $\|V\|$, where $\|U\|$ and $\|V\|$ are orthogonal matrices, and $\|S\|$ a diagonal matrix of which elements are noted $s_1, s_2, \ldots$ and called "singular values" of matrix $\|M\|$:

$$\|M\| = \|U\| \otimes \|S\| \otimes \|V\|^t \quad \text{with } \|S\| = \begin{Vmatrix} s_1 & 0 & \cdots \\ 0 & s_2 & \cdots \\ \cdots & \cdots & \cdots \end{Vmatrix}$$

Among the singular values we select the maximum and the minimum ones, called s_{max} and s_{min}. If one or some s_i are null, then $s_{min} = 0$.

Now we have two interesting properties linking the singular values and the norm of a matrix $\|M\|$: the norm of $\|M\|$ is equal to s_{max}, and the norm of $\|M\|^{-1}$ is equal to $1/s_{min}$.

The quotient s_{max}/s_{min} is called the condition number of $\|M\|$ and noted $\kappa(\|M\|)$. It can go from 1 to ∞, depending mainly of the value of s_{min}. The larger $\kappa(\|M\|)$ the worst conditioned is said matrix $\|M\|$ to be, which means it will be difficult to invert. Consequently it appears that s_{min} is a very important information regarding the structure of a data matrix.

The purpose of this study is still to study the variances of the formula coefficients or more exactly to investigate the origin of the high variance computed for some coefficients, when this origin is due to multi-collinearities – and not to a large scattering of the cost values.

The variances of the coefficients are given by the diagonal elements of the matrix (the other terms being equal to the covariances between the coefficients):

$$\operatorname{var}(\vec{b}) = \hat{S}^2 \times (\|^+x\|^t \otimes \|^+x\|)^{-1}$$

The set of the coefficients we are looking for, when looking for a linear relationship between the cost and the parameters, is here defined as a vector $\vec{b}$. In the case of our example, this vector will have five components, the first one being the "intercept" (a constant value), the other ones being related to the different cost drivers.

The computation of this vector is postponed to Part III. What we do here is to analyze the variance of each component of this vector; the result of the analysis is

[8] See Ref. ([31], p. 70). The SVD decomposition of a matrix is rather difficult to obtain; the interesting thing is that the procedure does not need the matrix inversion, which means that the SVD always exists.

Variance proportion for

Explained by	Intercept	Mass	Components	Connections	Boards	Condition index
s_1	0.034	0.18	0.523	0.162	0.471	1
s_2	0.058	0.16	40.044	0.048	6.91	3.906
s_3	5.976	12.836	8.727	3.527	41.884	7.599
s_4	92.899	3.673	36.505	0.168	36.766	12.981
s_5	1.034	83.15	14.2	96.095	13.967	17.601

Formula	R^2
Connections = 102.565 + 180.897 × Mass	0.874

Figure 6.16 Breaking down the variances according to the singular values.

presented in Figure 6.16 (proportions of the variances are given in percentages): for instance 83.15% of the variance of the coefficient of the mass is related to the singular value s_5 and this singular value is "responsible" for 96.1% of the variance of the coefficient of the number of connections!

When, on a single line, more than 50% of two or more variances are related to the same singular value, then it is interesting to see how much the related variables are correlated. A regression analysis made on these two variables shows that the value of one could be forecasted with a good accuracy from the other one.

It is generally recognized that a condition index higher than 10 signals a potential problem; the higher this value, the more serious is the problem.

6.2.3 What Are the Solutions?

In the presence of multi-collinearities, several solutions are possible, between which the cost analyst must decide.

Doing Nothing

This is of course the easiest solution! You may do it if the damages are small, which means that the level of collinearities (quantified by the Bravais–Pearson correlation coefficients) are small.

This decision means that you want – for any reason – to keep the parameters in the list of the variables on which the formula of the dynamic center is built; the first price you agree to pay is a loss of accuracy of the estimates you will do in the future.

The second price you will have to pay when using the specific model built on the formula is that you cannot choose the values of the parameters as if they were independent: when quantifying these values you must take into account the fact they are, at least a bit, correlated. In particular do not change the value of one parameter without changing the values of the parameters which are correlated to it.

This solution must however be strictly avoided if the level of correlation is too high.

Adding a New Data Point

This might be difficult in the domain of cost, as generally speaking, data are not easy to find. However, a short discussion about this possibility will add some light on the SVD algorithm. This discussion anticipates on the principal component analysis (PCA) which is the subject of Section 6.2.3.

The SVD, by analyzing the data matrix, quantifies in a very detailed way the damages caused by multi-collinearities. That is fine, but what can you do about it? Fortunately enough the singular values are closely related to the eigenvalues of the matrix: more exactly the singular values are the squares of these eigenvalues.

The PCA defines new axes, or new variables; to each axis is attached an eigenvalue (also called a "root value"). When the eigenvalue related to an axis is small, it shows that this axis is not precisely defined, due to the multi-collinearities effect as determined by its singular value.

Consequently if a new data point is added it should, if it is possible (other way it will not really help solve the problem), located along the axis for which the eigenvalue is the minimum. Johnston [34] demonstrated that, in such a case, it is effectively possible to increase the value of the relevant singular value without modifying the other ones. That would theoretically solve the problem but we shall not discuss this problem any further as its application is so rare (rather inexistent) in the domain of cost.

Deleting One Parameter

After all, saying that two parameters are correlated means they quantify about the same thing.

As it is the case, you may delete one of them and build a formula without it.

When you will estimate the cost of a future product with the specific model built on the new formula, you better check that the correlation you observed in the sample is still true. It may really happen that this correlation is just an artefact and does not exist in the whole population: this is the reason why you must always check if the correlation found in the sample is "statistically significant". This concept is studied in Chapter 15.

Create a New Variable

Let us explain the subject on one example.

We had in the past to prepare a model for an optical sensor. Engineers who worked on the subject were convinced that three parameters are sufficient to "explain" its cost:

1. the sensitivity;
2. the aperture (in what solid angle could an object be detected without moving the sensor?);
3. the accuracy (what should be the angular distance between two objects we want to separate?).

They gave us data about several sensors built in the past, at different periods.

The analysis of the data showed that these three parameters were highly correlated, thing that the engineers could not understand, because technically they were clearly different. The solution could not be found until we realized that, in the past, due to the technological progress, these three parameters always move together. As the engineers – as they all do – always tried to get the best from the sensors, they always chose what the technology could offer and the progress for the three parameters went hand in hand.

The solution was then to create a new variable – here the technological year of the sensor – in order to replace the three previously considered.

As it often happens in this type of analysis, the solution worked well for the sample, but we are never sure that the same rate of change will go on in the future when the relationship will be applied.

This solution may be a good one, but sometimes we want to get something more detailed.

Changing the Coordinates System

It may happen that you absolutely want to keep all the cost drivers. In such a situation, a first solution is given by changing the coordinate system.

The purpose of changing the coordinates system is to keep all the variables, and to look for combinations of these variables which would not be correlated. It means looking for new variables – which would be linear combinations of the natural variables – which could be used for building a formula with no collinearity problem, as they will be uncorrelated.

It may look artificial, but it is not. It is exactly what does the PCA which will be the subject of Section 6.3.

There is a small difference between the classical PCA and what we are trying to do here:

- The purpose of the standard PCA is to analyze the set of the causal variables only, without the cost, in order to understand the shape of this set: how scattered it is? Are there any outliers?
- The purpose of what we are trying to do here is to find new, uncorrelated variables.

In other words the classical PCA does not worry too much about the cost, which is dealt with afterwards, whereas in this section the cost is just a variable as the other ones.

Let us carry the study on our example.

First of all we make a PCA on the five "old" variables centered and scaled: cost (Y), mass (V_1), components (V_2), connections (V_3) and boards (V_4). This analysis[9] delivers five new vectors which are defined as a linear combination of the old ones:

$$U_1 = +0.517Y + 0.506V_1 + 0.179V_2 + 0.520V_3 + 0.418V_4$$
$$U_2 = +0.094Y - 0.027V_1 + 0.875V_2 + 0.013V_3 - 0.474V_4$$
$$U_3 = -0.042Y - 0.492V_1 + 0.396V_2 - 0.215V_3 + 0.744V_4$$
$$U_4 = +0.621Y - 0.669V_1 - 0.208V_2 + 0.278V_3 - 0.216V_4$$
$$U_5 = -0.580Y - 0.234V_1 + 0.050V_2 + 0.778V_3 + 0.011V_4$$

[9] EstimLab™ does that within a couple of seconds.

of which extensions (the term is defined in Section 6.3.3) are 70.9%, 22.1%, 4.6%, 1.6% and 0.8%. This shows that the first two new variables "explain" 93% of the cost. Consequently the formula giving the dynamic center could be written: $U_3 = U_4 = U_5 = 0$.

The reader may easily check that these new vectors are normalized (their Euclidian norm is equal to 1) and orthogonal as it can be, for instance, controlled on the product:

$$\vec{U}_1 \otimes \vec{U}_2 = \|0.517 \quad 0.506 \quad 0.179 \quad 0.520 \quad 0.418\| \otimes \left\| \begin{matrix} 0.094 \\ -0.027 \\ 0.875 \\ 0.013 \\ -0.474 \end{matrix} \right\| = 0$$

It may be thought that this solution solves the multi-collinearities problem, because, as we saw it, the new variables are uncorrelated, etc. But this is not exactly correct: when we will use the formula for estimating the nominal cost of a new product, the previous relationships will have to be taken into account.

The Ridge Regression

The Ridge regression, introduced by Hoerl and Kennard, is based on a simple idea: the problem when dealing with variables more or less correlated comes from the fact that the matrix:

$$(\|^{+}x\|^{t} \otimes \|^{+}x\|)$$

is, as we saw it in Section 6.2.2, "ill conditioned" and cannot be easily inverted because its determinant is too close to 0 (inverting a matrix needs to use this determinant as a divisor).

One way to solve the problem is to invert the matrix:

$$(\|^{+}x\|^{t} \otimes \|^{+}x\| \oplus k\|1\|)$$

where a small element (the matrix $\|1\|$ is a matrix having all its elements equal to 0, except for the elements of the main diagonal which are equal to 1) was added in order not to get the determinant of this matrix too close to 0. Then the matrix can be easily inverted.

The price to pay (there is always one!) is a slight bias in the value of the coefficients. For this reason a factor k was introduced in the matrix: the job of the cost analyst is then to choose a value of k as small as possible to make the matrix invertible with the minimum bias of the coefficients.

This Ridge regression is presented in more details in Part III.

6.3 Visualization of the Data

Visualization of the data is important: the eye is a very powerful sensor for transmitting to the brain a lot of information, if this information is presented in a suitable way. When only one or two variables are involved, the usual graph is a very good tool; when several variables coexist, something must be prepared in order to make the relevant information readable.

Three tools can be proposed:

1. The "star diagram", the simplest one.
2. The step-by-step analysis, one of the most powerful way to visually analyze the data.
3. The PCA, a classical tool in data analysis.

6.3.1 The Star Diagram

It is the simplest way to present, on one piece of paper, a synthetic view[10] of the available information. It deals with quantitative parameters. The relationship between the cost and the parameters does not necessarily have to be linear.

It requires two steps.

Preparation of the Graph

The support of the information is a circle. For preparing the graph, a set of $J + 1$ (as many as there are parameters, plus the cost) diameters are drawn on the circle; each diameter is dedicated to one variable. In order to make full use of the circle:

- the minimum and the maximum value of each parameter is computed;
- and each diameter is graduated from one end to the other one (not from the circle center), from the minimum to the maximum.

When this is done, a symbol representing the value of each product is displayed on the relevant diameter: Figure 6.17 illustrates.

In our example, we have four parameters, plus the cost: consequently, five diameters are drawn.

Needless to say, the figure is difficult to interpret: it is just displayed here in order to explain how the diagram is prepared.

This is the reason why a second graph has to be drawn.

Synthesis of the Information

On this second graph each product is represented by one point which is the center of gravity (or barycentre) of the $J + 1$ points displayed on the previous graph. Figure 6.18 illustrates.

[10] The star diagram, as it is presented here, has been proposed by Xavier Apolinarski, engineer in the CEA (Commissariat à l'Energie Atomique) in France.

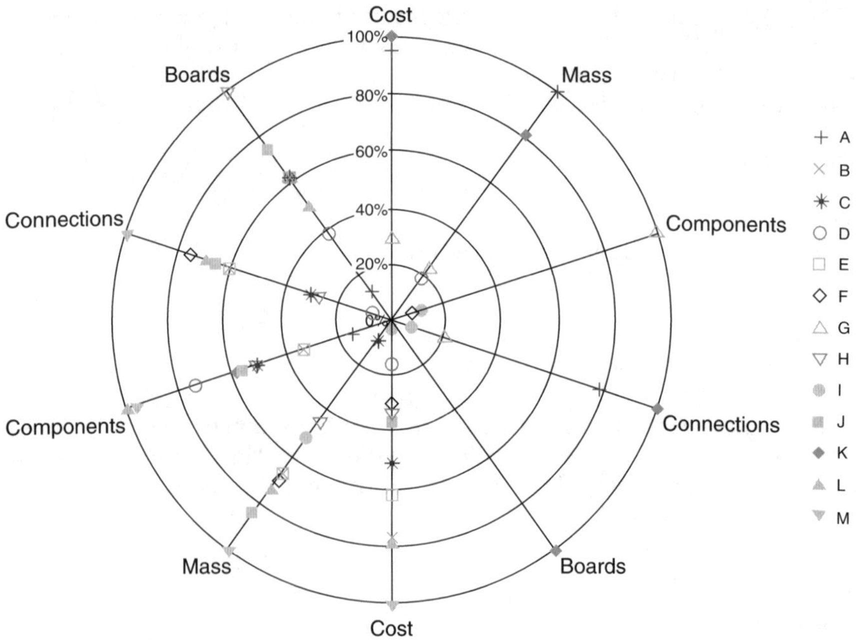

Figure 6.17 Preparation of the star diagram.

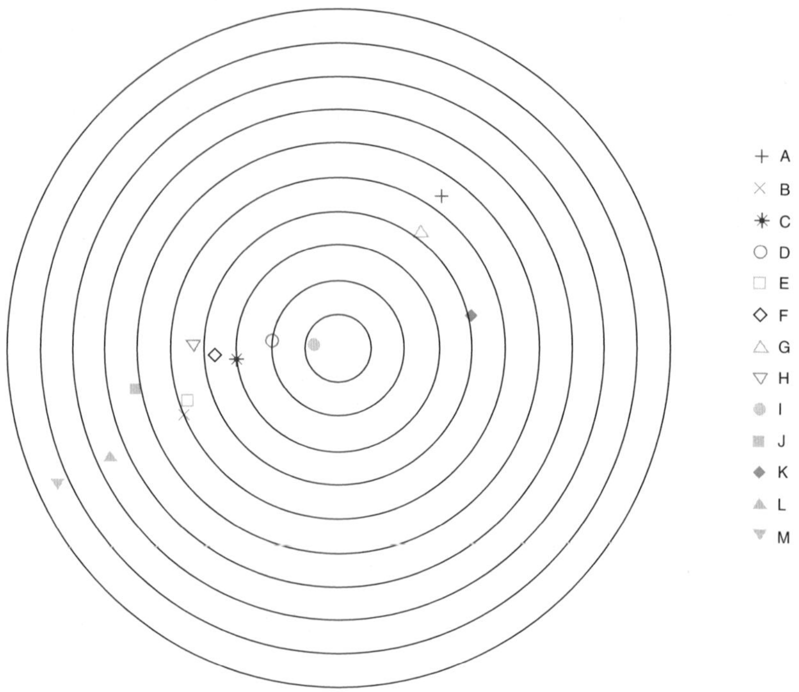

Figure 6.18 The star diagram.

The main purposes of this graph are as follows:

1. To look for outliers, if the outliers previously detected were kept, or if new ones can be discovered. In our example, product K clearly appears as an outlier. This product was detected as a possible outlier with the variances–covariances matrix analysis.
2. To confirm if the hypothesis of a linear relationship between variables, which is the basis of this whole chapter, is confirmed or not. Looking at our example, we have no reason to reject this hypothesis.
3. To have a broad view of the scattering of the data. In our example the data appear a bit scattered, but not too much if product K is eliminated.
4. To detect if the product family we are working with is composed or not of two sub-families. In our example, it does not seem so, even if products H and J seem globally slightly different from the family; none of these products was detected as a potential outlier. It would however be interesting to compute a dynamic center of the data without these products, in order to see if their absence improves the quality of the model.

6.3.2 The Step-By-Step Analysis

As we cannot see anything in a multi-dimensional space, the idea here is to present several two-dimensional diagrams, each diagram being dedicated to one variable. The basic idea, if we are interested in variable V_j, is the following one:

1. We search first how well all the variables but V_j "explain" the dependent variable Y. The best way to quantify this explanation is to look at the residuals of a regression of this dependent variable Y and these $J-1$ variables. These residuals are the values of the dependent variable "cleaned" from the influence of these $J-1$ variables: they are the "cleaned Y".
2. We look now to variable V_j. As this variable may be correlated to the other causal variables, we decide to also "clean" V_j from the influence of these variables. A good way to do that is to regress V_j to these variables: the residuals of this regression constitute the "cleaned V_j".
3. Now we search how the "cleaned V_j" can explain the "cleaned Y". The best way to do it is to regress both set of residuals and to look on a graph at the results: if the residuals of this third regression are small, this means that the cleaned variable V_j explains very well the dependent and cleaned variable Y.

This procedure may look a bit difficult at the first approach, but it is a very powerful tool for analyzing a set of data: it is therefore recommended. The following of this section details it.

Let us start with V_1. We start by making two linear regressions:

- The first one regresses (the algorithm is the standard linear regression which will be explained in Part III) the cost Y against all the variables *except* V_1, the result of this regression is the value of the dynamic center of the data, when variable V_1 is deleted. The equation giving this dynamic center can be written, the notation [1] reminding the reader that variable V_1 was removed from the computation:

$$\hat{y}_{[1]} = b_{0[1]} + b_{2[1]}x_2 + b_{3[1]}x_3 + \cdots$$

From this equation, the residuals can be computed; they are labeled, for product i:

$$u_{i[1]} = y_i - b_{0[1]} - b_{2[1]}x_{i,2} - b_{3[1]}x_{i,3} - \cdots$$

These residuals can be called (Theil [56], p. 183): "values of the dependent variable corrected for the effect of all other variables" (all variables less V_1). As, in the linear regression, the sum of the residuals is equal to 0, the arithmetic mean of $u_{n[1]}$ is equal to 0.

- We want now to investigate the relationship between variable V_1 and the other variables. Therefore, we regress V_1 against all the other variables (note that the cost is not involved in this regression). The equation giving the result of this regression can be written:

$$\hat{x}_1 = c_0 + c_2x_2 + c_3x_3 + \cdots$$

The residuals can be as usual computed; they are labeled, for product i:

$$v_{i,1} = x_i - c_0 - c_2x_{i,2} - c_3x_{i,3} - \cdots$$

For the same reason as mentioned upwards, the arithmetic mean of the $v_{i,1}$ equals 0.

We have now two sets of residuals which are labeled:

1. Values $u_{i[1]}$ of the dependent variable corrected ("cleaned") from the effect of all other variables.
2. Values of the causal variables V_1 corrected ("cleaned") from the effect of all other causal variables: $v_{i,1}$.

As the set of residuals $u_{i[1]}$ does not take into account the effect of V_1, it is interesting to check if the use of this variable V_1 could not reduce these residuals. However, we should not use V_1 itself because there is a relationship between V_1 and the other variables, and these other variables have already been taken into account in the computation of $u_{i[1]}$: we should not count twice the influence of the other variables.

In order to see if variable V_1 can help explain $u_{i[1]}$ we must use the value of V_1 corrected for the effect of the other variables; we already know this value: it is the v_i.

Consequently, the first thing we can do is to display the couples $(v_{i,1}, u_{i[1]})$. This is done, for our example, in Figure 6.19.

What can we observe on this figure?

1. About the vertical axis, giving the values of the dependent variables corrected for the effect of all other variables, one can see that the range which was for the cost about 600 is reduced to 180 (from 100 to −80). The "other" variables explain 70% of the cost variation, which is positive.
2. About the horizontal axis, the original range of 5.6 is now reduced to 2.5: the other variables explain 55% of the variable V_1. This clearly shows there is some correlation between V_1 and these other variables.
3. Now the important question is: would the inclusion of V_1 reduce the remaining range of variation of the cost? This is rather unlikely and can be seen making a

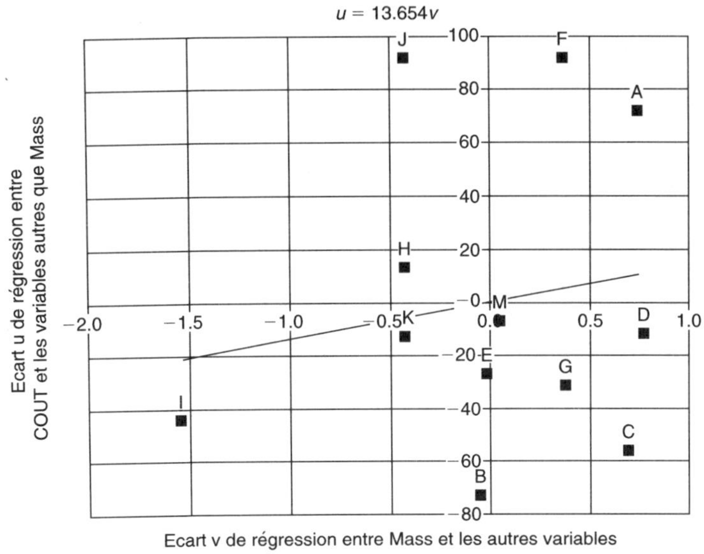

Figure 6.19 Partial regression of $v_{i,1}$ against $u_{i[1]}$.

regression between the residuals u_i and v_i. The equation of the regression is given by:

$$u_i = 13.654 \times v_i$$

which is displayed on the graph and the corresponding straight line (which has to go through the point [0, 0] because both arithmetical means of these variables are 0) is also plotted on the graph. The slope of this equation is small (this is obvious on the graph) and, consequently, we cannot expect inclusion of variable V_1 to have a great power to reduce the remaining residuals (after correction by the other variables) of the cost.

Let us examine now the "power" of the number of connections (Figure 6.20).

Note that the range of $v_{i,3}$ is small compared to the range of the number of connections (1148); this is due to the correlation between this number of connections and the other variables. Also note that the slope of the regression line is great: this variable, even cleaned from the influence of the other variables is an interesting cost driver.

This analysis should be done on all the variables.

A Useful Remark

An interesting point of this step-by-step analysis is that the coefficient of the regression of the residuals $v_{i,j}$ on $u_{i[j]}$ (0.393 in the example of Figure 6.20) is exactly equal to the coefficient of variable V_j in the regression of Y on all the variables. It very clearly indicates the origin of this coefficient, which otherwise seems to come from just the result of a computation: this remark shows the logic of it.

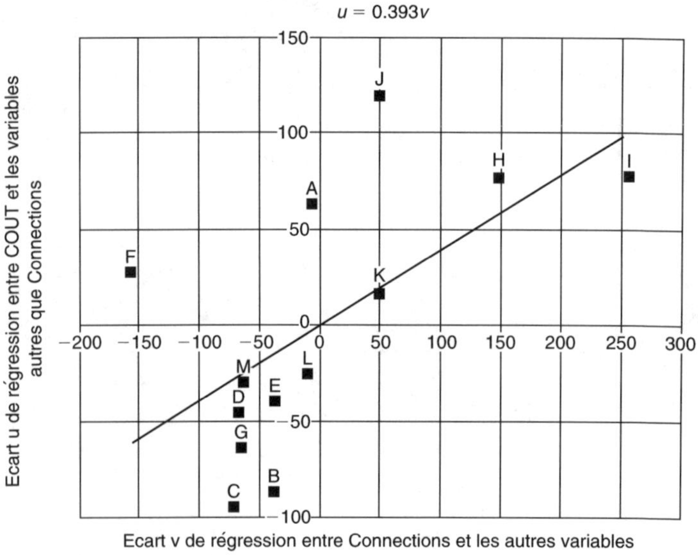

Figure 6.20 Partial regression of $u_{i[3]}$ against $v_{i,3}$.

This step-by-step analysis – also called "partial regressions" – is therefore a powerful visual tool for understanding and interpreting the data.

6.3.3 The PCA (Theil [56], p. 46)

The PCA is a very powerful tool for displaying a lot of information about data. Its origin is easy to explain on an example: look at Figure 6.21 and try to answer the question: What is this equipment? It could be a ladder for the firemen, an artistic candle, or anything else.

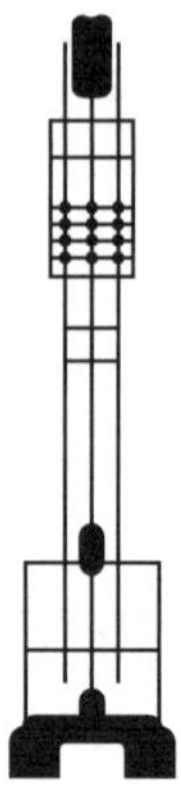

Figure 6.21 What is this machine? (What is this equipment?)

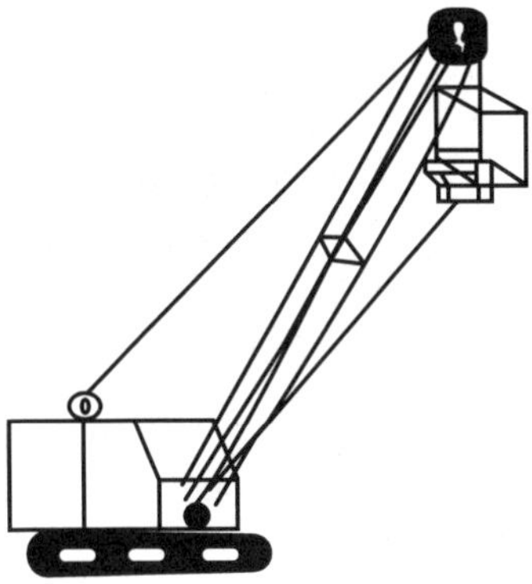

Figure 6.22 What is this machine?

Look now at Figure 6.22: it is the same machine and we are sure you recognize it immediately.

Why do you recognize the second one and probably not the first one? Because, on this second one, the equipment is represented by a projection on a plane, the plane being chosen to get the maximum extension of the picture. Another example? You cannot – unless you are an expert in fishing – recognize a fish when you see it in front, and it is certainly easier to recognize it when you see it laterally, under its maximum extension.

In a three-dimensional space, we are accustomed to see pictures of an object seen from three different places: this is the basis of industrial drawing. Images of the object are projections of this object on three different planes. The planes should be, as much as possible, chosen in order to see the object under its maximum extensions.

This can be extended to a several dimensions space (the number of variables) and this is the basis of the PCA.

How can this idea be implemented? Let us discuss this question with only two causal variables V_1 and V_2, plus of course the cost Y: using only two causal variables makes it possible to represent the set of the data points (labeled E) in the three-dimensional space (Figure 6.23).

The Starting Point

Figure 6.23 represents this set of data, the cost being plotted on the vertical axis.

On Which Variables Must We Carry Out a PCA?

There are two types of PCA:

1. The "full" PCA which works with all the variables, including the dependent variable (the cost). In such a case, we works with the set E.
2. The "partial" PCA which works with the causal variables only. The relevant set is then E'. Once this set is analyzed, the dependent variable can then be added and correlations between this dependent variable and the new variables U'_1 and U'_2 are then looked for.

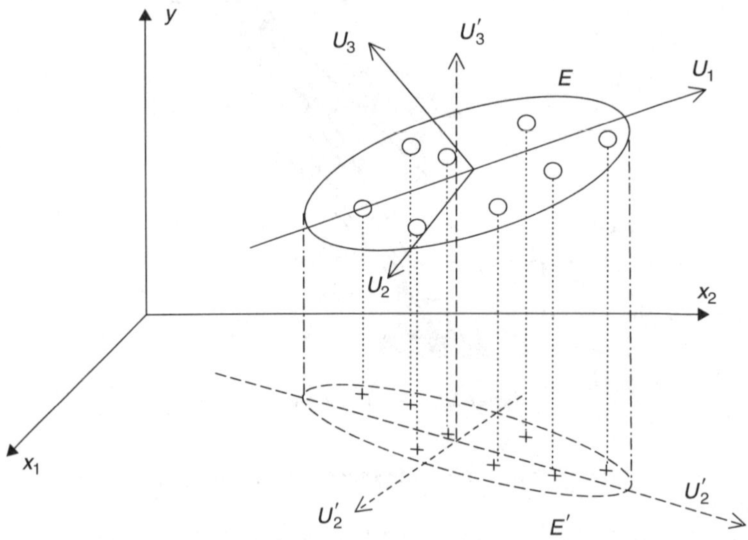

Figure 6.23 Representation of the data in the three-dimensional space.

The choice is let to the cost analyst. However, in the cost domain, the first procedure is not recommended. The reason for that comes from the fact, explained below, that we are looking for new variables that will be linear combinations of existing variables: linear combinations between cost and causal variables do not generally help the cost analyst. However, if other subjects that cost are studied, the "full" PCA may sometimes help.

Do We Have to Prepare the Values?

A PCA is always made on the centered values: this means that the origin of the coordinates system is always transferred to the center of the data set.

Therefore the choice is between:

- Making a PCA on the values only centered: the advantage is, on the graphs, to correctly represent the data, but to make the visual search of the correlations between the variables nearly impossible, due to the different scales which are used for these variables.
- Making a PCA on the values centered and scaled (which means that the values are divided by their standard deviation; this eliminates the scaling problem). Then the variables are not correctly represented, but the correlations are much easier to perceive.

The solution we prefer is the second one, in order to be able to look at these correlations. But both solutions must be available.

The data matrix, once the data are centered and scaled, will be represented here by $\|{}_{cs}x\|$. This matrix gives the centered and scaled values of the information relevant to each product (as the data are centered, there is no need for a

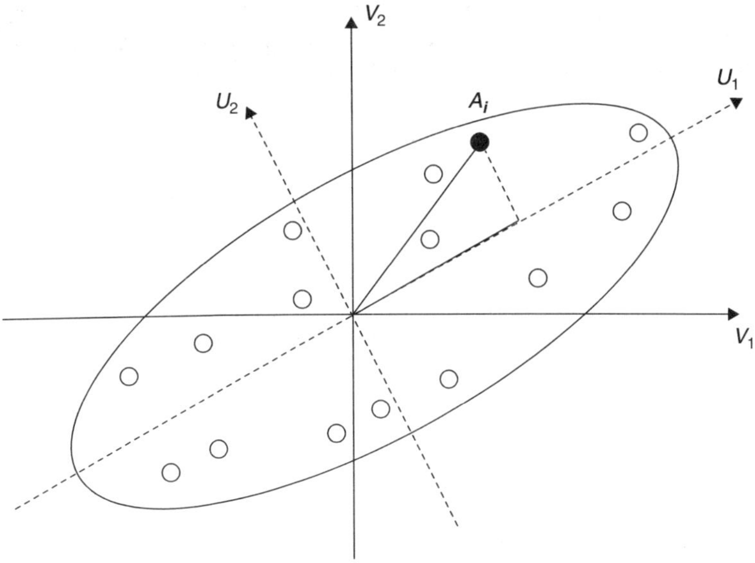

Figure 6.24 Representation of data defined by two causal variables.

column of 1):

$$\|_{cs}x\| = \begin{Vmatrix} \dots & \dots & \dots \\ \dots & \dfrac{x_{i,j} - m_j}{s_j} & \dots \\ \dots & \dots & \dots \end{Vmatrix} = \begin{Vmatrix} \dots & \dots & \dots \\ \dots & {}_{cs}x_{i,j} & \dots \\ \dots & \dots & \dots \end{Vmatrix}$$

where m_j and s_j are the arithmetic mean and the standard deviation of all values in the column *j*. As a product *i* is described by all the figures belonging to a row, the set of its values can be represented by the (row) $\|_{cs}x_{i,:}\|$ vector.[11]

The Computations

The computations will be shortly described in a two-dimensional space (Figure 6.24): this means that each product is defined by its values for two causal variables. The procedure can then easily be extended to any number of dimensions (the same number as the causal variables).

Let us consider the product A_i. In the "old" system of variables, it is defined by the two values of row *i* of the data matrix $(x_{i,1}, x_{i,2})$ which is the vector $\|x_{i,:}\|$.

The objective is to find out new variables, here called $U_1, U_2, \dots$ that will replace the existing variables $V_1, V_2, \dots$ These new variables will be linear combinations of the existing ones. Each new variable can then be represented by a vector, for instance $\vec{U}_1$.

[11] Using the colon notation is common in the literature about matrices.

As previously said, the procedure consists in searching variables for which the data set has the maximum extension. Let us start by finding the vector $\vec{U}_1$: the "extension" of data point i on $\vec{U}_1$ is given by the scalar product.[12]

$$||x_{i,:}|| \otimes \vec{U}_1$$

of which length is given by the square of this product, conveniently represented by:[13]

$$\vec{U}_1^t \otimes \left\| x_{i,:} \right\|^t \otimes \left\| x_{i,:} \right\| \otimes \vec{U}_1$$

Consequently $\vec{U}_1$ will be given by maximizing the sum of the projections of all the products:

$$\sum_i \vec{U}_1^t \otimes \left\| x_{n,:} \right\|^t \otimes \left\| x_{n,:} \right\| \otimes \vec{U}_1 = \vec{U}_1^t \otimes ||\vec{x}||^t \otimes ||x|| \otimes \vec{U}_1$$

given the constraint $U_1^t \otimes U_1 = 1$ because we want also all these vectors to have a length equal to 1.

The problem of maximizing a quadratic form given a quadratic constraint is solved by the use of the Lagrange multipliers. Consequently $\vec{U}_1$ is an eigen vector of the symmetrical matrix $||x||^t \otimes ||x||$. The same is true for all other $\vec{U}_j$.

The set of new variables is therefore the set of the eigen vectors of this matrix. As this matrix is symmetrical, we know – it is a theorem in the matrix algebra – that all the vectors of this set are orthogonal: they, together, constitute a basis of the space.

The fact they are orthogonal is very important: it means that the new variables are NOT correlated at all. This result will be used later on.

Another Interesting Feature

To each eigen vector is associated an eigenvalue, called $\lambda_1, \lambda_2, \ldots$ of which sum is equal to the trace[14] of the matrix $||x||^t \otimes ||x||$. It can be shown that each ratio such as $\lambda_p/trace$ represents the **extension** of the data set along the vector $\vec{U}_p$; it is generally given in percentage.

This result will help interpret the results of the PCA.

The Results[15]

Let us return to the example in order to illustrate the results provided by the PCA. Many interesting data and graphs are provided by the PCA: they give a lot of light on the structure of the data.

[12] We assume that the space is equipped with an Euclidian metric. Do not forget that the first term is a row vector, the second a column vector and that, consequently, the product is consistent with the rules of matrix multiplication.

[13] The transpose of a product is equal to the product of the transposes in the opposite order.

[14] The trace of a matrix is equal to the sum of its diagonal terms.

[15] All the pictures are extracted from EstimLab™.

The First Thing to Look at

Figure 6.25 gives:

- the extension along the four new axes,
- the data (centered and scaled) given in the new coordinate system.

Extensions explained for each axis in %
65.490
27.293
5.784
1.433

Centered and scaled data coordinates on new axis					
Copy		Axis 1	Axis 2	Axis 3	Axis 4
1	A	2.698	0.211	−0.749	−0.109
2	B	−0.945	0.100	0.250	−0.053
3	C	−0.129	−0.142	−0.483	−0.200
4	D	0.556	−0.745	−0.577	−0.203
5	E	−0.945	0.100	−0.250	−0.053
6	F	−0.985	0.801	0.637	−0.309
7	G	1.205	2.599	0.201	−0.171
8	H	−0.852	0.181	−0.793	0.308
9	I	0.045	0.978	0.210	0.619
10	J	−1.387	−0.212	0.047	0.146
11	K	3.821	−1.574	0.769	0.116
12	L	−1.197	−1.211	0.037	0.004
13	M	−1.886	−1.086	0.200	−0.096

Figure 6.25 Extension and new values of the data.

The extension table (Figure 6.25) shows that three new axes are enough to correctly represent the data: they represent 98.6% of the data, whereas the fourth axis has an extension of only 1.4%: it can therefore be neglected. Considering the three major axes, it is easy to see that, viewed from the appropriate direction of space (the one which is defined by the new axis), the data set has the form of an elliptic pancake (the extension of the first two axes are 65.5% and 27.3%) with a small thickness (5.8%). It means that the first two axes only are able to represent 92.8% of the data!

The values of the data on the new axes are given for information only: their interest is limited and the following graphs are easier for interpretation.

What Do Represent the New Variables?

As said earlier, the new variables are linear functions of the old ones; these functions are easily displayed on a screen (Figure 6.26).

It can be seen on this graph that the new variable U_1 is a mix, with about the same proportions of the mass, the number of connections and the number of boards. It can therefore be interpreted as a "size" of the equipment. The second important new variable, U_2, represents about the same thing as the number of components; it may be interpreted as the "complexity" of the equipment.

New variables	Mass (cs)	Components (cs)	Connections (cs)	Boards (cs)
U_1	0.591	0.031	−0.512	−0.623
U_2	0.175	0.900	0.384	−0.105
U_3	0.602	0.069	−0.227	0.762
U_4	0.507	−0.428	0.734	−0.143

Figure 6.26 The composition of the new variables. "cs" means "centered and scaled".

The Projection of the Data on the Planes U_1 and U_2

The purpose of the PCA being to represent the data, one can try to project these data on the planes built by the axes. On Figure 6.27, the plane made with axes U_1 and U_2 is used.

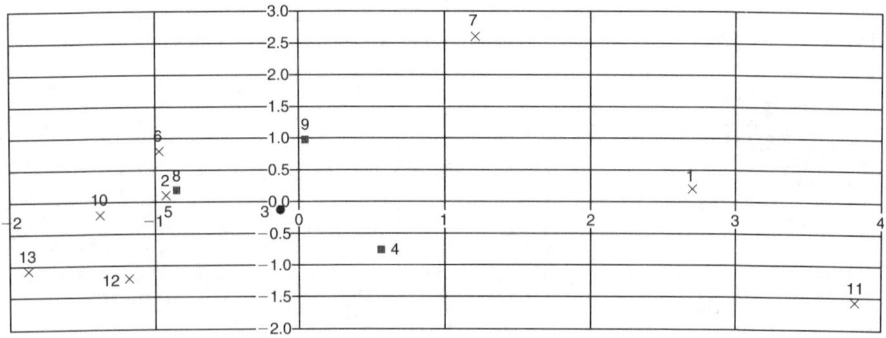

Figure 6.27 Graph showing projection of the data on the planes U_1 and U_2.

Data are represented by different symbols which express how far from the plane are the data: the distance is computed for a data A as the angle α between line OA with the plane (on the figure, crosses indicate that this angle is less than 30°, squares that this angle is between 30° and 60°, and circles that this angle exceeds 60°). This helps the interpretation of the figure: for instance point 3 which looks close to the center of the graph, may be, in fact, far away on axis U_3 (Figure 6.28).

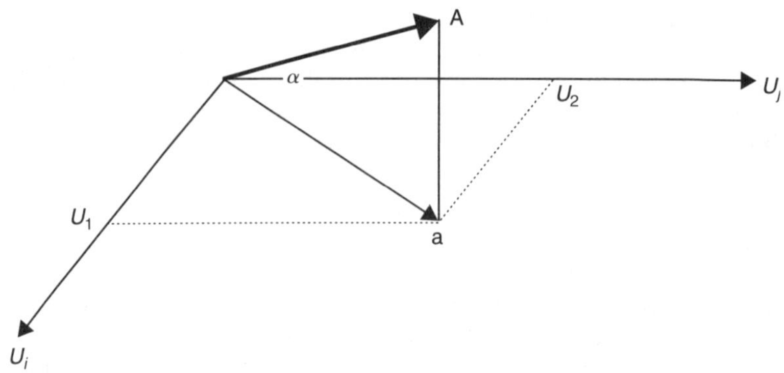

Figure 6.28 Projection of the data on the planes U_1 and U_2.

The distances between these data points and the planes can really be viewed on a projection of the data on the planes U_1 and U_3 which is perpendicular to the previous ones (Figure 6.29).

The purpose of these graphs is always the same:

- Looking for possible outliers (here data number 11 seems far away from the bulk of the data – it corresponds to data K – and therefore the analyst should pay attention to it).

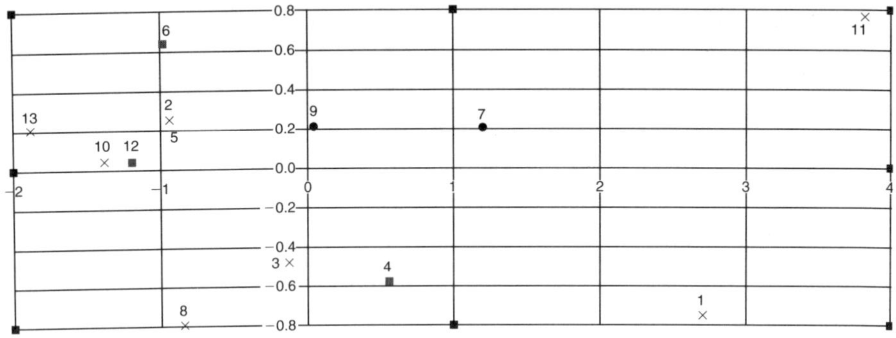

Figure 6.29 Projection of the data on the planes U_1 and U_3.

- Looking to the possible existence of two sub-families inside the family (it does not seem the case here).
- Looking to the extension of the values (here axis U_1 is far more extended than axis U_2 or U_3) and searching why (here it comes primarily from one data point).

Correlations Between the Two Set of Variables

Figure 6.30 displays the composition of the new variables from the old ones.

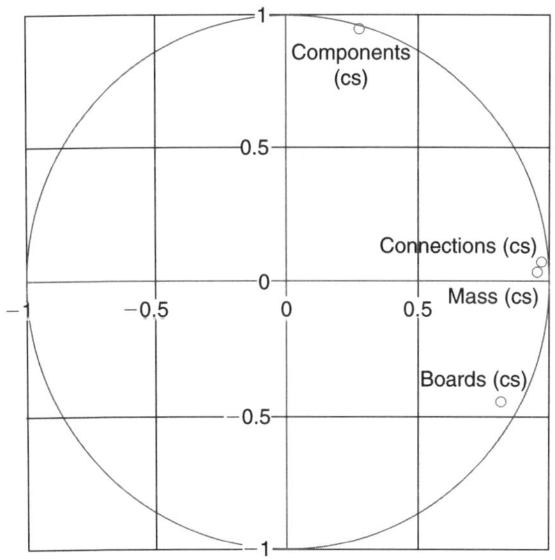

Figure 6.30 Correlation circle on planes U_1 and U_2.

Another graph may be useful to understand the links between the new and the old variables; it is called the "correlation circle". The idea is to display the projection of the old variables – of which composition in terms of the new variables can be computed – on the new variables: Figure 6.30 illustrates.

Name	Correlation with U_1		Correlation with U_2		Sum of squares (%)
	Value	Square	Value	Square	
Mass (cs)	0.957	0.916	0.032	0.001	91.723
Components (cs)	0.283	0.08	0.941	0.885	96.52
Connections (cs)	0.974	0.95	0.072	0.005	95.473
Boards (cs)	0.821	0.674	−0.448	0.2	87.417

Figure 6.31 Correlation table on planes U_1 and U_2.

On this figure we get the representation of an old variables on the planes U_1 and U_2: if these two new variables completely represent an old variable, its dedicated symbol will be on the circle (the name of the graph comes from this fact). Otherwise another variable, such as U_3, may be needed. On the figure it is clear that U_1 correctly nearly (we say "nearly" because a third component may have to be added) represents both the mass and the number of connections, as well as the number of boards but to a lesser extent (another new variable is required for describing it completely); U_2 is a partial representation of the number of components.

It is possible to quantify the visual perception: a value in the table below gives the correlation between the values of the data in the old coordinates system and the values computed in the new coordinates system.

In Figure 6.31 one can see, on a look to the "sum of squares", that the observed plane represents well – as expected from the extensions seen upwards – most of the old variables. This is an interesting point which shows that the two variables U_1 and U_2 would be sufficient to build a rather good cost-estimating relationship (CER)!

It is also possible to represent the cost on the same plane, as it is done on Figure 6.32.

It is clear from this picture that the new axis U_1 gives a very good representation of the cost, as the cost is collinear with it.

What to Do Afterward?

When the information provided by the PCA has been studied, a decision may be taken: should we go on and compute a formula for the dynamic center of the data with the data themselves or compute a formula based on the new variables?

We may here compute such a formula[16] on the new variables.

A First Method

The first method is rather standard: it uses the result of an analysis in principle components (ACP) made on all the variables but the cost, as explained on Figure 6.33: the ACP is made on the space E'.

[16] Note that this computation is completely different from what we did in section "Changing the coordinates system" ("rotating the axis"). In this section we made a PCA on all the variables, including the dependent variable which provided directly a formula. Here we have to make a regression on the new variables.

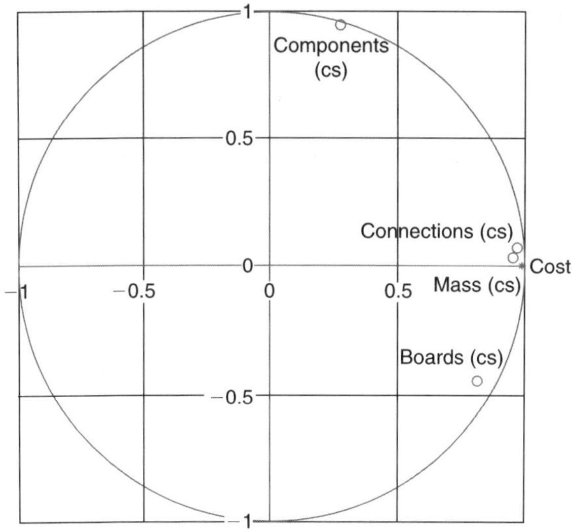

Figure 6.32 Correlation circle, with the cost added, on planes U_1 and U_2.

The first step to do that is to redo a PCA now on the variables not scaled, but only centered.[17] The values of the database now expressed with the new variables are given in Figure 6.33; the cost information, which is not transformed, has been

Couples	Correlation coefficient
U_1–U_2	−0.087
U_1–U_3	0
U_1–U_4	6.77E−05
U_2–U_3	3.84E−05
U_2–U_4	0.001
U_3–U_4	−0.002

Name	Cost	U_1	U_2	U_3	U_4
A	1278	363.57	509.84	−2.12	0.74
B	724	−29.43	−218.06	0.89	−0.04
C	809	−207.73	0.07	−1.28	0.69
D	920	−469.98	189.14	−1.19	0.77
E	772	−29.43	−218.06	0.89	−0.04
F	877	457.15	−410.11	2.88	0.37
G	1064	1719.83	−119.89	0.34	0.38
H	865	−209.49	15.75	−4.43	−0.38
I	961	606.06	46.82	−1.54	−1.53
J	856	−324.18	−185.68	−0.50	−0.43
K	1293	−146.92	748.00	4.44	−0.46
L	717	−867.17	−89.94	0.44	−0.09
M	648	−862.29	−267.90	1.18	0.03

Figure 6.33 Data (not scaled) transformed in the new coordinates system.

[17] This is not compulsory but helps understand the process.

repeated for convenience. It is easy to check that the correlations between these new variables are negligible (as displayed values are limited in precision, an exact zero cannot be all the times expected).

It clearly appears on the table in Figure 6.33 that the axis 4 (U_4) practically conveys no information and that axis 3 (U_3) conveys so little that it can be discarded.

The cost analyst may then decide to build a formula with the first two variables only. The result is:

$$\text{cost} = 906.462 + 0.144 \times U_1 + 0.513 \times U_2$$

For using this formula, it is now necessary to rebuild it with the old variables. The analysis on these centered data gives the following answer:

$$U_1 = 0.001{}_cx_1 + 0.978{}_cx_2 + 0.209{}_cx_3 + 0.000{}_cx_4$$

$$U_2 = 0.005{}_cx_1 - 0.209{}_cx_2 + 0.978{}_cx_3 + 0.100{}_cx_4$$

In these formulae the centered values appear; these centered values are defined as:

$${}_cx_1 = x_1 - 3.299$$

$${}_cx_2 = x_2 - 1015.154$$

$${}_cx_3 = x_3 - 699.385$$

$${}_cx_4 = x_4 - 7.231$$

After all the computations (which can easily be automated) we get:

$$\hat{y} = 493.1064 + 0.0026x_1 + 0.0336x_2 + 0.5318x_3 + 0.0513x_4$$

It is clear on this example that the mass and the number of boards are not really "cost drivers"! If we use the following set of data:

- $x_1 = 4$
- $x_2 = 1000$
- $x_3 = 500$
- $x_4 = 7$

the estimated cost is equal to 793, whatever the unit.

How does this formula differ from the one that would be computed with the standard linear regression on the old variables? We get:

$$\hat{y} = 479.275 + 13.654x_1 + 0.052x_2 + 0.393x_3 + 7.547x_4$$

which is rather different: due to the multi-collinearity effect, the weight given to the various cost drivers is very different. The algorithm does its best to "explain" the cost but is unable to properly discriminate between the variables.

The cost computed with the same set of data is equal to 835. The difference is not so large because the set of data corresponds about to the bulk of the data. Nevertheless the first result, for reasons already explained, should be preferred.

A Second Method

The second method starts with a PCA on all the pre-selected variables, including the cost.

From the analysis of the data we may decide to select the number of components V_2 and the number of connections V_3. We therefore make a PCA on the variables Y, V_2 and V_3 centered and scaled; this PCA provides us with three new variables U_1, U_2 and U_3.

The extensions along these new axes are, respectively, 72.482, 26.069 and 1.449. This means that the first two axes represents about 98.5% of the data. Consequently the formula will be given by writing that:

$$U_3 = 0.719\,_{cs}y - 0.072\,_{cs}x_2 - 0.691\,_{cs}x_3 = 0$$

Returning to the normal values this formula can be written:

$$\hat{y} = 486.297 + 0.02996x_2 + 0.5573x_3$$

An application of this formula to this set of data: $x_2 = 500$, $x_3 = 500$ gives an estimated value of 780.

The standard linear regression gives:

$$\hat{y} = 500.977 + 0.033x_2 + 0.532x_3$$

which, for the same set of data, gives an estimated cost of 783. The difference here is rather small because the correlation coefficient between the cost and the selected variable is quite good (0.958), but could be higher if the data were more scattered.

The difference here does not really come from the collinearity between both cost drivers (their Bravais–Pearson coefficient of correlation is only 0.319), but to the bias introduced by the linear regression. This bias is discussed in Chapter 9.

6.4 Quantification of the Perceived Relationships

Several correlations coefficients can be computed. We start, in the next section, by this Bravais–Pearson correlation coefficient.

6.4.1 Quantification Between the Couples of the Causal Variables

This section considers all the couples of variables and quantifies the correlations inside each couple. It considers only the Bravais–Pearson correlation coefficients.

One could very well use the results obtained in Chapter 5 about the correlation coefficients between two variables, but it is easier to generalize the process. This generalization is described here.

As we are here only concerned by the causal variables, we start here with the matrix $\|x\|$ of the causal variables, each variable being described by the values of one column (whereas a line is dedicated to a product). Note that we are not using in this section the matrix $\|^{+}x\|$.

The result we want to get is a symmetrical matrix called $\|R\|$ containing all the Bravais–Pearson correlation coefficients for all the couples of variables:

$$\|R\| = \begin{Vmatrix} 1 & r_{1,2} & \cdots & r_{1,k} & \cdots & r_{1,J} \\ r_{2,1} & 1 & \cdots & r_{2,k} & \cdots & r_{2,J} \\ \cdots & \cdots & \cdots & \cdots & \cdots & \cdots \\ r_{j,1} & r_{j,2} & \cdots & r_{j,k} & \cdots & r_{j,J} \\ \cdots & \cdots & \cdots & \cdots & \cdots & \cdots \\ r_{J,1} & r_{J,2} & \cdots & r_{J,k} & \cdots & 1 \end{Vmatrix}$$

This matrix is symmetrical because $r_{j,j} = 1$ and $r_{j,k} = r_{k,j}$ by definition, all $r_{j,k}$ being defined by:

$$r_{j,k} = \frac{\frac{1}{I}\sum_i (x_{i,j} - \bar{x}_{\bullet,j})(x_{i,k} - \bar{x}_{\bullet,k})}{\sqrt{\frac{1}{I}\sum_i (x_{i,j} - \bar{x}_{\bullet,j})^2}\sqrt{\frac{1}{I}\sum_i (x_{i,k} - \bar{x}_{\bullet,k})^2}} = \frac{s_{jk}}{s_j s_k} = \mathrm{cov}({}_{cs}V_j, {}_{cs}V_k)$$

with the following notations:

- $\bar{x}_{\bullet,j} = \sum_i x_{i,j} / I$ represents the average value of variable V_j on the I products.
- $s_j = \sqrt{1/I \sum_i (x_{i,j} - \bar{x}_{\bullet,j})}$ the standard deviation of variable V_j.
- ${}_{cs}V_j$ represents the centered and scaled variable V_j of which values are given by the vector, derived from the corresponding column of matrix $\|x\|$:

$$\begin{Vmatrix} \dfrac{x_{1,j} - \bar{x}_{\bullet,j}}{s_j} \\ \dfrac{x_{2,j} - \bar{x}_{\bullet,j}}{s_j} \\ \cdots \\ \dfrac{x_{i,j} - \bar{x}_{\bullet,j}}{s_j} \\ \cdots \\ \dfrac{x_{I,j} - \bar{x}_{\bullet,j}}{s_j} \end{Vmatrix}$$

How is it possible to define ||R|| from $||x||$?

In order to do that, we compute first a vector $||g||$ or $\vec{g}$ defined as the coordinates, in the I dimensional space, of the center of "gravity" of the variables $V_1, V_2, \ldots, V_j, \ldots, V_J$. Its components are:

$$\vec{g} = \left\| \begin{matrix} \bar{x}_{\bullet,1} \\ \bar{x}_{\bullet,2} \\ \cdots \\ \bar{x}_{\bullet,j} \\ \cdots \\ \bar{x}_{\bullet,J} \end{matrix} \right\| = ||x||^t \otimes \left\| \frac{1}{I} \right\| \otimes \vec{1}$$

where $||1/I||$ is defined as a diagonal matrix $\Re^{I\times I}$ of which the diagonal elements are all equal to $1/I$ and $\vec{1}$ a vector $\Re^I$ of which all components are equal to 1.

From this vector it is possible to get the matrix $\Re^{I\times J}$ of the centered data:

$$||_c x|| = ||x|| - \vec{1} \otimes \vec{g}^t$$

Now the matrix $\Re^{J\times J}$ of the variances and covariances *of the data* can be computed as:

$$||V|| = ||_c x||^t \otimes \left\| \frac{1}{I} \right\| \otimes ||_c x||$$

If we define now $||_{(-1)}s||$ the diagonal $\Re^{J\times J}$ matrix of which the diagonal elements are given by the inverse $1/s_j$ of the standard deviations of the variables, one can write:

$$||R|| = ||_{(-1)}s|| \otimes ||V|| \otimes ||_{(-1)}s||$$

Illustration

Let us illustrate these computations with our example. First of all the coordinates of the center of gravity is computed:

$$\vec{g} = \left\| \begin{matrix} 3.299 \\ 1.015 \times 10^3 \\ 699.385 \\ 7.231 \end{matrix} \right\|$$

and then the matrix of the centered data:

$$\|{}_{c}x\| = \begin{Vmatrix} 3.531 & 248.846 & 574.615 & 2.769 \\ -1.119 & 16.846 & -219.385 & -1.231 \\ 0.501 & -203.154 & -43.385 & -1.231 \\ 1.251 & -499.154 & 86.615 & 0.769 \\ -1.119 & 16.846 & -219.385 & -1.231 \\ -1.189 & 532.846 & -305.385 & -1.231 \\ 1.371 & 1.707\times10^{3} & 242.615 & -1.231 \\ -0.489 & -208.154 & -28.385 & -4.231 \\ -0.749 & 582.846 & 172.615 & -1.231 \\ -1.619 & -278.154 & -249.385 & -2.231 \\ -3.001 & -300.154 & 700.615 & 11.769 \\ -1.319 & -829.154 & -269.385 & -0.231 \\ -2.049 & -787.154 & -442.385 & -1.231 \end{Vmatrix}$$

The symmetrical matrix of the variances and covariances of the data is given by:

$$\|V\| = \begin{Vmatrix} 2.957 & 287.894 & 534.996 & 4.416 \\ 287.894 & 4.16\times10^{5} & 6.844\times10^{4} & -294.728 \\ 534.996 & 6.844\times10^{4} & 1.108\times10^{5} & 895.757 \\ 4.416 & -294.728 & 895.757 & 13.87 \end{Vmatrix}$$

and eventually, from this matrix, the matrix of the linear correlations between all parameters is computed:

$$\|R\| = \begin{Vmatrix} 1 & 0.26 & 0.935 & 0.689 \\ 0.26 & 1 & 0.319 & -0.123 \\ 0.935 & 0.319 & 1 & 0.723 \\ 0.689 & -0.123 & 0.723 & 1 \end{Vmatrix}$$

6.4.2 Quantification of the Other Correlations Inside the Couples

There is no way to compute directly from the matrix of the data the various correlations coefficients which were described in Chapter 5 dedicated to the analysis of a pair of variables. The only solution is to apply the algorithms to all the pairs of variables.

Here are the **Spearman correlation** coefficients for all couples (Figure 6.34).

Most of the Spearman correlation coefficients are rather similar the Bravais–Pearson ones, with the exception of the couple "connections–boards". A simple look at

Couples	Coefficient correlation
Mass–components	0.280
Mass–connections	0.934
Mass–boards	0.407
Components–connections	0.346
Components–boards	−2.209
Connections–boards	0.396

Figure 6.34 The Spearman correlation coefficients.

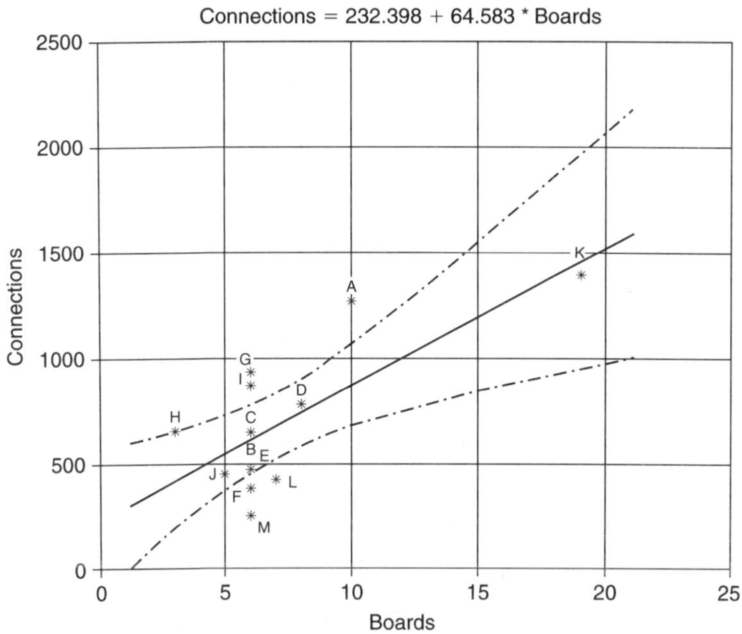

Figure 6.35 The correlation between the number of boards and the number of connections.

the diagram between these variables (Figure 6.35) explains this result: there is a – small – correlation between these variables, but it clearly appears that the correlation between the ranks is not high.

The Kendall correlation coefficients may also be computed (Figure 6.36).

Couples	Coefficient correlation
Mass–components	0.000
Mass–connections	1.000
Mass–boards	0.250
Components–connections	0.000
Components–boards	0.316
Connections–boards	0.250

Figure 6.36 The Kendall correlation coefficients.

6.4.3 Partial Correlations

The previous sections studied the correlation coefficients between two variables, independently of the other variables. Let us call $r_{i,j}$ the Bravais–Pearson correlation coefficient between variables V_i and V_j. For instance we found for our example $r_{1,2} = r_{\text{mass,components}} = 0.26$ whatever the values of the other variables. The question may be asked if this coefficient remains the same if another variable is kept constant.

The reason for investigating this point is that the correlation between V_1 and V_2 may be due to another variable, although V_1 and V_2 are not really correlated.[18]

Let us assume we know the value of variable V_3: What is then the correlation coefficient between V_1 and V_2? It can be demonstrated[19] that:

$$r_{1,2|3} = \frac{r_{1,2} - r_{1,3} \times r_{2,3}}{\sqrt{(1 - r_{1,3}^2) \times (1 - r_{2,3}^2)}}$$

This can be generalized to more known variables.

Numerical Application

In the example, we write:

$$r_{1,2|3} = \frac{0.26 - 0.935 \times 0.319}{\sqrt{(1 - 0.935^2) \times (1 - 0.319^2)}} = -0.114$$

which means that the correlation between V_1 and V_2 now changes its sign (but is so low that it can probably be inferred that there is no correlation at all)!

It is always interesting to have a look at these partial correlations in any study involving several quantitative variables. If it happens that two variables are highly correlated but that this correlation disappears – or, more exactly, nearly disappears – when a third variable is kept constant, then the best solution may be to delete these two variables and to keep only the first one.

This may happen for instance when two characteristics of an equipment (such as the sensitivity and the accuracy of a sensor) evolve at about the same – slow – pace with time, due to engineering efforts. In such a case both variables appear as correlated, but this correlation disappears when the time is kept constant. This is rather frequent for high-technology items, as the technical progress often goes simultaneously for several characteristics. The solution can be to delete both variables and to keep only the time (for instance the year of design).

[18] A well-known example is the study of the three variables "average personal income", "sales of personal computers" and "alcoholic consumption". The study may show that "sales of personal computers" and "alcoholic consumption" are very well correlated! However, if we recompute this correlation coefficient keeping the "average of personal income" constant, then the correlation disappears: it was due to the fact that when the personal income increases, then both the "sales of personal computer" as well as the "alcoholic consumption" increase.

[19] The notation $r_{1,2|3}$ means: correlation between V_1 and V_2, given V_3.

6.4.4 Multiple Correlations Between Variables

Up to now we investigated only the correlations between two variables, plus the possibility that this correlation may be due to another variable.

It is also possible to investigate the correlation between one variable and two other ones, in order to see if one variable can be explained by the simultaneous values of these two variables. The corresponding coefficient is called "multiple correlation coefficient" and is often represented by a capital *R*, the variables concerned being set as indices. For instance the multiple correlation coefficient between variable V_1 and the set of V_2 plus V_3 is named $R_{1.23}$, the dot in the index being used for clarity.

This multiple correlation coefficient is easily computed from the individual correlations:

$$R_{1.23} = \sqrt{\frac{r_{12}^2 + r_{13}^2 - 2r_{12}r_{13}r_{23}}{1 - r_{23}^2}}$$

6.4.5 Multiple Linear Correlation

The multiple correlation coefficient looks at the link between the dependent variable and the *J* causal, quantitative, variables.

It is computed (Saporta [50], p. 139) as the greatest value of the correlation between this dependent variable and any linear combination between the causal variables.

$$R_m = \sup_{\forall a_i} r\left(y; \sum_{j=1}^{J} a_j x_j \right)$$

This formula implies that $0 \leqslant R_m \leqslant 1$. We will find again this value when the linear regression will be studied; this formula will be obtained when the intercept is forced to 0.

If we call:

- $\|_c x\|$ the data matrix with centered values,
- $\|_c y\|$ the vector of the centered values of the dependent variable:

$$R^2 = \frac{\|_c y\|^t \otimes \|_c x\| \otimes (\|_c x\|^t \otimes \|_c x\|)^{-1} \otimes \|_c x\|^t \otimes \|_c y\|}{\|_c y\|^t \otimes \|_c y\|}$$

7 Working with Qualitative Variables

Summary

The introduction of qualitative variables is a bit disturbing, because everything we have been doing up to now was dedicated to quantitative variables.

However, it is possible to use the same logic once a slight change is made to the cost. This change implies a preliminary computation of the influence of the qualitative variables.

Once this is done the same data analysis can be carried out on the quantitative cost drivers and the modified cost:

1. search for outliers,
2. investigating the possible collinearities,
3. visualization of the data, in order to understand their structure,
4. quantification of the perceived relationships. Nothing has to be changed for the quantitative variables. One can also investigate the correlation between quantitative and qualitative variables, and even between qualitative variables. This will help understand the structure of the data and prepare future computations.

Using qualitative variables is a very important concept in cost estimating: the cost analyst should therefore be able to interpret them.

The quantitative variables are for instance the cost and the product size, the qualitative variable(s) being anything as previously indicated, such as the material, or the manufacturer, or the quality level, etc.

We are now going to illustrate the concepts with the following data; these data refer to electrical engines: column 1 gives the product name, column 2 its price (the unit is irrelevant), column 3 the mass (kg) and column 4 a qualitative variable which describes the way these engines were designed; here the change in design is given by the number of poles (0 means 2 poles, 1 means 6 poles). Values 0 and 1 should not be confused with quantitative variables; one could have named them "engine with 2 poles" and "engines with 6 poles" (Figure 7.1).

The analysis of the data, in the presence of qualitative variables, is about the same as with quantitative variables only.

However, most of the algorithms which were developed for the search of outliers and for the study of multi-collinearities, use the results of linear algebra, algebra

Name	Price	Mass	Poles
a	819	4.5	0
b	934	5.5	0
q	934	5.0	1
r	983	5.5	1
c	1038	6.5	0
d	1212	9.0	0
e	1403	10.0	0
g	1643	13.0	0
h	2049	16.0	0
i	2474	21.0	0
s	2807	27.0	1
j	2830	25.0	0
t	3540	39.0	1
k	3726	37.0	0
u	4315	46.0	1
l	4710	42.0	0
v	5462	54.0	1
m	6883	76.0	0
w	7211	79.0	1
n	8742	85.0	0
x	9877	93.0	1
o	10402	95.0	0
p	12357	120.0	0
y	12739	145.0	1

Figure 7.1 An other example.

which is only able to handle quantitative values. Consequently some preliminary work has to be done in order to use its procedure.

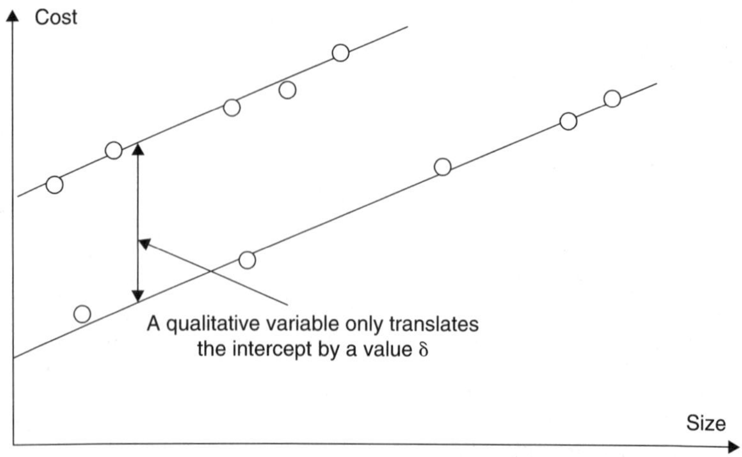

Figure 7.2 The role of a quantitative variable in the case of an additive formula.

What can be proposed is the following treatment. This treatment anticipates on the results of Chapter 10. We will see in this chapter that the qualitative variables only "translates" the formulae parallel to the cost axis by a value δ as illustrated in Figure 7.2

in the case of an additive formula with two modalities only (the result is also true for the other types of formulae). The idea is then to compute, for each modality of the qualitative parameter, the amount of translation and to remove this amount from the cost of the relevant product. This procedure provides "adjusted" costs which are not anymore influenced by the qualitative parameters: it is then possible to proceed with the quantitative variables only.

If outliers or other problems are detected on these adjusted costs and if you decide that a product or even a variable must be deleted, the whole process must be redone without this product or this variable. Of course it is a little more time consuming that the treatment of pure quantitative variables, but the treatment is so quick that it is still worth doing it (it can be – and has been – automatized).

7.1 Looking for Outliers

The procedures described in the previous chapters can all be used. The result depends on the presence or the absence of the qualitative variable: column 2 gives the list of the potential outliers if the analyst does not care about the qualitative variable (no correction is made for the change of intercepts), whereas column 3 adjusts the cost data according to this change (Figure 7.3).

Procedure	Potential outliers detected	
	With no qualitative	With qualitative
Looking for residuals	y	y
Looking to the "HAT" matrix	p and y	y
Changes of the covariance matrix	o and y	o, x and y

Figure 7.3 Potential outliers detected.

The potential outliers are about the same, but some differences can be noticed. Our preference goes to the third column results, as we consider that qualitative variables should be taken into account.

7.2 Dealing with Multi-Collinearities

The procedure is exactly the same as described earlier. No example is given here due to the presence of one quantitative variable only.

7.3 Visualization of the Data

The visualization of the data should not take into account the qualitative variable, in order to clearly see their interest. This interest immediately appears

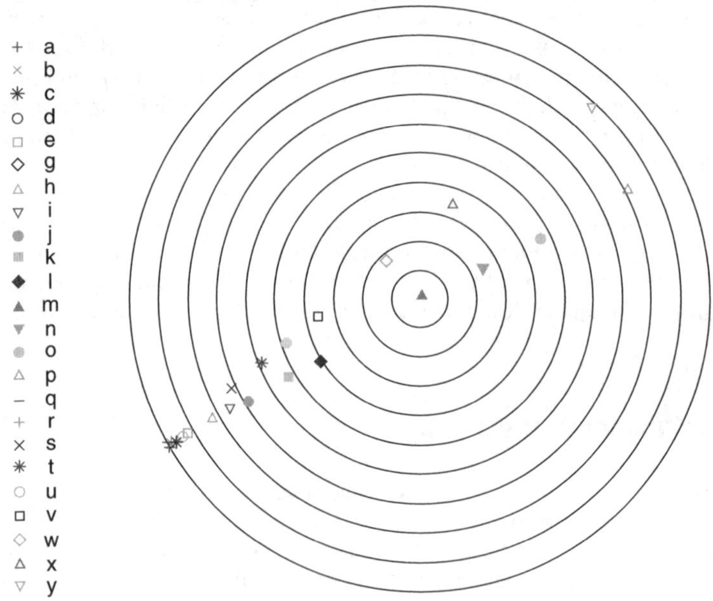

Figure 7.4 The star diagram.

on the "star diagram" as illustrated: two trends are seen which shows that qualitative information should be taken into account when preparing the formula (Figure 7.4).

The question is slightly different for the principal component analysis (PCA): the standard use of the PCA is only interested in the relative position of the quantitative variables. So there is no use of it for the qualitative variable, of which unique function is to adjust the cost.

However, new visualization can be introduced in order to get a representation of the independence or the relationship between one quantitative and one qualitative variable, or even between two qualitative ones.

Visualize One Quantitative Variable and a Qualitative One

The solution which can be proposed in order to give to the eye a general picture of the data is illustrated on Figure 7.5: using a graph on which both axis refer to the quantitative variable. Then the data points are all on the bisector of the graph. In order to distinguish the qualitative variable a different symbol is used depending on the value of the modalities.

For the example given, no relationship seems to exist: the modalities seems to be randomly distributed among the values of the quantitative variable.

Visualize Two Qualitative Variables

It is possible to visualize two qualitative variables simultaneously in order to look at the possible correlation between these variables.

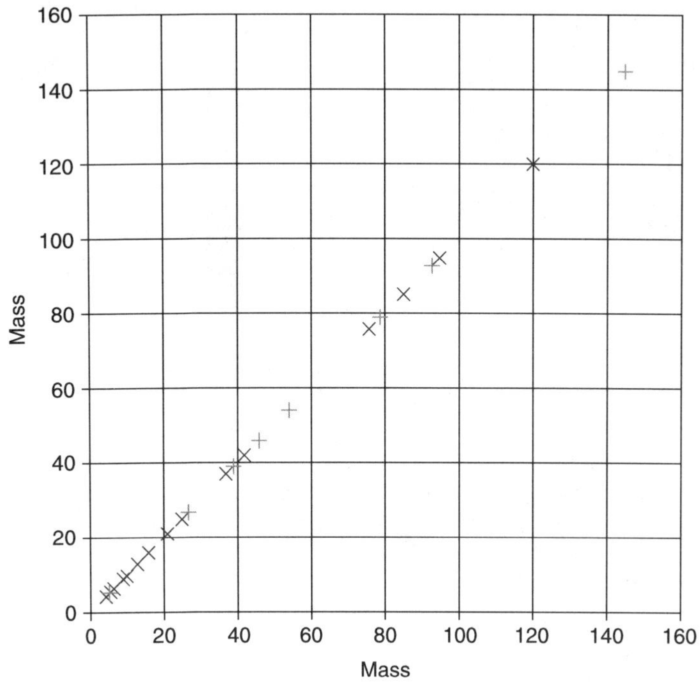

Figure 7.5 Visualization of the relationship between a quantitative variable and a qualitative one.

The following example illustrates the idea; it displays the possible relationship between two qualitative variables used by Barry Boehm (Ref. [8], p. 496):

- *Type of software*: business application, process control, human–machine interactions, scientific application, support software, system software.
- *Programing language.*

It immediately appears that:

- *Cobol* is practically only used for business applications (this was expected).
- *Fortran* and high-order language were (COCOMO database refers to software made in the 1970s) the most used languages.
- *Fortran* was primarily used for scientific applications (this was expected too).
- Human–machine interactions applications are the software with the greatest diversity of language (Figure 7.6).

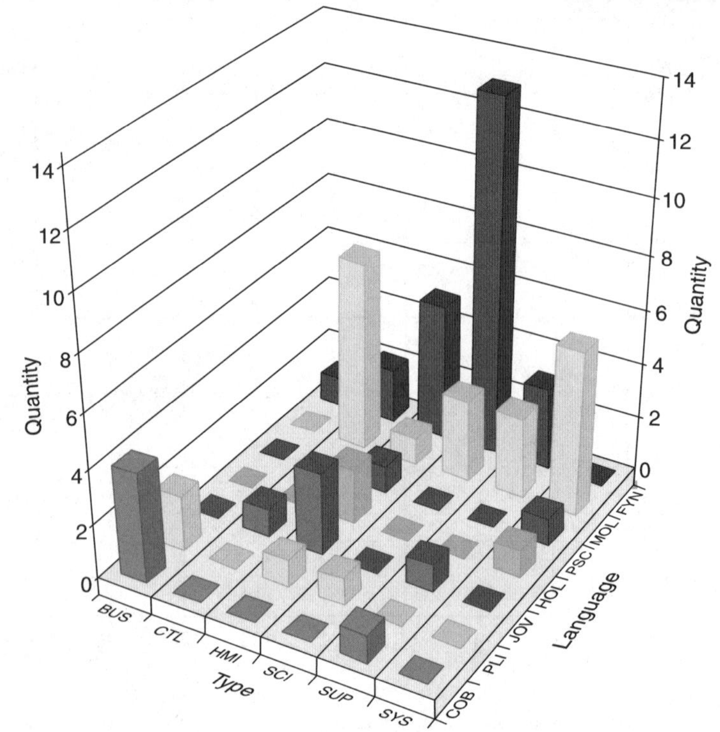

Figure 7.6 Visualization of the relationships between a type of application and programing language.

7.4 Quantification of the Perceived Relationships

This section deals only with the analysis with at least one qualitative variable.

7.4.1 Correlation Between One Quantitative Variable and One Qualitative

This section analyzes the possible correlation between a quantitative variable (which may be the cost or any quantitative parameter: it will be noted z to remain general) and a qualitative one, inside the sample, the objective being to see if both variables express or not the same idea.

The sample includes I products, for which the average value of the quantitative variable is $\bar{z}$ and the standard deviation s_z.

Let us assume that the qualitative variable has M modalities, numbered from 1 to M. This allows splitting the data into M subsets, each subset corresponding to one modality. Inside each subset m:

- the number of products is called I_m,
- the average value of the quantitative variable is called $\bar{z}_m$ (this is the arithmetic mean).

The correlation coefficient between both variables is given by Saporta ([50], p. 148):

$$\eta = \frac{\frac{1}{I}\sum_{m=1}^{M} I_m \times (\bar{z}_m - \bar{z})^2}{s_z^2}$$

It is clear that $\eta^2 = 0$ if $\bar{z}_1 = \bar{z}_2 = \cdots = \bar{z}_m = \cdots = \bar{z}_M = \bar{z}$ (there is absolutely no correlation between both variables) and that $\eta^2 = 1$ if all the products presenting the same modality have the same value of z (the correlation is the perfect). η^2, therefore has the properties of a correlation coefficient.

Example

For the example given the correlation coefficient takes the value 0.222: this is practically no correlation between the mass and the number of poles, result which confirms the visual perception we had about these variables.

7.4.2 Correlation Between Two Qualitative Variables[1]

Let us consider two qualitative variables:

- the first one has M modalities noted from 1 to M,
- the second has L modalities noted from 1 to L.

The objective is always to see if they express the same idea or not.

In order to answer this question, the "contingency table" is computed: it gives in cell l, m (row l, column m) the number of products which have both modalities l and m:

	1	2	m	M	
1	$n_{1,1}$	$n_{1,2}$	$n_{1,m}$	$n_{1,M}$	$n_{1,+}$
2	$n_{2,1}$	$n_{2,2}$	$n_{2,m}$	$n_{2,M}$	$n_{2,+}$
l	$n_{l,1}$	$n_{l,2}$	$n_{l,m}$	$n_{l,M}$	$n_{l,+}$
L	$n_{L,1}$	$n_{L,2}$	$n_{L,m}$	$n_{L,M}$	$n_{L,+}$
	$n_{+,1}$	$n_{+,2}$	$n_{+,m}$	$n_{+,M}$	I

Totals in line and in row are given by (the "+" sign which replaces an index means that the values rated to this index were added):

$$n_{l,+} = \sum_m n_{l,m} \text{ and } n_{+,m} = \sum_l n_{l,m}$$

[1] See Saporta ([50], p. 150).

The contingency coefficient is given by K. Pearson[2]:

$$\eta = \left(\frac{d^2}{I + d^2} \right)^{1/2}$$

with

$$d^2 = \sum_m \sum_l \frac{\left(n_{l,m} - \dfrac{n_{l,+} n_{+,m}}{I} \right)^2}{\dfrac{n_{l,+} n_{+,m}}{I}}$$

It is clear that $d^2 = 0$ when both variables are independent.

Example

Returning to the data displayed in Figure 7.6, one computes a value for the contingency coefficient equal to 0.74.

This quantification shows that, as it is appeared in a qualitative way on the graph, that languages are not uniformly spread on all types of software, and that programers who work for different applications have a trend to select a language as a function of the software (and the contingency coefficient quantifies this trend). This is certainly obvious for people in the software industry, although the importance of the trend may not be known, but probably not for the cost analyst who discovers this business.

[2] See Saporta ([50], p. 154).

Part III

Finding the Dynamic Center of a Multi-Variables Sample

Part Contents

Once the data analysis has been properly carried out, it is time to investigate the purpose of this book, which is the construction of a specific model, specific meaning "dedicated to a proper family".

As explained in Part I, the procedure for doing so is to replace the complex (because many variables may be involved) distribution of the cost by something simpler. This is achieved by removing from the cost values the influence of all the variables. What remains once this has been done is a simple distribution which does not depend anymore on any variable.

The remains are called the "residuals": it is what is left over when the influence of all the variables has been removed.

The purpose of this part is to find out the "best" way to do this removal. As a matter of fact you will discover there is no best way! Several ways are possible and cost analysts use one or the other depending on the data or their personal preference.

But no solution is perfect; the cost analyst should therefore know them (more exactly several of them, because there is potentially an infinite number of ways) in order to select the one which is adequate for the data he/she has to work with.

These ways are called "metric", for a reason which will appear in this chapter: it is the way the "distance" between two values is defined and computed.

Several metrics will be investigated for two reasons:

1. The first reason is to convince the cost analyst that he/she has to make a choice, because no metric is really perfect under all circumstances.
2. The second is to let him/her know the advantages and the inconveniences of all of them.

An important consequence of this choice is that two cost analysts, working with the same data, may propose two different formulae, which may differ more especially as the data are more scattered: there is nothing such as a unique formula.

8 Finding the Center of the Cost Distribution for Choosing a Metric

Summary

This chapter is an important one of this book. It is an introduction to the algorithms that have to be used for finding a specific model.

In this chapter we study the distribution of the cost without using any other variable; the purpose is to see how the center of this distribution can be found.

Generally speaking the distribution of a quantitative variable can be described (see Chapter 2) as a set of two things:

1. The value of the "center" of this distribution.
2. How the "residuals" (the differences between the values of the variable and the center just found) are scattered around this center.

Splitting a distribution into these two terms is extremely useful and is the basis of the work the cost analyst has to carry out. When trying to extract from the data a specific cost model for forecasting the cost of future products, the center of the distribution will lead to what can be called the cost-estimating relationship (or CER) from which the "nominal cost" of these future products will be computed.

The "residuals" are what is left from the data when the value of the center has been removed from the cost values.

This part deals only with the search of the center as the definition of the metric to be used is a prerequisite toward this end. Part IV will study the residuals.

The question of finding the center of a distribution is completely independent of the formula type that can be chosen by the cost analyst. The answer to it can be used whatever this type: this is the reason of its importance.

Choosing the algorithm which will be used for finding the center of a distribution is one of the most important decision that the cost analyst has to make (the second one being the formula type). For this reason the different algorithms which can be used will be explored in order for the cost analyst to understand the advantages and disadvantages of each. The reader will discover that **no one solution is theoretically better than the other ones** and imposes itself to the cost analyst.

The distributions which are studied here use one variable only. Consequently the "centers" are quantified with one value only. When we study the distribution involving two or more variables, the center cannot be one value anymore: we will introduce, in this Part III, the concept of the "dynamic center". This concept will use all the developments made in this chapter.

8.1 Introduction

This chapter introduces the most important concepts for using the data for cost estimating.

Let us start with the concept of "distribution"; it is a very important one and the reason is obvious: we are looking for the cost of a new product. This product belongs to a product family. Consequently its cost will be known if the distribution of the cost of all the products belonging to this product family is known.

Inside a product family several figures (cost and parameters, quantitative or qualitative) are attached to a product.

The distribution of the cost data inside a product family is the set of all the costs, each one being related to one product. As the number of products is potentially infinite, this distribution may be extremely complex and unmanageable.

The idea which was found in order to manage these data is *to replace this distribution* by the couple made of:

1. a "**center**", which contains, hopefully, most of the information included in the data.
2. distribution of the **residuals** around this center.

The interest of this idea is to replace something which could be extremely difficult to handle by something very manageable. In other words, a multi-variable distribution is replaced, once the center has been found, by a one variable distribution, the distribution of the residuals around the center.

The center will give birth to the formula which we are looking for. The distribution of the residuals – which is, for the cost analyst, as important as the formula itself – will be used to provide some useful information about the quality of the formula in order to answer the question: "don't we loose too much information when we replace the distribution of the cost by a simple formula?"

The price to pay for going from complex to simple is twofold: it requires some work for:

1. establishing the value of this center,
2. studying the residuals around this center, this subject being postponed to Part IV.

This price is far from negligible: many statistical books have been written on these subjects, without giving a definite answer, at least in the domain of cost. The reason for that, as the reader will discover, is that there cannot be a definite answer. He/she will realize that it is impossible to define one unique center. The cost analyst has to choose the type of center he/she wants to use for his/her own purpose. Practically this means that two persons, working with the same data, will generally (unless the data perfectly fit with the formula type he/she selects) obtain different values for this center, and the reader will notice that the differences increase with the spread of the data.

This creates a very serious problem, because it means that these two persons, although working with the same data, will not estimate the same cost for the same new product. For this reason, if you have to share formulae with somebody else, it is highly recommended that the formulae will be accompanied by the procedure which was used to compute it.

At this stage it must be remembered that:

There is nothing such as a unique formula for a CER even if the formula type is imposed because there is nothing such as a unique center for a distribution.

As it is a very important subject for cost estimating, this whole chapter is dedicated to it. It must be considered as an introduction to the subject.

As an introduction, this chapter deals with one variable only (generally speaking the cost). It is not expected – unless you want to carry out special studies such as the costs, proposed by several manufacturers, for the same product, or the specific cost (cost per size unit, such as the cost per kilogram or per square meter) for a product family – that you will work often with one variable only. The purpose of this chapter is not to suggest that you should do it, but to introduce on simple distributions some very important concepts; all these concepts will be used when "serious" matters will be considered: building a formula for preparing a CER.

Searching for the center generally supposes that some preliminary work has been performed. This preliminary work, which is described in the previous chapters, is supposed to have been performed.

One Example

For illustrating the discussion of the different approaches, we will use the following example (which is the same as the one used in Chapter 2):

3.4
4.2
6.3
8.2
9.9
16.3
21.2
35.9
46.5
64.5
84.5

This example uses one variable only. It may be the cost of the same product, or any other observations of the same phenomenon (for instance the speeds of vehicles which cross a signal at a given moment of the day). Of course these data are rather scattered, but they help understand what the problems are.

We will make an extensive use of such a distribution when we study the distribution of the residuals around the distribution center. The present example was selected for this purpose: it appears rarely when studying a phenomenon, but it is a representative example of what may happen for the residuals.

The first important characteristic of any distribution is its center; it has therefore to be determined with care. In Chapter 2 a preliminary discussion was carried out about this center; we need now to have a full discussion of it. You may have an intuitive perception of what the center of a distribution is. This is nice, but in order to work with it (calculate it and use it), we need a formal definition of what the center is. You will then discover that the concept of center is far from obvious and that it is impossible to define the center in an unambiguous way: *there are as many centers as you may think of*, and each center has a special purpose.

The general **definition** of the center of a distribution – in the present case when dealing with just one variable – can be the following one: the center is "**a value**

which is as close as possible to all values present in the database". This value will be generally represented by $\hat{y}$; left indices (such as ${}_b\hat{y}$ or ${}_m\hat{y}$) can be used for representing different types of centers; however the very common notation $\bar{y}$ will be kept in order to represent the arithmetic mean.

A First Attempt: Trying a Global Approach

The global approach was introduced in Chapter 2 because it was needed as an introduction to the center of a distribution. It is reminded here; it will be shown in the following pages of this chapter that this approach is just a particular case of more general metrics.

The global approach is rather straightforward. It consists in saying that the center $\hat{y}$ of the distribution will be the value computed in the following ways.

Using the Differences

If the differences are used (two data are said to be close to each other if their difference is close to 0), the center is the value for which the sum of all the differences between this center and all the data is equal to zero:

$$\sum_i (\hat{y} - y_i) = 0$$

The value can be then immediately computed as:

$$\hat{y} = \frac{1}{I}\sum_i y_i$$

which is called the "arithmetic mean". Let us remind the reader that, because it is largely used, it is generally just called the "mean" and receives a particular symbol: $\bar{y}$. For the example given, $\bar{y} = 27.355$.

Using the Ratio

If the ratios are used (two data are close together if the ratio of their values is close to 1), the center is the value for which the product of all the ratios between the center value and all the data is equal to 1:

$$\prod_i \left(\frac{y_i}{{}_g\hat{y}} \right) = 1$$

Here also the value is immediately computed as:

$${}_g\hat{y} = \sqrt[I]{y_1 \times y_2 \times \cdots \times y_i \times \cdots \times y_I}$$

which is called the "geometric mean", generally represented by the symbol ${}_g\hat{y}$. For the example given, ${}_g\hat{y} = 16.39$.

These values are rather different! But both can legitimately be called the "center" of the distribution. It is a first perception of the fact that the center is a value which results from conventions, or from a deliberate choice.

For our purpose this procedure for finding the center is a bit "rough", as – due to the sign effect in the first case, to the quotient effect in the second case – values may compensate each other. We would like to define the center with more flexibility in order to adjust it to the particular problem we have to solve.

8.2 Defining the Distance Between Two Values: Choosing a Metric

The definition of the center given in the introduction to this chapter is perfect if we are able to define what we mean by "close". In order to make such a definition operational, we have to decide on the way the closeness will be measured, which means we have to choose a "metric".[1]

Without starting from the mathematical axioms, we can simply say that a distance between two numbers a and b or $d(a, b)$ is a number attached to the two values a and b having the following properties:

1. $d(a, b) \geqslant 0$
2. if $a = b$, then $d(a, b) = 0$
3. $d(a, b) + d(b, c) \geqslant d(a, c)$.

There are obviously many different ways to define such a distance; to start with, one can consider the following metrics:

- Using the difference: we can say two values are close together if their difference is in the vicinity of 0. In order to express this difference, we will use $d(a, b) = |a - b|$; for adding some flexibility in this definition, we prefer to use $d(a, b) = |a - b|^{\alpha}$, α being a value chosen by the cost analyst.
- Using the ratio: two values are close together if their ratio is in the vicinity of 1. The ratio itself cannot be used as a distance, because property 2 is not satisfied. The distance must then be defined as:

$$d(a,b) = \left|\frac{b}{a} - 1\right|, \text{ or preferably } d(a,b) = \left|\frac{b}{a} - 1\right|^{\alpha}.$$

These two ways (using the difference or the ratio) express of course the same idea: saying that a is close to b means that $a - b \approx 0$ which means that $a \times (1-(b/a)) \approx 0$

[1] According to the "Webster's New Collegiate Dictionary", a metric is defined either as a standard of measurement or a mathematical function that associates with each pair of elements of a set a real non negative number constituting their distance and satisfying the conditions that the number is 0 only if the two elements are identical, the number is the same regardless of the order in which the two elements are taken, and the number associated with one pair of elements plus that associated with one member of the pair and a third element is equal or greater than the number associated with the other member of the pair and the third element. This is exactly the definition we use here.

<table>
<tr><td colspan="3">“Global” approach</td></tr>
<tr><td colspan="2">Using the difference</td><td>Using the ratio</td></tr>
<tr><td colspan="3">“Local” approach</td></tr>
<tr><td rowspan="2">No weight</td><td>Using the difference</td><td>$\left|\hat{y} - y_i\right|^{\alpha}$</td></tr>
<tr><td>Using the ratio</td><td>$\left|\frac{\hat{y}}{y_i} - 1\right|^{\alpha}$
$\left|\frac{y_i}{\hat{y}} - 1\right|^{\alpha}$
$\left|\log \frac{\hat{y}}{y_i}\right|^{\alpha}$</td></tr>
<tr><td>Weighted</td><td colspan="2">Using the biweight</td></tr>
</table>

Figure 8.1 The main distances generally used for searching the center $\hat{y}$.

and therefore $(b/a) \approx 1$. But the mathematical computations – and some of the results – are different.

Instead of the ratio itself, we could use its log and define $d(a,b) = |\log(b/a)|^{\alpha}$ which now fulfills property 2. This is an interesting metric.

At this stage it is not possible to decide what is the best choice between these different metrics: they have to be investigated. Figure 8.1 gives the list of the different metrics we will investigate.

This list is not exhaustive: the number of algorithms which could be used for quantifying the distances is potentially infinite. The list gives the major ones.

The “global” approach is the one we first considered. We saw that this “global” approach is too rough for our purpose; then we turn now to the “local” approach. The rest of this chapter is dedicated to the investigation of the metrics which were just defined.

About the weighted approaches: giving a “weight” to each product of the sample data may help solve several problems which appear for searching for a center; for example, it allows to decrease the influence of outliers that we prefer (because their values are considered, at least partly, as reliable) not to delete. Such weights could be manually entered for each product, but one can prefer to have an automatic weighting process. Several processes have been proposed by different authors, the most interesting being the biweight: so the discussion about the weighting process will be limited to this procedure.

Abbreviations and Other Metrics

The difference $y_i - \hat{y}$ is sometimes called the "additive error" (although we do not like the term "error" as it is used here: we prefer to use the term "residual") and noted e_{+i} because $y_i = \hat{y} + e_{+i}$.

The metric based on the squared differences $|\hat{y} - y_i|^2$ is generally called the Euclidian metric, or the "ordinary least squares", or "OLS". When the difference of the log is used instead, it becomes the "OLS on logs" or "OLSL".

The ratio $y_i/\hat{y}$ is often called the "multiplicative error" and noted $e_{\times i}$, because $y_i = \hat{y} \times e_{\times i}$.

The quantity $((y_i/\hat{y})-1) = (y_i-\hat{y})/\hat{y}$ is often called the "percentage error" (it can be multiplied by 100 without changing anything) and noted $e_{\% i}$. The metric based on its square $((y_i-\hat{y})/\hat{y})^2$ is then called the "minimum percent error" or "MPE".

One can also define other metrics which are similar to the previous one, **plus a constraint.** For instance one may want to use the metric defined by $((y_i-\hat{y})/\hat{y})^2$ with the constraint $\sum_i((y_i-\hat{y})/\hat{y}) = 0$. As such a sum is sometimes called the "bias" (although the term is not correct, as the bias is generally defined as an asymptotic property), such a metric is sometimes called "MPE with 0 bias" or "ZMPE".

One can immediately note here that the OLS automatically generates a 0 bias, because, as we will see it, for this metric $\sum_i(y_i-\hat{y}) = 0$.

8.3 A First Approach: Using the Differences

8.3.1 Definition

The distance (this is sometimes called the Minkowski's distance) between y_i and $\hat{y}$ is given by:

$$d(y_i, \hat{y}) = |\hat{y} - y_i|^\alpha$$

Figure 8.2 presents, on an example ($y_i = 20, \hat{y} = 30$, or the opposite because the formula is symmetric), the value of the distance as a function of α the range of α will be limited from 1 to 2, which is the generally accepted range (the distance still decreases when α becomes lower than 1).

As expected, the distance becomes greater and greater when α increases. This is intuitive, and just need to be reminded.

Figure 8.3 shows, for three given values of α, how this distance changes when $\hat{y}$ goes from 10 to 30. The curves are symmetrical around $\hat{y} = 20$ – as is the formula. Two points must be noticed:

1. When $\hat{y}$ is close to y_i by about $\pm 10\%$, the value of the distances are rather similar, whatever the value of α.
2. But this distance grows extremely fast with α when $\hat{y}$ is at more than 20% of y_i. This explains that, when using $\alpha = 2$ in a sum including small and high differences, the small differences have a very little "weight" in the sum, which is dominated by the large differences: the center is "attracted" by large values.

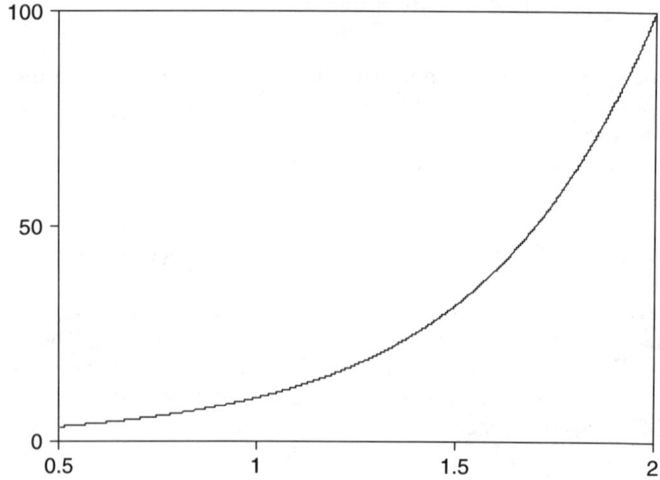

Figure 8.2 How does the distance change with α computations made with $y_i = 20, \hat{y} = 30$ and α from 1 to 2.

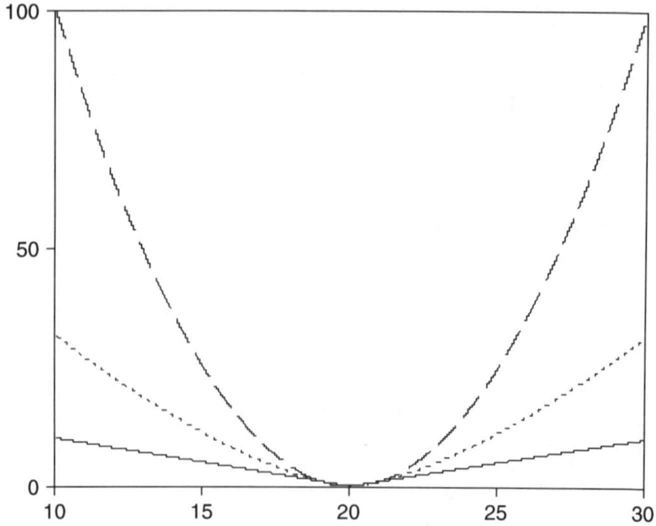

Figure 8.3 How does the distance change with $\hat{y}$ (from 10 to 30) for three different values of α: $\alpha = 1$ (full line); $\alpha = 1.5$ (dotted line) and $\alpha = 2$ (dashed line).

8.3.2 Computing the Center According to This Metric

In order to find the center $\hat{y}$ of this distribution, let us try to compute the value which minimizes the sum – which is a function of α and $\hat{y}$ – of the distances:

$$\text{sum}(\alpha, \hat{y}) = \sum_i |\hat{y} - y_i|^\alpha$$

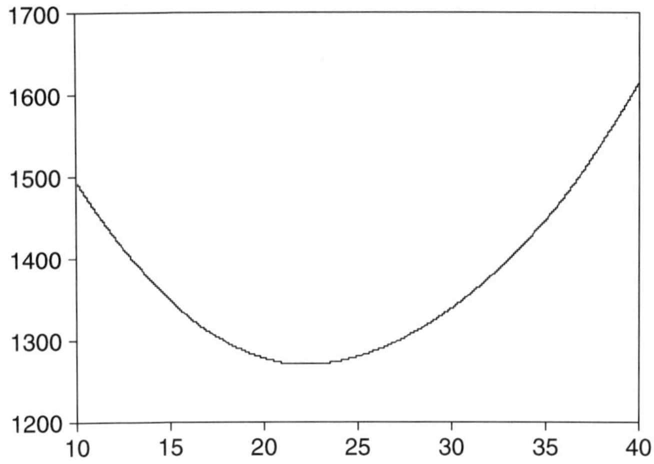

Figure 8.4 Searching $\hat{y}$ for minimizing the sum for $\alpha = 1.5$.

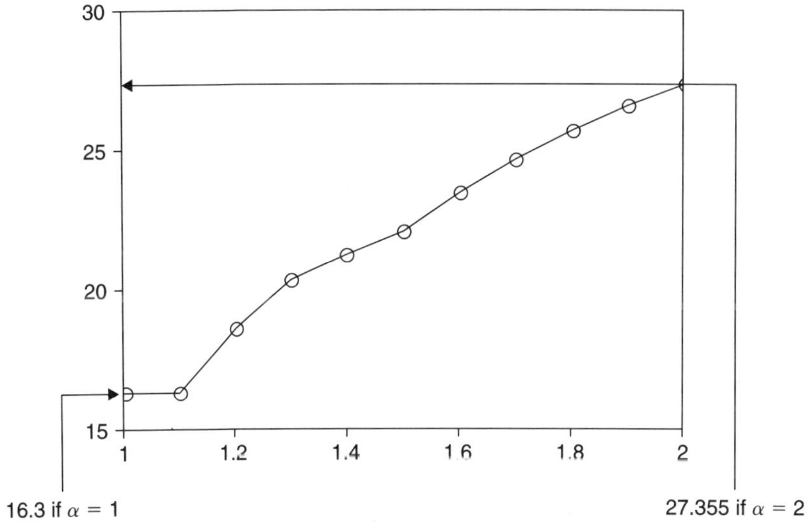

Figure 8.5 How does this center $\hat{y}$ change with α?

The computation must be made step by step: one chooses a value for α, computes the sum for various values of $\hat{y}$ and determine the minimum values. This is represented in Figure 8.4 for $\alpha = 1.5$. With the set of values listed on this figure, the minimum is obtained for $\hat{y} = 22.076$ which is the center of the distribution for this metric.

An interesting feature of this curve is that the minimum is rather well defined: the curve is not very flat in the vicinity of the minimum.

How does this value change with α?

The same type of computation allows to find the change of the center when α is changed. Results are given in Figure 8.5. Two important values must be noted: $\hat{y} = 27.355$ for $\alpha = 2$ (this the Euclidian metric), $\hat{y} = 16.3$ for $\alpha = 1$. The first value corresponds to the usual "arithmetic mean" or simply the **mean** $\bar{y}$, the second to the **median** $\tilde{y}$. This result shows that these values are just particular cases of using the Minkowski's distance.

Let us briefly demonstrate the first one. We already saw the mean when working on the global approach; we work now with a second definition. The mean $\bar{y}$ can also be defined as the value which is the closest to all values according to the Euclidian metric: the value sum $= \sum_i(\bar{y} - y_i)^2$ will reach a minimum for $(\partial\text{sum}/\partial\bar{y}) = 2\sum_i(\bar{y} - y_i) = 0$ which gives the usual result $\bar{y} = (1/I)\sum_i y_i$.

The mean therefore has interesting properties, as it is the center of the distribution either when it is computed globally, or locally (with $\alpha = 2$). This is one of the reasons of the success of this characteristic.

Using the concept of distances as the difference between the values affected by an exponent enables to find out the two frequently used centers of a distribution: the **mean** and the **median.** Other values of α are practically never used (but they could be!).

Sensitivity Analysis

How do these values compare together? First of all it is clear that the higher α, the more the center is "attracted" by the largest values (this is logic). Consequently **the mean** – although the most used value for the center of a distribution – **has one serious drawback:** it is sensitive to the extreme values (and consequently to the outliers if there are some). We say it lacks "robustness". The mean goes down to 27.35 if the 3.4 value is changed to 1, and to 28.76 if the 84.5 is changed to 100.

This obviously comes from the fact that the difference $|\hat{y} - y_i|$ becomes large when y_i is very different from $\hat{y}$, the result being symmetrical.

The major advantage of **the median** as the center of a distribution is its relative insensitivity to extreme values: the value of one outlier may change dramatically without any change of the median; in the given example, the value 84.5 could change from 17 to infinite without changing the median. The median is a "robust" characteristic; this is an interesting quality when the data are rather scattered.

8.3.3 Study of the Influence

The previous section studied the different values which can be computed when the differences are used as a base for determining the distances.

A small sensitivity analysis was carried out.

A more general study of a metric can be done using the concept of influence. This idea is to see what happens to the center when a new data is added to the existing data set: how does this new data modifies the value of the center?

In order to explore this concept, we will add a new data of which value can change from -100 to $+200$, this large interval being used to investigate the influence of the outliers. The procedure is straightforward:

- A new value is added.
- The new center is computed.
- The way the center changes is displayed on a curve, called the "influence curve".

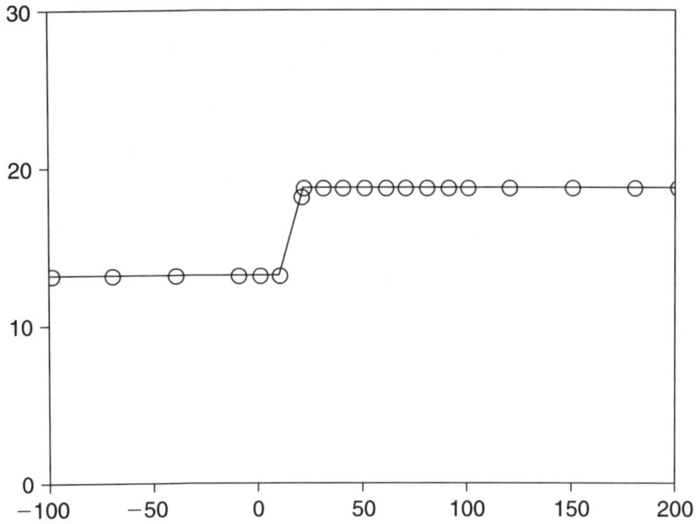

Figure 8.6 How a supplementary data influence the median.

Using the Median

When the added value is present, the "center" takes the value given in Figure 8.6, as a function of this value.

The result is straightforward, as it can be expected: the median does not change as long as the added value is lower than 9.9, then climbs smoothly to 18.75 when the new value reaches 21.2 and does not change afterwards.

As expected the "center" is completely insensitive to outliers: it is a very robust characteristic of the distribution.

Note

Be careful if you try to draw this graph with a simple algorithm working with iterations, because the minimum value of the sum computed to get the median presents a flat minimum, as it can be seen in Figure 8.7: the minimum of the sum is not defined between 9.9 and 16.3. The algorithm may then display any value in the flat range, depending on the approximate value of the center you start from.

Using the Mean

It is clear from Figure 8.8 that the mean is sensitive to new data, but in a rather moderate way.

The relationship between a new data is normal: the mean changes linearly with its value.

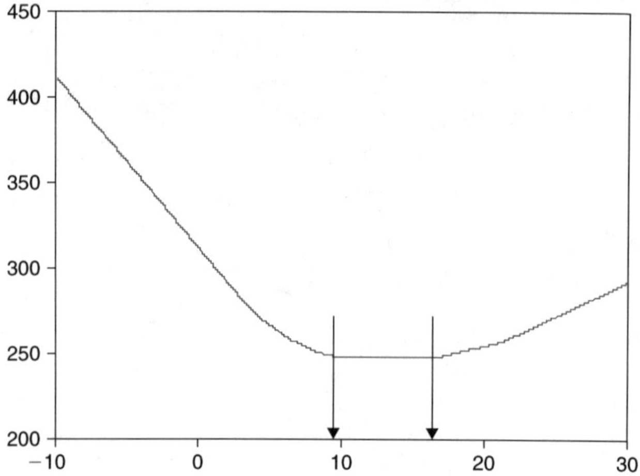

Figure 8.7 Value of the sum of which the minimum is looked for in order to get the median (in this example, the value added is – 10).

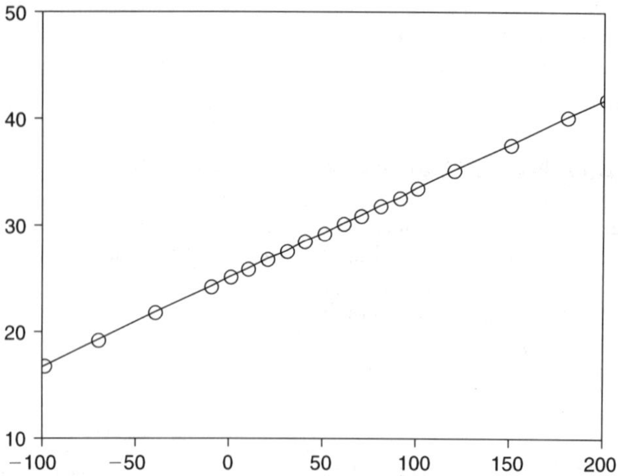

Figure 8.8 How a supplementary data influence the mean.

8.4 Using the First Type of Ratio: The Center Appears as the Numerator

8.4.1 Definition

We turn now to the distance given by:

$$d(y_i, \hat{y}) = \left|\frac{\hat{y}}{y_i} - 1\right|^{\alpha}$$

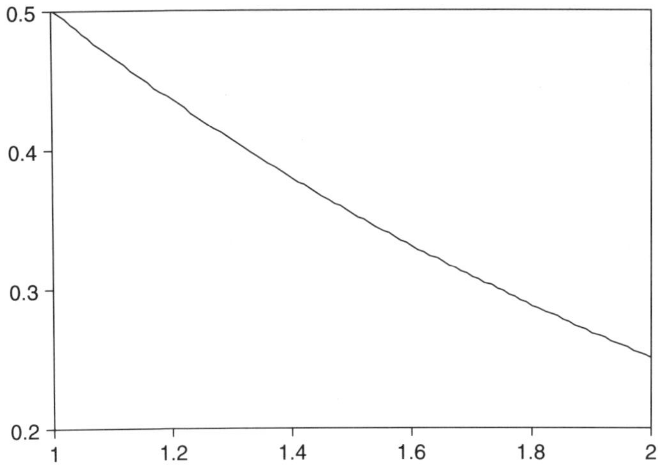

Figure 8.9 How does the distance change with α computations made with $y_i = 20, \hat{y} = 30$ and α from 1 to 2 ($\hat{y}$ is the numerator of the fraction).

When $\hat{y}$ is at the numerator of the formula, the change in the distance between y_i and $\hat{y}$ with α is given by Figure 8.9.

Two points must be noted:

1. the distance seems to – and does in this example – decrease with α. This comes from the fact that, in this example, $(\hat{y}/y_i) - 1 = 0.5 < 1$ is raised at a power larger than 1. If we had chosen $\hat{y} = 2y_i$ we would have found that the distance does not change with α; if we had chosen $\hat{y} > 2y_i$, we would have drawn an opposite conclusion.
2. the distance does not change so much if the values of y_i and $\hat{y}$ are not too far away: it is far less sensitive to α than with the metric using the difference.

If we look now at the way the distance changes when $\hat{y}$ is modified, we get the results in Figure 8.10. The interesting point is that the distance does not change too much with α, except of course if $\hat{y}$ becomes very large.

The point to be noticed here is that the change in the distance is symmetrical around y_i.

8.4.2 Computing the Center According to This Metric

The center is now given by the value ${}_{r1}\hat{y}$ which minimizes the sum:

$$\text{sum}(\alpha, y_c) = \sum_i \left| \frac{{}_{r1}\hat{y}}{y_i} - 1 \right|^{\alpha}$$

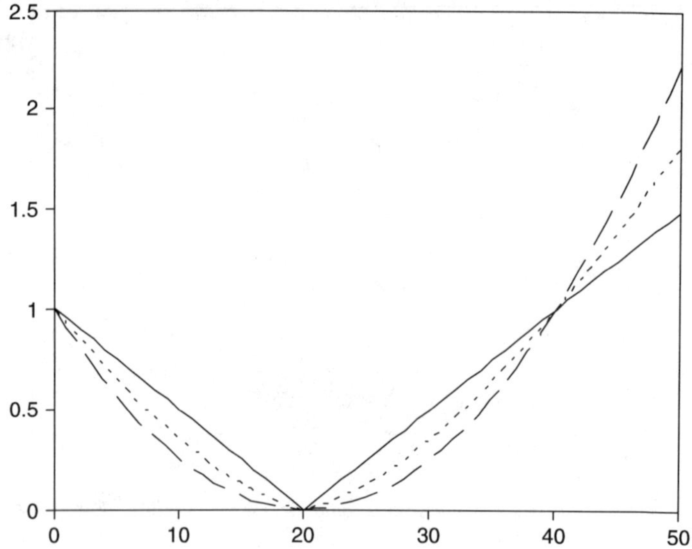

Figure 8.10 How does the distance change with $\hat{y}$ (from 10 to 50) for three different values of α: $\alpha = 1$ (full line); $\alpha = 1.5$ (dotted line) and $\alpha = 2$ (dashed line).

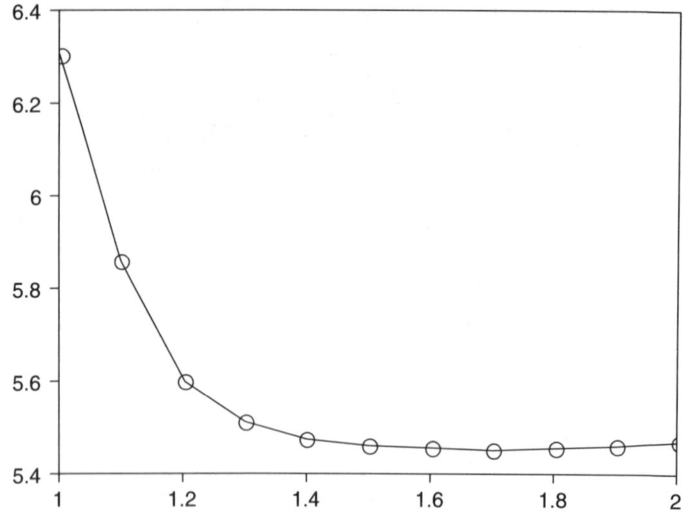

Figure 8.11 How does this center ${}_{r1}\hat{y}$ change with α?

Minimizing this sum leads, in our example, to values of ${}_{r1}\hat{y}$ which depend on α; the results, according to the values of α, are plotted in Figure 8.11.

We still used here the same example as given in the beginning of this chapter; it is reproduced here for easiness.

There is always a minimum for each value of α and this minimum does not change so much with α.

3.4
4.2
6.3
8.2
9.9
16.3
21.2
35.9
46.5
64.5
84.5

Two important comments have to be mentioned:

1. The center value is much smaller than the mean and even the median (of course the "much" comes to the fact that the data are intentionally rather scattered in order to make the characteristics of this metric clearly appear). It can be said that, **for this metric, the center is "attracted" by the low values.**
2. This center does not change a lot with the value of α. For this reason, the value **$\alpha = 2$ is generally adopted** by cost analysts who want to use this metric. We will now limit the investigation to this value.

Sensitivity Analysis

Starting from ${}_{r1}\hat{y} = 5.466$ (for $\alpha = 2$), the center goes down to 1.619 if the 3.4 value is changed to 1, and to 5.458 if the 84.5 is changed to 100. This shows the "attraction" exerted by low values but it is not representative of the quality of the formula because 0 is a singular point.

8.4.3 Study of the Influence

For studying the influence curve, we add to the previous set of data one new data and observe what happens when this new value goes from -100 to 200 (Figure 8.12).

The center is nearly completely insensitive to the new data point, as long as the new data point value is clearly different from 0; this is understandable.

Where does that come from? The sum we are now investigating is given by ("new" represents the data which is added)

$$\text{sum}(\alpha, y_c) = \sum_i \left| \frac{{}_{r1}\hat{y}}{y_i} - 1 \right|^\alpha + \left| \frac{{}_{r1}\hat{y}}{\text{new}} - 1 \right|^\alpha$$

with ${}_{r1}\hat{y}$ rather small; as soon as the new value becomes higher than ${}_{r1}\hat{y}$, the last term of the expression becomes negligible and does not influence the value of the sum.

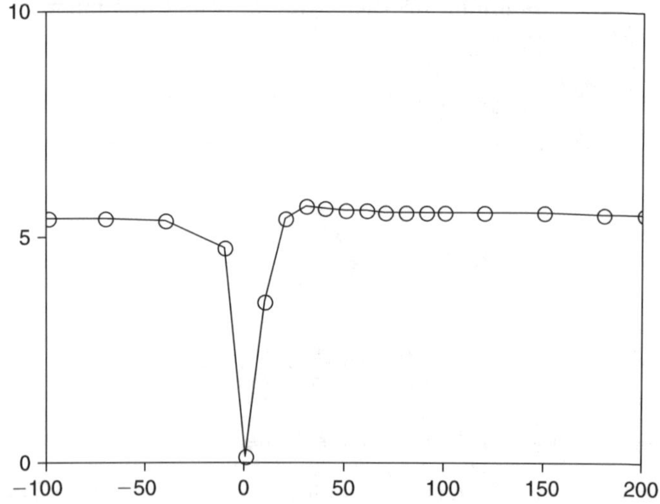

Figure 8.12 How does this center ${}_{r1}\hat{y}$ change according to the value of a new data?

8.5 Using the Second Type of Ratio: The Center Appears as the Denominator

8.5.1 Definition

The metric now is:

$$d(y_i, \hat{y}) = \left|\frac{y_i}{\hat{y}} - 1\right|^{\alpha}$$

The difference between the first ratio may look negligible, because we think that y_i and $\hat{y}$ play the same role. This is not true for our purpose because, in our computations, y_i will remain constant (it will be represented by our data, which are fixed), whereas $\hat{y}$ will be the center we are looking for. It means that we will have to modify $\hat{y}$ until we find its value: it is not the same if the value we change is at the numerator or the denominator of the fraction.

Figure 8.13 illustrates what happens when the value at the denominator is smaller than the value at the numerator.

The conclusions are the same as in the previous section.

Here again the value which is raised at the power α is equal to $0.333 < 1$. This explains the shape of the curve.

Let us have a look now on how this distance changes with the value $\hat{y}$ (which, let us remind it, appears at the denominator of the fraction). The result is given in Figure 8.14.

The graph is now not symmetrical of both sides of $\hat{y}$: distances are computed greater when y_i is larger than $\hat{y}$ than it is in the opposite situation. This dissymmetry will have to be taken into account.

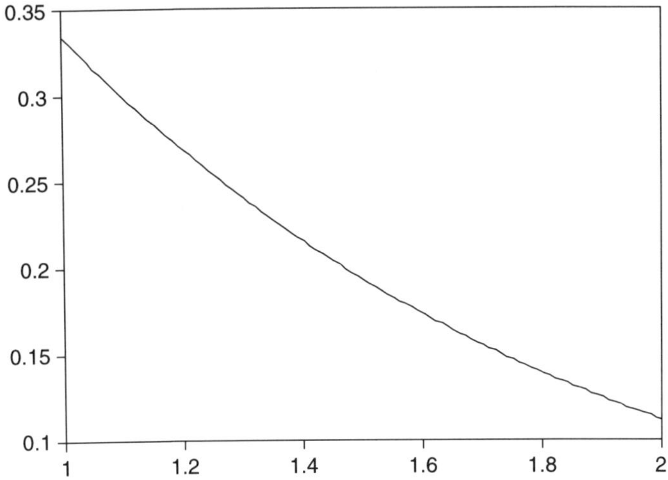

Figure 8.13 How does the distance change with α computations made with $\hat{y} = 20, y_i = 30$ and α from 1 to 2 ($\hat{y}$ is the denominator of the fraction).

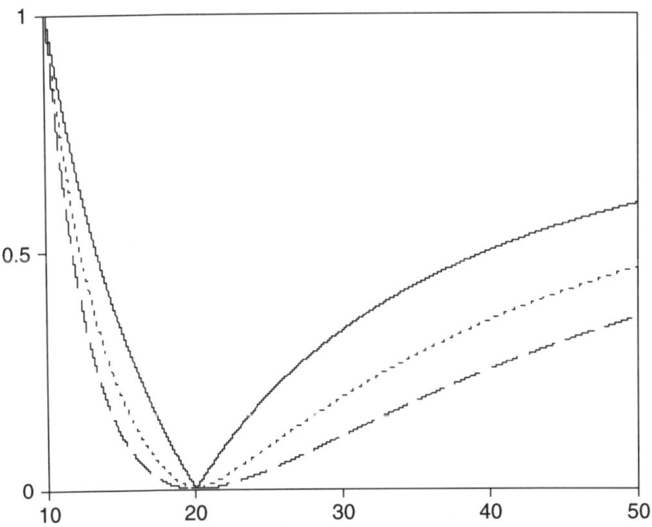

Figure 8.14 How does the distance change with $\hat{y}$ (from 10 to 50) for three different values of α: $\alpha = 1$ (full line); $\alpha = 1.5$ (dotted line) and $\alpha = 2$ (segmented line).

8.5.2 Computing the Center According to This Ratio

The center of the distribution will now be given by the value ${}_{r2}\hat{y}$ which minimizes the sum:

$$\mathrm{sum}(\alpha, y_c) = \sum_i \left| \frac{y_i}{{}_{r2}\hat{y}} - 1 \right|^{\alpha}$$

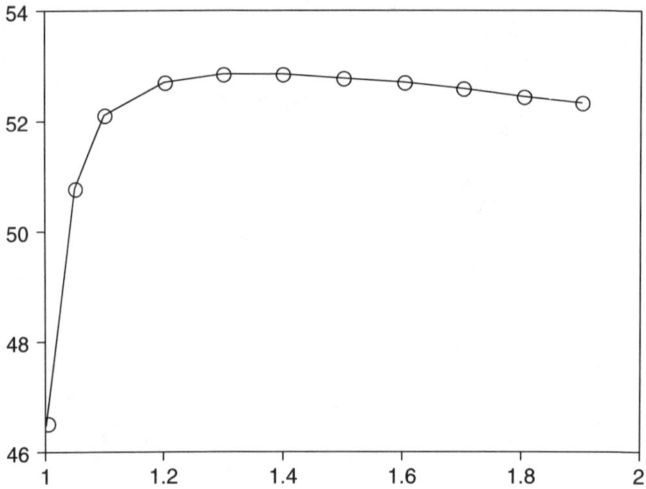

Figure 8.15 How does this center ${}_{r2}\hat{y}$ change with α?

Minimizing this sum leads, for our example, to values of ${}_{r2}\hat{y}$ which depend on α; the results, according to the values of α, are plotted in Figure 8.15.

One immediately notices that the position of the center is now completely different from the position which was computed when using the previous ratio: about 53 compared to 6! The reason for that is clear: from the definition of the center, it can be seen that a large value of ${}_{r2}\hat{y}$ will – up to a certain point – decrease the value of the sum.

For $\alpha = 2$ the center has, with this metric, a value of 52.18, which is this time much higher than the mean, and considerably higher than the previous metric. **The center is now "attracted" by the high values.** Very clearly this metric has a behavior opposite to the previous one.

It can also be noticed here that the value of the center does not really depends on α as soon as $\alpha > 1.1$.

Sensitivity Analysis

Starting from ${}_{r1}\hat{y} = 52.18$ (for $\alpha = 2$), the center goes down to 52.564 if the 3.4 value is changed to 1, and to 58.662 if the 84.5 is changed to 100. Two comments can be made:

1. This metric is rather insensitive to low values (which confirms the previous remark). Notice nevertheless that the change, although it is small, goes in the direction opposite to what could be expected: decreasing a data increases the value of the center.
2. But it is sensitive to high values and the difference is not negligible.

8.5.3 Study of the Influence

The sum to be minimized is now given by:

$$\text{sum}(\alpha, \hat{y}) = \sum_i \left| \frac{y_i}{{}_{r2}\hat{y}} - 1 \right|^{\alpha} + \left| \frac{\text{new}}{{}_{r2}\hat{y}} - 1 \right|^{\alpha}$$

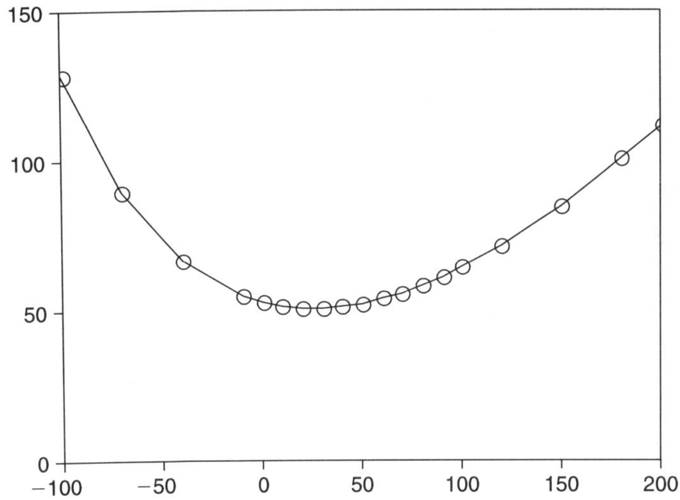

Figure 8.16 How does this center ${}_{r2}\hat{y}$ change with the new value?

The interesting fact here is that the value of the center always goes up as soon as the new value becomes different from the original center, whatever the change: this algorithm is very sensitive to outliers and should therefore be used when the outliers have been properly detected and possibly eliminated (Figure 8.16).

Another interesting fact is that the presence of a new data, if it is inside the range of the other data, does not really change the value of the center.

8.6 Using the Log of the Ratio

8.6.1 Definition

The formula is now given by

$$d(y_i, y_c) = \left|\log\frac{\hat{y}}{y_i}\right|^{\alpha}$$

where y_i and $\hat{y}$ have a symmetrical role: it is not therefore necessary to study the log of $y_i/\hat{y}$.

As usual, Figure 8.17 displays the change in the distance as a function of α. Due to the log effect, the value of the distances are small, but this is not important: what is important is the change in the distances.

Here we find again, for the same reason, that the distance decreases when α grows. Should we have taken $\hat{y} = 20$ instead, the result would have been different (a horizontal line) and of course also for $\hat{y} > 20$.

Figure 8.18 displays, for three different values of α how the distance between two points changes as a function of their values.

Distances are now slightly "biased" toward values of y_i larger than values of $\hat{y}$, but this bias is limited.

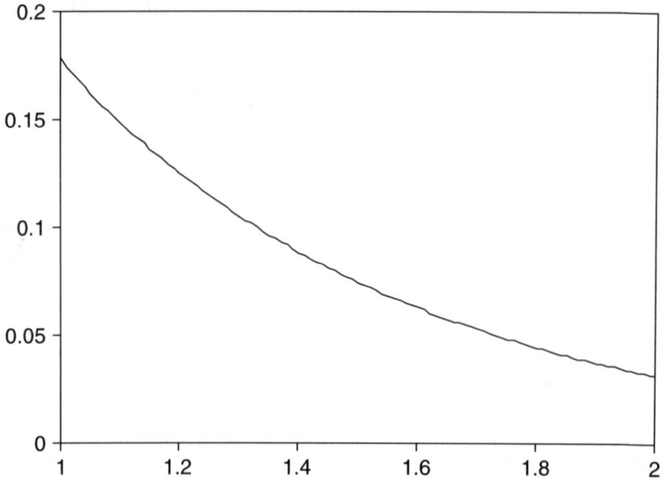

Figure 8.17 How does the distance change with α computations made with $y_i = 20, \hat{y} = 30$ and α from 1 to 2.

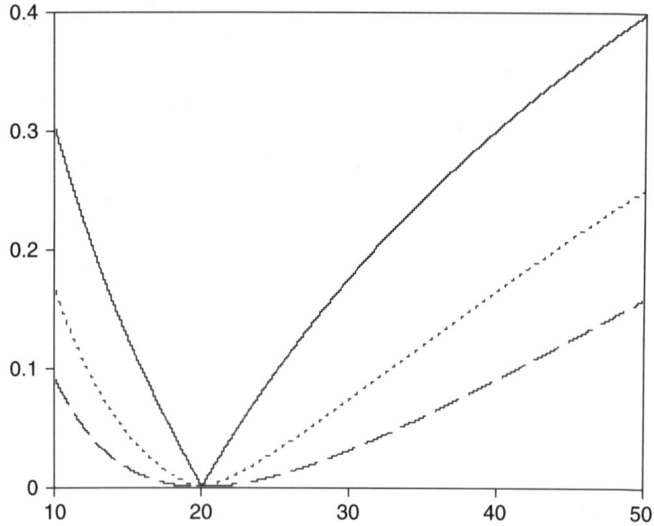

Figure 8.18. How does the distance change with $\hat{y}$ (from 10 to 50) for three different values of α: $\alpha = 1$ (full line); $\alpha = 1.5$ (dotted line) and $\alpha = 2$ (line).

8.6.2 Computing the Center According to This Metric

We are now looking for a center ${}_{l}\hat{y}$ which minimizes the sum

$$\text{sum}(\alpha, {}_{l}y_c) = \sum_i \left| \log\left(\frac{{}_{l}\hat{y}}{\text{new}} \right) \right|^{\alpha}$$

Figure 8.19 displays the value of this center as a function of α.

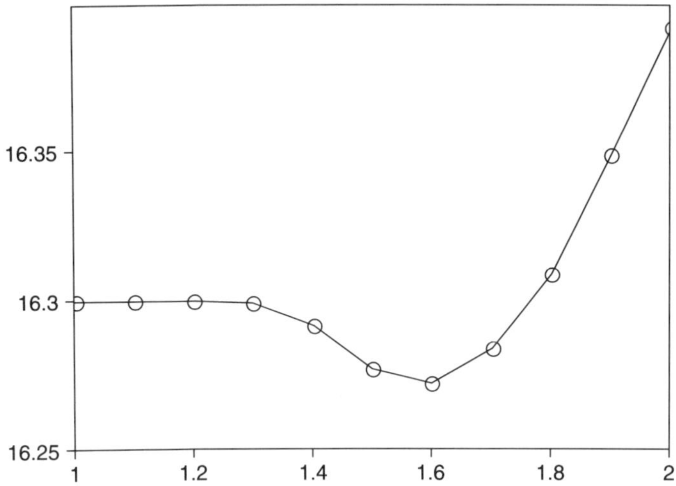

Figure 8.19 How does this center $_{l}\hat{y}$ change with α?

It is interesting to note that this center changes very little with the value of α. Furthermore the algorithm produces a value which is very close to the median and even equal to the median as soon as $\alpha < 1.3$.

This is an interesting fact and such a center should deserve more attention than it does.

Sensitivity Analysis

Starting from $_{r1}y = 16.392$ (for $\alpha = 2$), the center goes down to 14.666 (-10.5%) if the 3.4 value is changed to 1 (-70%), and to 16.645 ($+1.5\%$) if the 84.5 is changed to 100 ($+18.3\%$). Two comments must be made:

1. The changes go in the right direction: the center goes down if a value is lowered, it goes up if a value is increased.
2. This center is rather robust, as the changes are moderate. The level of robustness is comparable to the use of the differences.

8.6.3 Study of the Influence

We introduce now a new value and investigate – for $\alpha = 2$ – how the minimum of the sum:

$$\text{sum}(\alpha, {}_{l}\hat{y}) = \sum_i \left| \log\left(\frac{{}_{l}\hat{y}}{y_i}\right) \right|^{\alpha} + \left| \log\left(\frac{{}_{l}\hat{y}}{\text{new}}\right) \right|^{\alpha}$$

changes with the new value. The result is displayed in Figure 8.20.

It obviously appears that $new = 0$ is a singularity. If this value is eliminated, the value of the center grows slowly with the value of *new*: **having an outlier is not a problem** as its sensitivity is not really a problem.

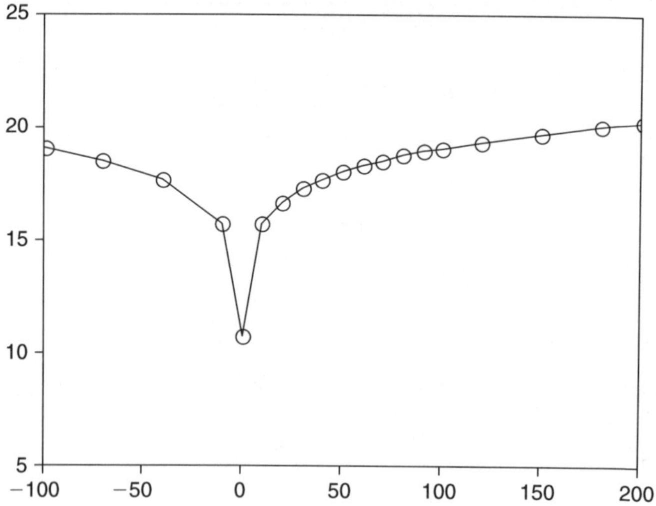

Figure 8.20 How does this center ${}_l\hat{y}$ change as a function of a new value?

Using this metric – for positive values – seems to be a satisfactory solution to the search for a center value.

8.7 Using the Biweight

8.7.1 Definition

Associating a "weight" to each data point is a common way to solve some problems. The problem we try to solve here is to reduce the influence, on the center, of data points which are far away from this center. The idea is that a formula should be based on the "bulk" of the data, without considering rare outliers, or, more exactly, to reduce gradually their influence as soon as their distance – here measured by something based on the differences – to the bulk increases.

In order to implement this gradual influence, the idea is to give to each data point a weight which decreases as soon as the data becomes distant from this bulk.

Immediately a problem appears: the weight depends on the distance of the center, but the center depends on the distances. Obviously a step by step process will be required.

Let us introduce the subject gradually. In this section the way the weight is defined is investigated.

Suppose we know the center of the distribution computed by the biweight[2] algorithm; let us call:

- ${}_w\hat{y}$, the formula giving the center.
- s, the standard deviation of the data around this center.

[2] The name comes from an abbreviation for "bisquare" weight.

- w_i, the weight attached to data point y_i.
- y_0, the cut-off value. This cut-off value is the distance from the center from which we want the data to have a 0 weight (which means completely removing the influence of the data which are farther than this value from the center). What could be the value of this cut-off value? It is let to the choice of the cost analyst, depending on how his/her data are scattered. Mosteller and Tukey recommend (Ref. [43], p. 353) to use 3 times the interquartile range, which corresponds to about 4*s* (for a normal distribution). The example will be carried out with this 4*s* value, because it is easier to compute than the interquartile range. The ratio $e_{+i}/y_0 = (y_i - {}_w\hat{y})/y_0$ could be called the "normalized" difference between the data y_i and the center ${}_w\hat{y}$.

The biweight procedure uses weights given by

$$w_i = \begin{cases} \left[1 - \left(\dfrac{e_{+i}}{y_0}\right)^2\right]^2 = \left[1 - \left(\dfrac{y_i - {}_w\hat{y}}{y_0}\right)^2\right]^2 & \text{if } \dfrac{e_{+i}}{y_0} \leq 1 \\ 0 & \text{elsewhere} \end{cases}$$

How does this weight changes with the difference e_{+i} between the center and the data value?

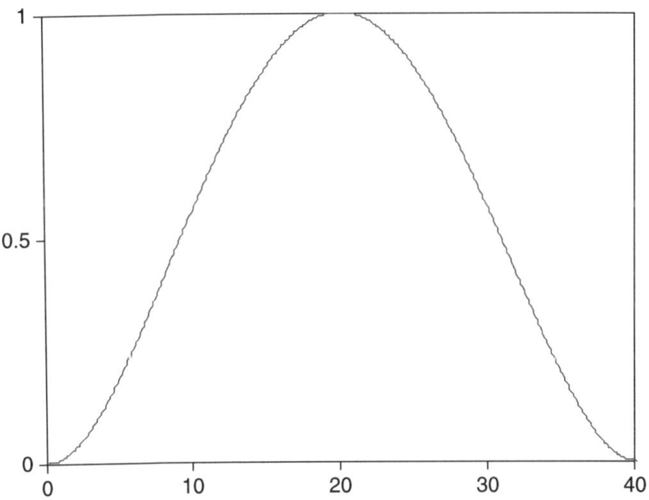

Figure 8.21 Weight computed for the biweight as a function of the difference between ${}_w\hat{y} = 20$ and y_i ($y_0 = 20$ corresponds to the cut-off value).

The answer to this question appears in Figure 8.21. The weight starts with a value of 1 when both values are close together and decreases slowly when the data becomes distant from the center, until it reaches the cut-off value (20 in the example): starting from this value the weight equals 0. The double square is needed to get a smooth transition with the weight 0 when y_i is close to the cut-off value (a cosine could also have been used).

This weight clearly satisfies our needs.

8.7.2 Computing the Center According to This Metric

There could be of course many ways to use the weights, depending on the metric which is selected. Due to the fact that the weight has no interest with the ratio of type 1 or with the log of the ratio (as the influence curves showed it) and that the type 2 ratio is rarely used, we limit here the discussion to the use of the metric based on the differences.

For this metric we know that if $\alpha = 1$, we have nothing to fear with the outliers. Consequently we limit here the discussion to the case $\alpha = 2$; the center of the distribution is then given by the arithmetic mean.

Once the cut-off value has been decided, the process starts with an estimated value of the center ${}_w\hat{y}^{(0)}$ (the arithmetic mean is a good starting point), the exponent (0) reminding that it is just a starting point of the iterations; from this value, the weights of all the values can be computed and a new center ${}_w\hat{y}^{(1)}$ is computed with the formula:

$$ {}_w\hat{y}^{(1)} = \frac{\sum_i w_i y_i}{\sum_i y_i} $$

which is the standard formula used to compute a mean with weighted values. From this value new weights are computed, which gives a ${}_w\hat{y}^{(2)}$ value for the center from which new weights are computed, etc. A set of values ${}_w\hat{y}^{(0)}, {}_w\hat{y}^{(1)}, \ldots, {}_w\hat{y}^{(k)}, \ldots$ is therefore established. The process stops when ${}_w\hat{y}^{(k+1)} = {}_w\hat{y}^{(k)}$ at a predefined level of precision. The last value ${}_w\hat{y}^{(k)}$ is the required center value.

Two approaches may be tested: the first one uses a cut-off value equal to $4s = 104.4$, the second one $2s = 52.2$.

Using the first value ($4s = 104.4$) produces a center of 23.289, whereas the standard mean – computed with all the weights equal to 1 – was equal to 27.355. This comes from the fact that the weights which correspond to the center are given by the vector:

$$: \begin{pmatrix} 0.929 \\ 0.934 \\ 0.948 \\ 0.959 \\ 0.967 \\ 0.991 \\ 0.999 \\ 0.971 \\ 0.903 \\ 0.712 \\ 0.429 \end{pmatrix} $$

From this vector, it appears that the two "strong" values 64.5 and 84.5 have decreasing weights, which decrease their influence for computing the mean and therefore move the new center toward lower values.

Using the second value ($2s = 52.2$) produces a center of 14.102. This comes from the fact that the weights which correspond to this center are now given by the vector displayed below. Now the last two values have a very small weight (the last one being 0) and even the value 46.5 has a weight equal to 0.378. All that "pushes" the center toward the low values.

It is quite possible that you think that the cut-off value is, in this last case, too small. You are probably right. The discussion was presented here to demonstrate the concept of the weighted values – of which the biweight is certainly the best implementation – in order to make it clear that the choice of the cut-off value is an important choice that you have to make. The "standard" value of $4s$ is probably here a more reasonable option.

We return to this point downwards.

$$\begin{pmatrix} 0.918 \\ 0.929 \\ 0.956 \\ 0.975 \\ 0.987 \\ 0.996 \\ 0.963 \\ 0.682 \\ 0.378 \\ 4.667 \times 10^{-3} \\ 0 \end{pmatrix}$$

The major interest of using this biweight is to have a much better focus of what may appear to be the real center of the distribution, the values too far away from it being eliminated or at least seeing their influence strongly reduced.

The biweight does accomplish, for the selected metric, the purpose for which it was created: the center being more representative of the focus of the distribution. Note that its value is not far from the median.

8.7.3 Study of the Influence

Now a new value is introduced in order to examine how the center is going to move when this new value goes from -200 to $+300$. The center of the distribution is still given by the weighted mean:

$${}_w\hat{y} = \frac{\sum_i w_i y_i + w_{\text{new}}\text{new}}{\sum_i y_i + \text{new}}$$

the weight being computed according to the same formula. As the weight still depends on the center and the center on the mean, iterations are necessary.

We examine in the following section how the influence curve changes as a function of the cut-off value y_0.

$y_0 = 4s$

The example will show that this cut-off value is not necessarily the best choice.

The center was computed for several values of the new data point; the results are given in Figure 8.22.

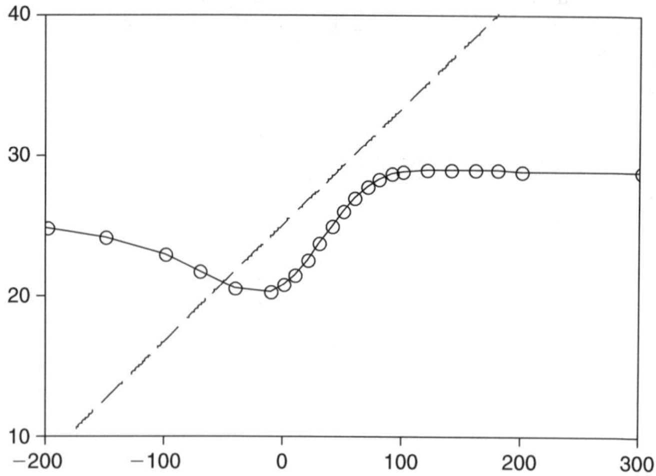

Figure 8.22 Center computed for the biweight as a function of the new value (cut-off value $= 4s$).

The result is not exactly what was expected for two reasons:

1. the high value of the new data point is still taken into account, even if it is larger than 100,
2. inside the range of the starting point, the center is slightly lower than the mean, represented by the dashed line.

The reason for that is that the new data changes substantially the value of the standard deviation which increases with its value: therefore its weight never goes to 0.

Let us try something else.

$y_0 = 4\text{MAD}$

MAD stands for "median of absolute deviations". It is computed the following way as soon as a center is computed: one calculates the deviations around this center, takes their absolute values and then computes their median. It is obvious, as demonstrated in Section 8.3.3, that this value is less sensitive to outliers.

The principle of the computation is about the same:

- a starting center ${}_w\hat{y}^{(0)}$ is chosen,
- the MAD is computed,
- the cut-off value is taken at $y_0 = 4\text{MAD}$,
- from this value the weights are computed,
- as well as the estimated center.

A new starting center is chosen as long as the estimated center is not equal to this starting point: the iteration is rather long to do it manually.

The influence curve of the center as the new data moves from −200 to +200 is given in Figure 8.23.

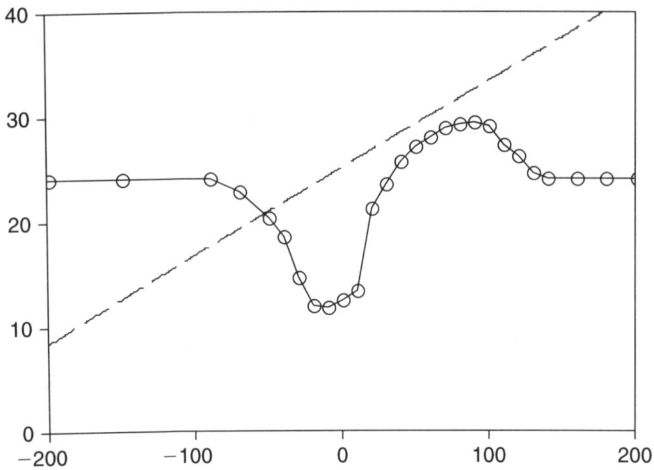

Figure 8.23 Center computed for the biweight as a function of the new value (cut-off value = 4MAD).

The result is improving for one part (when the new value is too high the center "returns" to the previous value), but not for the other part (the changes in the center is too steep in the vicinity of the center).

y_0 = 9MAD

The cut-off value is now higher.

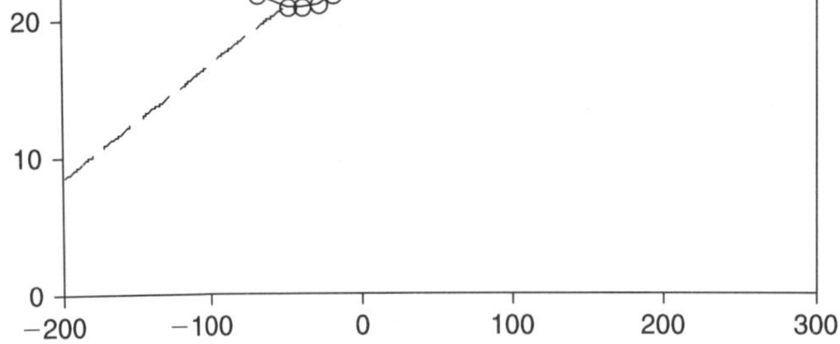

Figure 8.24 Center computed for the biweight as a function of the new value (cut-off value = 9MAD).

The result is satisfactory: outliers have no influence on the center and, inside the range of the previous data, the change of the center follows about the change of the mean (dashed line).

8.7.4 Conclusion

If you decide to use the biweight, you should avoid using a cut-off value related to the standard deviation: this measure of distribution spread is not robust enough for this purpose.

Using 9 times the MAD appears to be, at this stage, the best choice.

8.8 What Is the Center of a Distribution?

Figure 8.25 summarizes the value of the center computed by different algorithms. Note that all values are – largely – different, although all values can pretend to represent the distribution center.

It is clear that there is nothing such as one and only one center for a distribution. The choice of the center value to be used depends first on the problem to look at: for instance for computing the center of a set of values when each one is computed from the previous one (such as growth rate computed per period), the geometric mean is the most appropriate.

It also depends on the properties of the value: the arithmetic mean has interesting properties – that will be discovered later on – which explains why it is so often retained as the value of the center. It is certainly the most appropriate when the data have a very low dispersion.

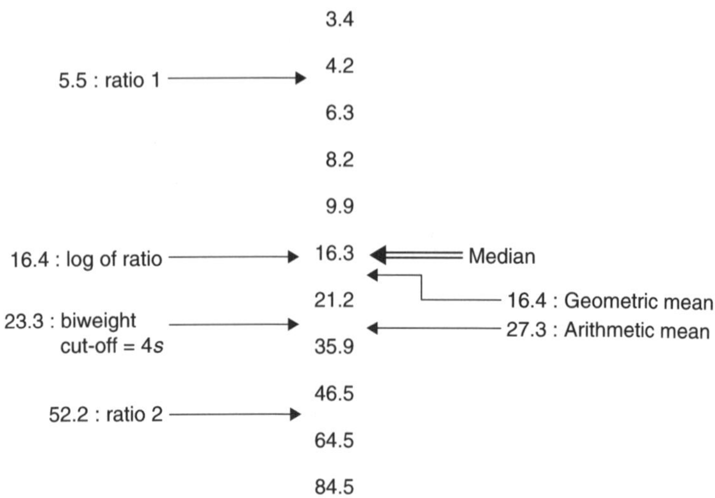

Figure 8.25 The set of center values for the example.

The major interest of the median is its insensitivity to outliers: it should therefore be preferred when the data are scattered.

Most of the values are in the set [16.3, 27.3], between the median and the mean, the difference being nearly equal to the MAD around the mean (12.1).

The ratios create a problem: ratio of type 1 is "attracted" by small values, whereas ratio of type 2 is attracted by large values. The reader must be interested by two points:

1. The difference between both centers computed by the ratios is rather large.
2. Both values are "far away" from the traditional centers.

Consequently both metrics must be used carefully.

Selecting a Value

The center of a distribution will be used in the following chapters to compute the "nominal" cost estimate of a new product: it is therefore very important to make a proper selection.

On what can be based this selection?

Suppose the values given in the example are the speeds of the last 11 vehicles observed on a road. You have this information and you are asked the question: what is your best guess for the speed of the next vehicle?

As you have no other information, you are certainly inclined to look for some value in the middle. Which value? Think about the subject and decide. Personally, I would probably choose the median: that there are as many vehicles with a speed lower than it, there are with a speed higher than it and therefore I think it is a good choice, when no other information is available of course.

My opinion would certainly change if I have other information: for instance if I know (this is not part of the experiment, but it can be a previous information I have) that the 3.4 and even the 4.2 are extremely rare because these vehicles probably had an engine problem, my decision would be different.

This example shows that **the knowledge of previous information or of previous experience is an important guide** for selecting the way the center should be computed. Returning to the cost, you may for instance know that, if you measure 11 different values for the cost of the same product, the distribution of these costs generally follows a normal curve: in such a case, you should probably prefer to use the arithmetic mean as the value of the center. If you know from experience that this distribution is always skewed to the right (higher values are more likely to occur than lower values), you should choose some kind of a ratio.

Of course the spread of the distribution used in the example is very large and we hope you will not meet such distributions when you have to make an estimate. The purpose here was to make the results clearly appear. The conclusions will remain the same if the spread is reduced by 2 or 3, but the result will be less visible.

Nevertheless such distributions may be found for the distribution of the residuals around the center of a cost distribution: as the nominal cost depend on minimizing some sum of these residuals, the knowledge of the various metrics is a must.

9 Looking for the Dynamic Center: The Bilinear Cases

Summary

In the previous chapter, we studied the information φ in a sample containing just one variable (the cost or more often the specific cost, which is the cost per unit of "size", such as the cost per kilogram or the cost per square meter) and established that this information can be conveniently represented by:

- a center, called $\hat{y}$,
- the distribution ψ of the "residuals" around this center.

For computing the value of the center, the concept of metric was introduced and several metrics were investigated. We saw that using different metrics leads to different values for this center, but that all of them could legitimately pretend to be the center.

If no other information is available, this center $\hat{y}$ will be used for estimating the cost of a new product belonging to the same family of product. This is what is done when the cost-estimating method based on the "ratios" is used.

For the present time the distribution ψ of the residuals (mainly its mean and its standard deviation) can be used for quantifying the quality of the cost estimates which can be done with this method: the larger the standard deviation, the less accurate will of course be a cost estimate based on $\hat{y}$ only.

If we have more information, we can try to reduce the standard deviation of ψ.

In this chapter we study information of a sample containing several variables, one of them being of course the dependent variable, the other ones being the causal variables.

We assume that the analysis of the sample data (according to the procedures described in Chapter 2) was carried out:

1. The potential problems related to possible outliers and/or multi-collinearities between the causal variables were discovered and solved.
2. An "interesting" *linear* correlation (measured by the Bravais–Pearson correlation coefficient) between the dependent variable and the (selected) causal variable(s) was found, which legitimates to devote some effort to go on with this sample.

This chapter tries to use this correlation to carry out the analysis of the sample, the purpose being to get a better understanding of the distribution φ. The presence of the correlation suggests that we can reduce the importance of the residuals by adopting a better value for the center.

The idea is rather simple: as there is a correlation between the dependent variable and the causal variable(s), it would be interesting, instead of using a "static center" $\hat{y}$ (the one we studied in the previous chapter) to use a center which would depend on the causal variable(s). **This new center will be called the "dynamic center" of the distribution** φ. It will be represented, for simplifying the notations, by the same symbol $\hat{y}:\hat{y} = f(x_1, x_2, \ldots, x_P)$; as this dynamic center changes with the values of the causal variables, the value it will take for a particular product, let us say P_i, will be called $\hat{y}_i$, the difference between $\hat{y}_i$ and the "observed" value y_i being called the "residual":

$$e_{+i} = y_i - \hat{y}_i$$

if it is defined as additive.

The purpose of this analysis must be clearly understood:

> We will compute a dynamic center of φ for the purpose of "improving" the distribution ψ of the residuals (which means here reducing its standard deviation).

Another way of expressing the same view is the following one: we want to replace the complex distribution of the y_i by the distribution of the e_i, this second distribution being "cleaned" from the "pollution" brought along by the causal variables; it will therefore be easier to handle.

The word "linear" was underlined. This chapter assumes that there is a linear correlation between the dependent variable and the causal variable(s): this means that we are looking for a dynamic center which will be a linear function of these variables.

In most statistical books, "linear" means that the function we are looking for in the sample, $y = f(\vec{b}, x)$, is a linear function of the coefficients $\vec{b}$; in other words a function such as:

$$y = b_0 + b_1 x + b_2 x^2$$

is said to be linear. This is not our definition in this chapter: what we are looking for is a relationship which is linear in x. It so happens here that the relationship will also be linear in terms of the coefficients: it must therefore be called "bilinear" and this is the reason of the title of this chapter.

This hypothesis will be deleted in Chapter 12.

There are several ways to establish this function, the most commonly used being called the "linear regression". For this reason it will be studied first. As it is largely developed in most statistical books, the discussion will be limited to its more important features, with no demonstration.

This "linear regression" is not the best tool which can be used in the domain of cost. Consequently other methods will be investigated and the results compared.

This chapter will investigate the use of one causal variable – or parameter – only. The next chapter will use several parameters.

9.1 The Classical Approach: The Ordinary Least Square or the "Linear Regression"

In this first section,

1. We use one causal variable only.
2. The metric is limited to the use to square of the differences; this means that the distance between two values y_a and y_b is defined by $|y_a - y_b|^2$. This metric is known as the "Euclidian metric" or the "ordinary least squares" (OLS). Other metrics will be studied in the following section.
3. The analysis of the sample revealed that the correlation between the dependent variable and the causal variable is linear, or about linear. More exactly, whatever the correlation, the cost analyst *decides* to investigate the interest of a linear relationship.
4. The residuals are defined to be additive.

As in the other chapters the two variables are named Y (for the dependent variable, generally the cost) and V_1 (for the causal variable).

Both variables can be studied independently as we did in Chapter 2: it is possible to look for the center of the distribution of each variable. However, in the presence of two variables, when the study (see Chapter 5, Paragraph 2.3) has shown that a relationship exists between both, it is reasonable to search if some improvement can be done. This chapter investigates such an improvement, when a linear relationship, quantified by the Bravais–Pearson correlation coefficient, has been found.

This chapter is therefore devoted to the use of the revealed correlation between both variables for improving the future cost estimates.

The text uses the logic described in Chapter 1.

In the previous chapter the set φ of the values $\{y_1, y_2, \ldots, y_i, \ldots, y_I\}$ was replaced by the value $\hat{y}$ of the center, plus the distribution ψ of the residuals around this center.

The distribution φ is now defined as a set of couples of two variables among which there exists a linear correlation $\{\langle y_1, x_1\rangle, \langle y_2, x_2\rangle, \ldots, \langle y_i, x_i\rangle, \ldots, \langle y_I, x_I\rangle\}$. The existence of a correlation between values y and x suggests that we can "clean" the distribution of the y_i by removing the influence of the causal variable; for this purpose, we will consider that the center $\hat{y}$ is now a function of x: this center now becomes a "dynamic" – with x – center; as no confusion may arise, the same symbol $\hat{y}$ will be used.

This chapter assumes that the correlation between y and x is bilinear. For this reason we are looking for a **dynamic center** defined by the following formula:

$$\hat{y} = b_0 + b_1 x$$

b_0 and b_1 being called the coefficients of the formula; b_0 is called the intercept (the value of the dynamic center when $x = 0$), b_1 being the slope (how much the dynamic center changes with x).

We do not expect (except in exceptional circumstances) that all y_i will be exactly equal to $\hat{y}_i$. The *difference* between both values is called the residual related to y_i. This is what is meant when we said that the residuals are defined in an additive way: y_i can therefore be written:

$$y_i = \hat{y}_i + e_{+i} = b_0 + b_1 x_i + e_{+i}$$

where the b_0, b_1 and all e_{+i} are, for the time being, unknown.

In this process, instead of studying the complex distribution of the y_i, we will only have to study the simpler, because it does not (hopefully) anymore depends on the x, distribution of the e_{+i}. Experience shows that we can win a lot by doing so.

An Example

In this chapter we will study the following data drawn from a sample (Figure 9.1).

Name	Cost	Mass (kg)
A	1278	6.83
B	724	2.18
C	809	3.80
D	920	4.55
E	772	2.18
F	877	2.11
G	1064	4.67
H	865	2.81
I	961	2.55
J	856	1.68
K	1293	6.30
L	717	1.98
M	648	1.25

Figure 9.1 The sample.

The characteristic of the distribution of each variable is given by:

- for the cost: arithmetic mean: 906.462, standard deviation: 192.979;
- for the mass: arithmetic mean 3.299, standard deviation: 1.720.

If we decide not to use any parameter, we would use, for estimating the cost of a new product – assuming we use the Euclidian metric, which delivers the arithmetic mean – the mean value of 906.462. The residuals around this center are then given in Figure 9.2.

Residuals
371.54
−182.46
−97.46
13.54
−134.46
−29.46
157.54
−41.46
54.54
−50.46
386.54
−189.46
−258.46

Figure 9.2 The residuals around the center $\bar{y}$.

The distribution ψ of these residuals is here simply described by:

- Its mean, equal to 0 (which is normal when using the mean as the center of a distribution).
- Its standard deviation, equal to 192.979, which is obviously – because, for going from the cost values to the residuals, we only translated the values – the same as the standard deviation of the cost values.

Due to this high level of the residuals we cannot expect to get an accurate estimate for the cost of a new product by just using this mean value of the cost.

The question is now: Can we reduce this amount of residuals by introducing a parameter? This parameter will be used to create, instead of the simple arithmetic mean, a dynamic mean.

The Matrix Form

The $\|{}^{+}x\|$ matrix (the matrix of the parameter) for this example will be written as:

$$\|{}^{+}x\| = \begin{Vmatrix} 1 & 6.83 \\ 1 & 2.18 \\ 1 & 3.80 \\ 1 & 4.55 \\ 1 & 2.18 \\ 1 & 2.11 \\ 1 & 4.67 \\ 1 & 2.81 \\ 1 & 2.55 \\ 1 & 1.68 \\ 1 & 6.30 \\ 1 & 1.98 \\ 1 & 1.25 \end{Vmatrix}$$

Why using a column of "1"? It is of course for finding the intercept. Let us explain why: the formula giving the dynamic center should be written:

$$\hat{y} = b_0 x_0 + b_1 x_1$$

where x_0 is the value of a variable V_0 which will take the same value for all the products: it is a constant which partly explains the cost. Any number could have been chosen for its value, for instance the arithmetic mean we used upwards; in such a case the value found for b_0 would have, of course, to be adjusted accordingly. The value of 1 is generally used, just because it is the simplest one!

The cost analyst is not forced to use such a matrix: he/she can simply use the matrix $\|x\|$ without a column of 1. As the general form used in regression analysis when dealing with cost includes such a column, we will say, in this case, that the intercept is "forced" to be 0.

9.1.1 Looking for the Center of the *Y* Distribution: The Concept of the Dynamic Center

Y is the dependent variable, the one we will try to forecast: generally for the cost analyst, it is the cost. Its values are denoted $y_1, y_2, \ldots, y_i, \ldots, y_I$ in the sample.

In a tentative to understand the behavior of this variable, and to prepare future forecasts, its distribution must be studied. As we did when it was the only variable, this study has to be split in two phases:

1. Finding the center of the distribution.
2. Examining the distribution of the (additive here) residuals $e_1, e_2, \ldots, e_i, \ldots, e_I$ around this center.

This section is devoted to the first phase, the study of the residuals being postponed to Part IV.

The center $\bar{y}$ of variable Y, considered alone, has here a limited interest! Correlation, studied in the previous chapter, reveals that the values y_i are influenced by the values x_i: the distribution of the values of Y is certainly "disturbed" by V_1. In other words studying this distribution without taking into account the influence of V_1 would be too complex.

The idea is then the following one: let us remove the disturbance generated by V_1 before studying the distribution of Y.

How can we do that?

When we studied the distribution of Y in Chapter 4 (Y considered alone), we started by finding out its center before we computed the other values which really describe it: variance – or standard deviation – skewness and kurtosis. The computation of all these values starts by removing the influence of the distribution center. Refer to all the formulae of this chapter: they are all based on the differences $y_i - \bar{y}$. Any analysis of a distribution is carried out by observing what happens "around the center".

The same logic can be used here. However, in order to remove the disturbance generated by V_1, it would be a good idea to remove not $\bar{y}$, but also preferably the "disturbing effect" of V_1. We will do that by removing not a center value, but a "**dynamic**" center, a center which moves with V_1! The idea is therefore to replace the fix center previously used by a center which is a function of V_1.

What is this function? There is no way to find out, by an algorithm, this function. It is then given *a priori*: it is the result of a *decision* made by the cost analyst.

Let us start with the easiest one (we will certainly not say the best one, because at this stage we do not know), which is the subject of this chapter: we assume a *linear relationship* between Y and V_1 and write that the "dynamic center" of Y, which is called $\hat{y}$, is linearly dependent on V_1:

$$\text{dynamic\_center\_of\_}Y = \hat{y} = b_0 + b_1 x$$

where b_0 and b_1 are constant – but presently unknown – values. They are called the "coefficients" of the relationship.

This idea is very simple, and will appear very powerful.

How can we find out b_0 and b_1? Exactly as we did in the previous chapter, the dynamic center will be defined by the b_0 and b_1 which together minimize, in this section, the sum of the distances between $\hat{y}_i$ and y_i. As we decide to use the metric defined by the square of the differences, we have to minimize:

$$\sum_i (y_i - \hat{y}_i)^2 = \sum_i (y_i - b_0 - b_1 x_i)^2$$

9.1.2 Computing the Formula Giving the Dynamic Center

No hypothesis is necessary to carry out the computations: the computation aims at finding b_0 and b_1 which together minimize the sum:

$$\text{Sum} = \sum_{i=1}^{I} (y_i - b_0 - b_1 x_i)^2$$

As it is well known the values of b_0 and b_1 will be given by the couple which equals to 0 the partial derivatives:

$$\frac{\partial \text{Sum}}{\partial b_0} = 0 \quad \text{and} \quad \frac{\partial \text{Sum}}{\partial b_1} = 0$$

These two equations are linear equations in b_0 and b_1 of which solutions is easily computed:

$$b_1 = \frac{\sum_i (x_i - \bar{x})(y_i - \bar{y})}{\sum_i (x_i - \bar{x})^2} = \frac{s_{xy}}{s_x^2} \qquad \text{this is called the slope}$$

$$b_0 = \bar{y} - b_1 \bar{x} \qquad \text{this is called the intercept}$$

Note that the value of b_1 is proportional to the covariance between x and y (this is rather logic), and inversely proportional to the variance of x; it means that for the same covariance, the slope will be higher when the spread of x is smaller (this is intuitive). About the units? If y is a cost in euros and x is the mass in kilograms, the unit of b_1 is in €/kg, the unit of b_0 being in €.

This procedure is generally known under the name of "least squares method" (because its purpose is to minimize the sum of the squares of the deviations between the dynamic center and the observed values), or, for a reason which will appear below, the "linear regression". The abbreviation OLS – which stands for "ordinary least squares" – is sometimes used.

These expressions can take many forms, as found in the literature. For instance when using the means:

$$b_1 = \frac{\sum_i x_i \cdot y_i - I \cdot \bar{x} \cdot \bar{y}}{\sum_i x_i^2 - I \cdot \bar{x}}$$

Changing the Scales

It is sometimes necessary to change the scales: What does happen if we change the scales of x and y?

- Let us change the scale of x (for instance expressing x in kilogram instead of tons): all x values are multiplied by a factor k_x – 1000 in the example. Then b_1 is divided by k_x. This is logic: if the value of x are multiplied by 1000, geometrically, the line representing the dynamic center will be much more horizontal than before. Quite obviously the intercept must be changed accordingly, in order for the dynamic center to pass through the data point $\{\bar{x}, \bar{y}\}$.
- If the scale of y is changed by a factor k_y (for instance if y is given in euros instead of thousand euros), then both the slope and the intercept are multiplied by k_y. This comes from the fact that the relationship between b_0 and b_1 on one hand, y on the other hand is linear as it can immediately be seen from the above relationships. This is interesting to remember if you have to change the currency unit in which the costs are given.

Consequently if both scales are simultaneously changed, the slope is multiplied by the ratio of the change (k_y/k_x) and if both scales are multiplied by the same factor, the slope does not change; this is geometrically obvious.

If the change of scale is just a translation, then the slope does not change, but the intercept does. A useful application of this is what happens if we use *centered data* (translation of x by $-\bar{x}$, of y by $-\bar{y}$ which is geometrically equivalent to transferring the center of the coordinates axes to the arithmetic mean values). Then the coefficients ${}_cb_0$, ${}_cb_1$ computed on the centered data will be given by:

$${}_cb_1 = \frac{\sum_i {}_cx_i \cdot {}_cy_i}{\sum_i {}_cx_i^2} = b_1 \quad \text{the value of the slope does not change}$$

$${}_cb_0 = 0 \quad \text{the value of the intercept does change}$$

Note that this expression shows that, when using the OLS procedure with an intercept not forced to be 0, the dynamic center passes exactly through the center of the data values. This is not the case when the intercept is forced to be 0.

If we now use centered and scaled values (the centered values are divided by the standard deviation s_x and s_y), then the slope ${}_{cs}b_1$ is multiplied by the ratio (s_x/s_y):

$${}_{cs}b_1 = \frac{\sum_i {}_{cs}x_i \cdot {}_{cs}y_i}{\sum_i {}_{cs}x_i^2} = \frac{s_x}{s_y} b_1$$

$$b_0 = 0$$

As the logic does not really change, it is sometimes easier to work with centered and scaled values instead of direct values.

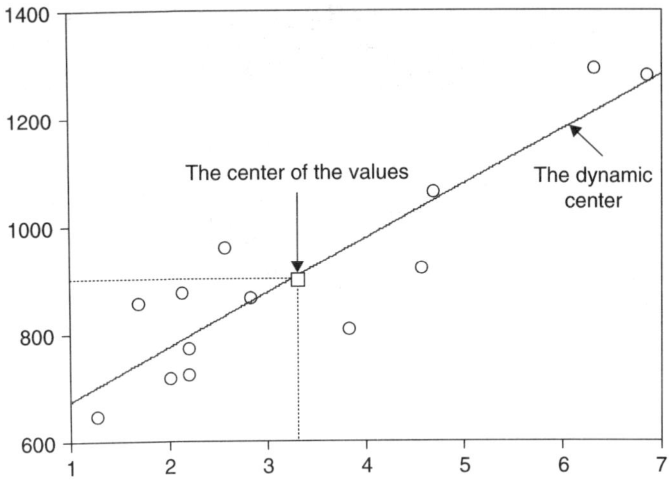

Figure 9.3 The dynamic center computed by the OLS.

If the OLSs is used on these centered and scaled values, one can return to the normal coefficients by using the first equation for the slope:

$$b_1 = \frac{s_y}{s_x}\, {}_{cs}b_1$$

The intercept can easily be computed from the fact that the dynamic center must pass through the center of the data:

$$b_0 = \bar{y} - b_1\bar{x}$$

Example

Using this metric on this sample, we found the formula giving the dynamic mean of this sample (Figure 9.3):

$$cost = 572.972 + 101.081 \times Mass$$

9.1.3 What Did We Win by Using This Dynamic Center?

The purpose of using the dynamic center – here the dynamic arithmetic mean which could be noted $\hat{\bar{y}}$ – instead of the mere arithmetic mean $\bar{y}$ was to reduce the amount of the residuals around this center.

Let us check on the example what we won by this process.

The distribution ψ around $\bar{y}$ was computed at the beginning of this chapter: we found that its standard deviation was 192.979.

The distribution ψ of the residuals around the dynamic center $\hat{y}$ is now given in Figure 9.4.

The mean value of these residuals is 0 (this is a property of the linear regression when we use an intercept different from 0) and its standard deviation 83.805.

Name	Residuals
A	14.644
B	−69.328
C	−148.080
D	−112.891
E	−21.328
F	90.747
G	18.980
H	7.990
I	130.272
J	113.212
K	83.217
L	−56.112
M	−51.323

Figure 9.4 The residuals around the dynamic center.

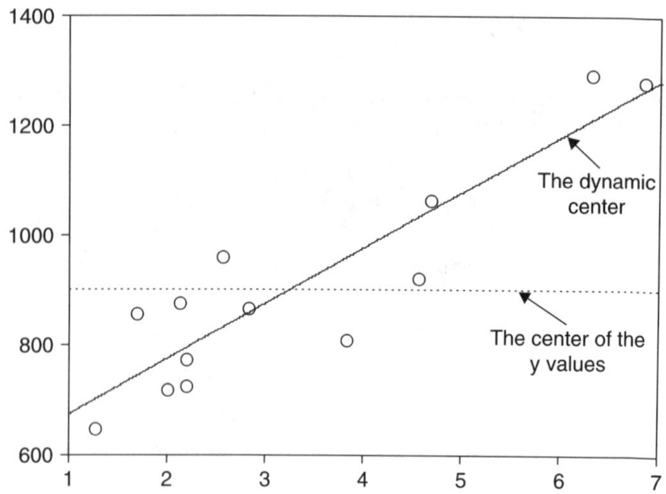

Figure 9.5 Comparing both centers.

The residuals are therefore considerably reduced as this clearly appears in Figure 9.5. Residuals are the distance (measured parallel to the y-axis) between the data points – represented by small circles – and either the horizontal line representing the distribution center of these data points or the inclined line representing the dynamic mean. This will allow to make much better estimates by using $\hat{y}$ instead $\bar{y}$. We won something!

9.1.4 Using the Matrix Notation

In order to prepare the more complex case of more than two quantitative variables, the matrix notation is developed here for the reader who is not familiar with this notation. It will of course give the same result, but in a more concise form.

The matrix we need derives from the $||x||$ matrix to which a column of 1 is added (in order to compute the intercept). In general terms, assuming the database includes I products, we have:

$$||^{+}x|| = \begin{Vmatrix} 1 & x_1 \\ 1 & x_2 \\ \dots & \dots \\ 1 & x_I \end{Vmatrix} \quad \text{and its transpose} \quad ||^{+}x||^{t} = \begin{Vmatrix} 1 & 1 & \dots & 1 \\ x_1 & x_2 & \dots & x_I \end{Vmatrix}$$

We also introduce a vector $\vec{b}$ of which components are

$$\vec{b} = \begin{Vmatrix} b_0 \\ b_1 \end{Vmatrix}$$

As we write (in order to conform with the matrix algebra because $\vec{b} \in \Re^{2\times 1}$ and $||^{+}x|| \in \Re^{I\times 2}$):

$$||y|| = ||^{+}x|| \otimes \vec{b}$$

this vector $\vec{b}$ is computed by the linear regression algorithm as:

$$\vec{b} = \left(||^{+}x||^{t} \otimes ||^{+}x||\right)^{-1} \otimes ||^{+}x||^{t} \otimes ||y||$$

Let us develop, just for once, this expression. First of all we have:

$$||^{+}x||^{t} \otimes ||^{+}x|| = \begin{Vmatrix} I & \sum x_i \\ \sum x_i & \sum x_i^2 \end{Vmatrix}$$

of which the inverse must be computed. In order to compute this inverse, we need the determinant "**det**" of this matrix:

$$\det = I \cdot \sum x_i^2 - \left(\sum x_i\right)^2 = I \times \left(\sum x_i^2 - I \cdot \bar{x}^2\right)$$

which is I times the denominator of the value previously computed for b_1. This determinant will play in the future an important role: it is clear that it has to be different from 0 and one of the major problems when dealing with several variables will be to make sure that it is different from 0.

Let us go on with the computations:

$$\left(||^{+}x||^{t} \otimes ||^{+}x||\right)^{-1} = \begin{Vmatrix} \dfrac{\sum x_i^2}{\det} & -\dfrac{\sum x_i}{\det} \\ -\dfrac{\sum x_i}{\det} & \dfrac{I}{\det} \end{Vmatrix} = \frac{1}{\det} \otimes \begin{Vmatrix} \sum x_i^2 & -\sum x_n \\ -\sum x_i & I \end{Vmatrix}$$

The product of this $\Re^{2\times 2}$ matrix by a $\Re^{2\times I}$ matrix will generate another $\Re^{2\times I}$ matrix which is given by:

$$\left(\|^{+}x\|^{t}\otimes\|^{+}x\|\right)^{-1}\otimes\|^{+}x\|^{t} = \begin{Vmatrix} \sum x_i^2 - x_1\cdot\sum x_i & \sum x_i^2 - x_2\cdot\sum x_i & \dots & \sum x_i^2 - x_I\cdot\sum x_i \\ -\sum x_i + I\cdot x_1 & -\sum x_i + I\cdot x_2 & \dots & -\sum x_i + I\cdot x_I \end{Vmatrix} \otimes \frac{1}{\det}$$

and the product of this $\Re^{2\times I}$ matrix by a $\Re^{I\times 1}$ matrix will produce a $\Re^{2\times 1}$ matrix which is the $\vec{b}$ vector:

$$\vec{b} = \frac{1}{\det}\otimes\begin{Vmatrix} \left(\sum x_i^2\right)\times\sum y_i - \left(\sum x_i\right)\times\sum x_i\cdot y_i \\ -\left(\sum x_i\right)\times\sum y_i + I\cdot\sum x_i\cdot y_i \end{Vmatrix}$$

Taking into account that $\Sigma x_i = I\cdot\bar{x}$ and $\Sigma y_i = I\cdot\bar{y}$ we can write for b_1 (the computation of b_0 can also be done but is more complex):

$$b_1 = \frac{I\cdot\sum x_i\cdot y_i - I^2\cdot\bar{x}\cdot\bar{y}}{\det} = \frac{\sum x_i\cdot y_i - i\cdot\bar{x}\cdot\bar{y}}{\sum x_i^2 - I\cdot\bar{x}^2}$$

which is, of course, exactly what was computed earlier.

This detailed computation will not be repeated anymore. It was made just once in order to become familiar with the matrix computations which are rather simple when you are used to it.

Computations for Example A

We get, with five significant digits:

$$\|^{+}x\|^{t}\otimes\|^{+}x\| = \begin{Vmatrix} 13 & 42.89 \\ 42.89 & 179.95 \end{Vmatrix}$$

$$\det = 499.81$$

$$\left(\|^{+}x\|^{t}\otimes\|^{+}x\|\right)^{-1} = \begin{Vmatrix} 0.36004 & -0.085812 \\ -0.085812 & 0.02601 \end{Vmatrix}$$

The product $\left(\|^{+}x\|^{t}\otimes\|^{+}x\|\right)^{-1}\otimes\|^{+}x\|^{t}$ gives a matrix with two rows and seven columns, of which only the first three ones are written below:

$$\begin{Vmatrix} -0.22606 & 0.17297 & 0.033951 & \dots \\ 0.091834 & -0.029111 & 0.013025 & \dots \end{Vmatrix}$$

and eventually

$$\left(\|{}^{+}x\|^{t} \otimes \|{}^{+}x\|\right)^{-1} \otimes \|{}^{+}x\|^{t} \otimes \|y\| = b = \begin{Vmatrix} 572.97 \\ 101.08 \end{Vmatrix}$$

This solution may appear complex (as indeed is the case) but it becomes so simple – compared to the normal computations – when there are several variables that it cannot be avoided.

9.1.5 A Word of Caution

The expressions are always valid, unless of course x has a constant value (the determinant "**det**" and therefore the denominator of b_1 is then equal to 0; but in such a case, it is useless to look for a dynamic center!).

Another related case may happen when the range of the x values is small; in such a case, the standard deviation of x may be very small, which may give an "enormous" value to b_1 (it may happen that this will produce an overflow of your computer). The result is mathematically correct, but is useless if your computer has an overflow!

Let us illustrate with one example. Consider the data (the standard deviation of x is 0.02) in Figure 9.6.

$$\|y\| = \begin{Vmatrix} 1000 \\ 1010 \\ 1020 \\ 1030 \\ 1040 \\ 1050 \\ 1060 \end{Vmatrix} \quad \text{and} \quad \|x\| = \begin{Vmatrix} 10.00 \\ 10.01 \\ 10.02 \\ 10.03 \\ 10.04 \\ 10.05 \\ 10.06 \end{Vmatrix}$$

Figure 9.6 Data values for Example B.

A dynamic center can be easily computed, and it is perfect (see Figure 9.7):

$$\hat{y} = -9000 + 1000 \times x$$

If the formula is not wrong, one would probably prefer (people who receive the formula may object to the high-negative intercept!) to get a formula such as:

$$\hat{y} = b_0' + b_1' \cdot \Delta x$$

with $\Delta x = (x - 10) \times k$. For instance taking $k = 100$ will generate the formula:

$$\hat{y} = 1000 + 10 \times \Delta x$$

which is more comfortable and will never produce an overflow.

This situation is rare when using a linear relationship, but may happen more frequently when using a "multiplicative" dynamic center. This will require to take the

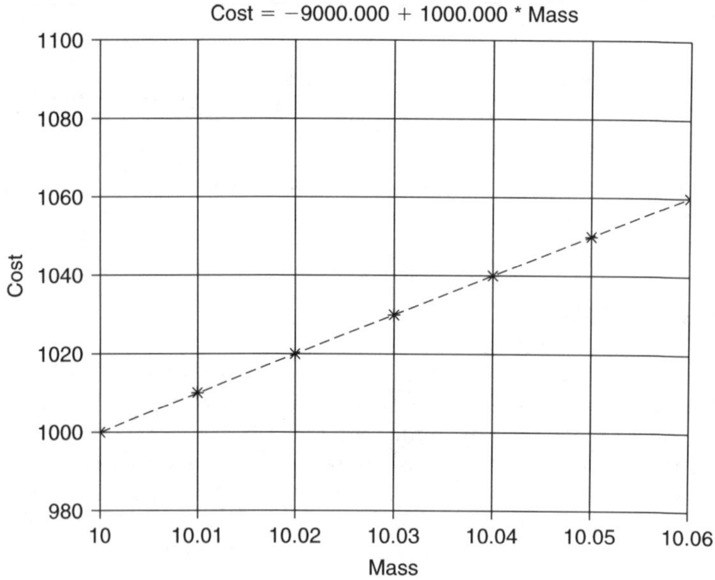

Figure 9.7 The dynamic mean for Example B.

logarithm of the x values: the standard deviation might, in such a case, be extremely small (because it is based on logarithms) and will certainly produce an overflow.

It is, of course, possible to automatically detect such a situation and to display a warning explicitly displaying the determinant which appears in the computations:

$$\det = I \cdot \sum x_i^2 - \left(\sum x_i\right)^2 = I \times \left(\sum x_i^2 - I \cdot \bar{x}^2\right)$$

when it is lower than 1. The user may go on or change the x variable accordingly.

9.1.6 The Characteristics of the Linear Regression

The linear regression has, at this stage[1], four important properties:

1. The dynamic center, when the intercept is not "forced" to 0, passes exactly through the center of the data, the center being defined as an hypothetical data point with values $\bar{y}$ and $\bar{x}$. This is obvious from the expression giving b_0:

$$b_0 = \bar{y} - b_1 \bar{x}$$

The dynamic center is then written as:

$$\hat{y} = (\bar{y} - b_1 \bar{x}) + b_1 x$$

and therefore $\hat{y} = \bar{y}$ when $x = \bar{x}$.

[1] Other characteristics – related to the variances of the coefficients – are dealt with in Chapter 15.

Note that is not true if we force the intercept to be equal to 0, which means we want to find a moving center such as $\hat{y} = b_1 x$. When minimizing the distances as they were previously used, we find the following solution:

$$\hat{y} = \frac{\sum_i x_i y_i}{\sum_i x_i^2} x$$

which will not gives $\bar{y}$ when $x = \bar{x}$.

2. The sum of the residuals is equal to 0: $\Sigma_i e_{+i} = 0$. It is sometimes said that this implies that the coefficients giving the dynamic center are not biased. But the concept of "bias" is something different; it will be studied in Chapter 15.
3. The value of b_1 is not symmetrical in x and y. This fact has important consequences which are developed in the next section.
4. b_0 and b_1 are strictly correlated. This is also obvious from the expression giving b_0: the intercept is a linear function of the slope.

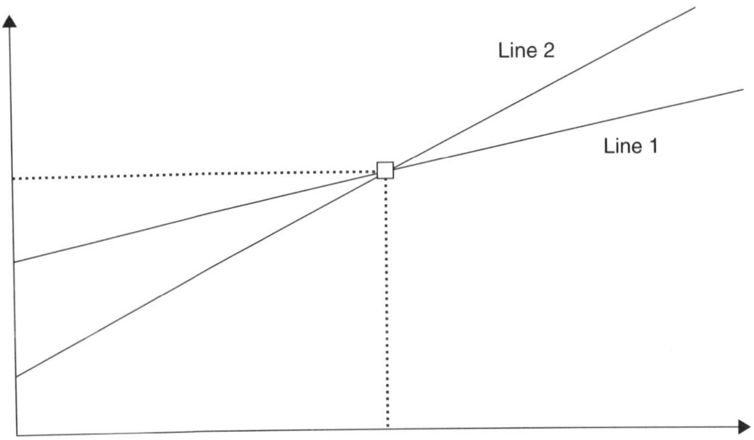

Figure 9.8 Correlation between b_0 and b_1.

Geometrically this result is obvious (Figure 9.8): starting from solution given by line 1, if another solution is found which increases the value of the slope, then the intercept has to decrease, for line 2 must also go through the center of the data.

The linear regression is extensively used by the community of cost analysts.

We would like, however, attract at this stage the attention of the cost analyst to two major problems that frequently occur when dealing with cost:

1. The first problem appears when the data are rather scattered (and it is generally the case with cost data, as previously explained): then the linear regression introduces a bias which can be very detrimental.
2. The second problem occurs when the range of costs exceeds – let us say – a ratio higher than 3: then the linear regression gives a high "weight" to large cost values, sometimes seriously disregarding the low-cost values.

Solutions to these problems will be presented.

9.1.7 Problems, When Dealing with Cost, with the OLS

This book is dedicated to cost estimating. It is therefore necessary to investigate if the OLS procedure is the best tool for this purpose.

Unfortunately it is not! The question is not at this stage to accept the hypothesis of a linear relationship between the causal variable and the dynamic center, but, when this hypothesis can be considered as correct, to look at potential problems. We do not pretend that the linear regression is imperfect in all the domains, but to limit this discussion to cost applications.

A First Serious Probem: The Linear Regression is Biased!

Assessing the Problem

This problem was well known to Karl Friedrich Gauss[2]: the name of "regression analysis", often given the OLSs procedure refers to this problem.

Let us introduce this question geometrically: in the plane x–y, the data can be represented by a set of points, all of them being included inside an ellipsis E (Figure 9.9). When you look for the dynamic center, you naturally *expect*, in the hypothesis of a not negligible linear correlation between x and y, to find the line represented by AB.

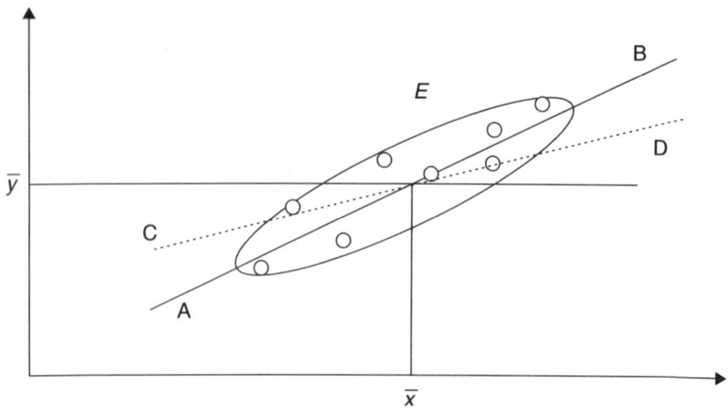

Figure 9.9 What do we mean by "regression"?

Is it the case? No in most of the situations! The line we found out is something such as CD (depending on your data, as it is quantified below): the **dynamic center "regresses" towards the arithmetic mean $\bar{y}$.**

An illustration of the phenomenon can be given in making two regressions: y to x on one hand, x to y on the other hand. You may consider that looking for the size x when the cost y is known does not really make sense: you are not going to estimate, but this is quite possible, the size of a product when you know its cost. But forget for the moment the logic and concentrate on the mathematical aspects of the question.

[2] He said that any other (different from trying to minimize the sum of the squares of the differences between the observed values and the values of the dynamic center) method for computing the coefficients of the straight line corresponding to the dynamic center would involve very complex computations. He was right; other methods can only be used practically thanks to the computer.

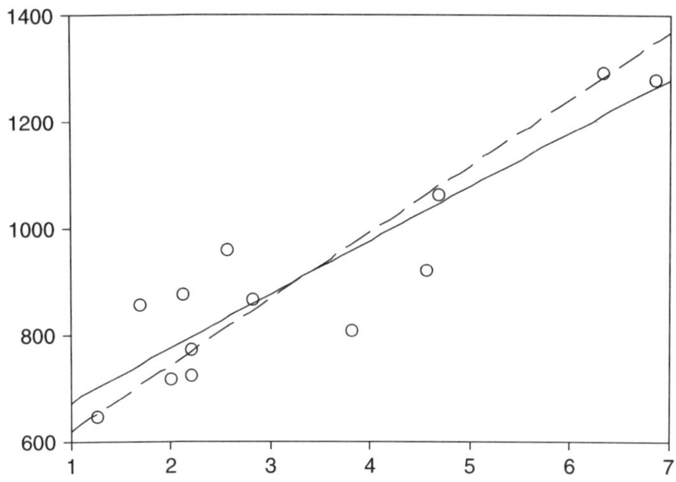

Figure 9.10 The two regression lines: the continuous line displays the regression of *y* to *x* the dotted line displays the regression of *x* to *y*.

Let us illustrate the point on the example given at the beginning of this chapter:

- The dynamic center line generated by making a regression of y to x is given by:

$$\hat{y} = 572.972 + 101.081 \times x$$

- Whereas the dynamic center line generated by making a regression of x to y is given by:

$$\hat{x} = -3.9774 + 0.00803 \times y$$

which gives

$$\hat{y} = 495.460 + 124.569 \times x$$

Both lines are displayed on Figure 9.10; the difference in the vicinity of the center $\langle \bar{y}, \bar{x} \rangle$ of the data is negligible but it becomes important when the x value is away from $\bar{x}$.

It is interesting to note that the product of both slopes is equal to the square of the Bravais–Pearson correlation coefficient: if one labels $b_{1(y/x)}$ the slope of the regression of y on x and $b_{1(x/y)}$ the slope of the regression of x on y, then it can be written that:

$$r_B = \sqrt{b_{1(y/x)} \times b_{1(x/y)}}$$

For the present example:

$$0.901 = \sqrt{101.081 \times 0.00803}$$

What Is the Origin of This Phenomenon? Algebraically this is caused by the fact, already mentioned in Section 9.1.2. of this chapter, that the value of b_1 is not symmetrical in x and y.

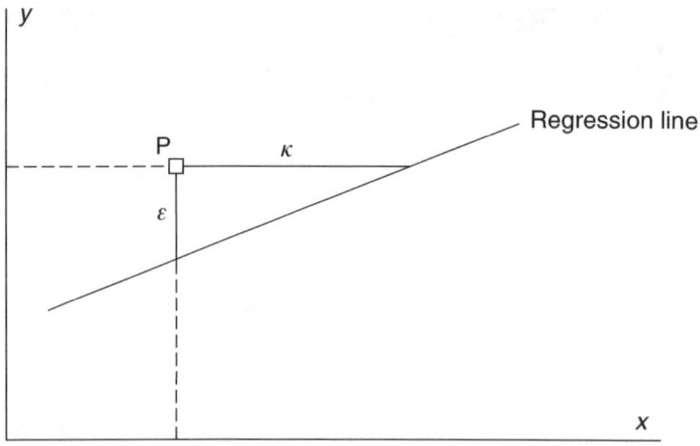

Figure 9.11 Geometric interpretation of the regression analysis.

Geometrically, it comes from the fact we are minimizing "deviations" and not "distances". Refer to Figure 9.11, where P represents a data point and the solid line the linear relationship we are interested in. When we make a regression of y to x we compute b_0 and b_1 by minimizing $\Sigma_i \varepsilon_i^2$, whereas when we make a regression of x to y we compute the coefficients by minimizing $\Sigma_i \kappa_i^2$. It is obvious that there is absolutely no reason to obtain the same values for these coefficients.

Some Comments You may find in the literature people who seem to be happy with the situation. An example is given by psychologists[3]: suppose, they say, that x represents the note in English, y the note in mathematics for the same student. Data were plotted as in Figure 9.12. Let us assume we have a student who got a note in English higher than the mean $\bar{x}$. What can we forecast for his note in mathematics? You could expect a note on the main axis of ellipsis. No, they say: this note would be too optimistic, we prefer to forecast a note closer to the mean $\bar{y}$: his note should "regress" towards this mean.

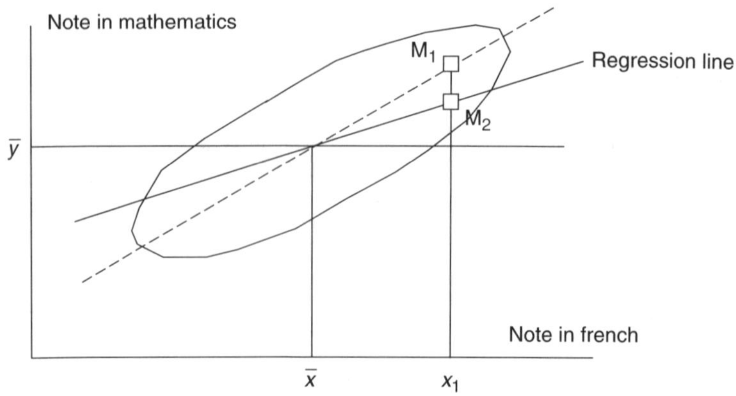

Figure 9.12 The two regression lines.

[3] See Thomas H. Wonnacott and Ronald J. Wonnacott. [60], 4th French edition. p. 556.

The logic of the comment is not clear to us.

Galton was also very pleased with the phenomenon: observing that tall fathers have tall sons but not so tall as their fathers (the same observation is true for small fathers), he concluded that the sons' sizes "regress" towards the mean size of the population (it seems he coined the word "regression").

This "regression" might be an advantage in some sciences, but in the domain of cost we have no reason to be pleased with this fact: after all the facts are there: a forecast which does not follow the facts presents a problem! We do not see any reason why the cost of a new product (maybe made by another manufacturer) should regress towards the mean of the cost of other products belonging to the same family, just because data are a bit scattered (as, if there is no scattering, there is no regression!).

Quantification of the Damage How serious is the damage?
First of all, it is clear that, if the residuals are all in the vicinity of 0, then line CD is very close to line AB and therefore the problem does not really exist. Unfortunately this is rather rare when dealing with cost.

Let us compute the damage: using the centered data (centering the data simplifies the computations without changing the slopes), the first regression line is given by:

$$_c\hat{y} = \frac{\sum {_c x_i} \cdot {_c y_i}}{\sum {_c x_i^2}} {_c x} = b_{1(y/x)} \times {_c x}$$

whereas the second one is given by:

$$_c\hat{x} = \frac{\sum {_c x_i} \cdot {_c y_i}}{\sum {_c y_i^2}} {_c y} = b_{1(x/y)} \times {_c y}$$

or

$$_c\hat{y} = \frac{\sum {_c y_i^2}}{\sum {_c x_i} \cdot {_c y_i}} {_c x} = \frac{_c x}{b_{1(x/y)}}$$

The ratio of the slopes is easily computed: it is equal to:

$$\frac{\left(\sum {_c x_i} \cdot {_c y_i}\right)^2}{\sum {_c x_i^2} \sum {_c y_i^2}}$$

which is nothing else than the square of the correlation coefficient r^2 (this value is now symmetrical in x and y). A correlation coefficient equal to 0.901, as it appears in the example (and which is rather good in the domain of cost) leads to a ratio of 0.811 between the slopes, about 20% (this can easily be checked by the figure which appears upwards); it is easy to imagine the error that will be done if one extrapolates the dynamic center far away from the arithmetic mean $\bar{x}$ of the causal variable!

One could object – and some authors do – to this reasoning that we interchange "causal" and "dependent" variables. This is true, but:

1. As Lothar Sachs says (Ref. [49], p. 393): "hypothesis on causation must come from outside, not from statistics". The work of statistics – as we do it here – is simply to quantify a relationship between two variables, without deciding on which causes the other one.
2. We are interested in the quality of predictions. Whatever can be said, a prediction based on line AB (Figure 9.9) is, in the domain of cost, better than one based on line CD.

Conclusion **The linear regression is often not the best solution when dealing with cost!**

Another comment about the Bravais–Pearson correlation coefficient: the angle (Figure 9.10) between both lines becomes exactly equal to 0 when $r = 1$. In such a case there is a perfect linear correlation between x and y. This is the reason why this correlation coefficient is said to quantify a linear correlation, as we mentioned it in Chapter 5.

Solutions

There are three ways to solve this problem:

1. Averaging the regressions of y to x and of x to y.
2. Using the euclidian distance.
3. Rotating the axes.

A First Solution: Averaging the Regressions *y/x* and *x/y* Let us go on with the example: averaging both formulae leads to the following formula for the moving center:

$$\hat{y}_i = 534.216 + 112.828 \times x_i$$

which is displayed as a mixed line of Figure 9.13.

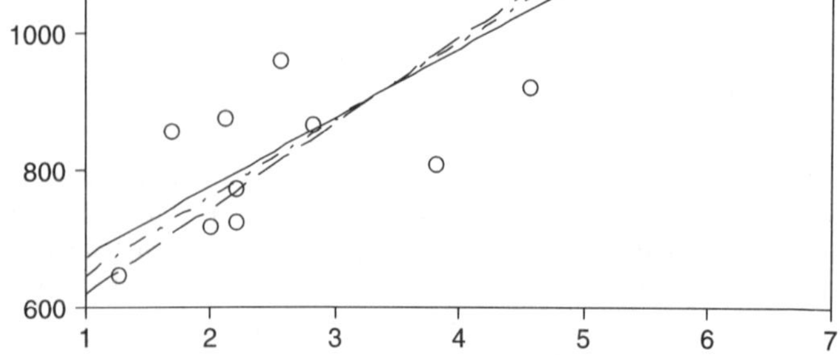

Figure 9.13 Averaging slopes.

What are the characteristics of the residuals related to this average formula. One can easily compute that the mean of the distribution ψ of residuals equals 7×10^{-5}, very close to 0, and its standard deviation equals 86.205, very close from what was computed for the standard regression analysis (83.805).

The conclusion at this stage is that, for a very limited loss of precision for the future estimates, we get a formula which will give much better values when the causal variable is away from its centered value $\bar{x}$.

This procedure can obviously be easily automated, and has been.

A Second Solution: Using the Euclidian Distance We saw in a previous section that the cause of the bias was the fact we tried to minimize $\Sigma_i \varepsilon_i^2$. We can instead try to minimize the sum of the Euclidian distances between the data points and the searched moving center.

The computation requires several steps.

The Euclidian distance between two points P_1 and P_2 defined by their coordinates (x_1, y_1 for data point P_1, x_2, y_2 for data point P_2) is given by:

$$\delta_{1,2}^2 = (x_1 - x_2)^2 + (y_1 - y_2)^2$$

Let now (Figure 9.14):

$$y = b_0 + b_1 \cdot x$$

be the equation of the line D we are looking for.

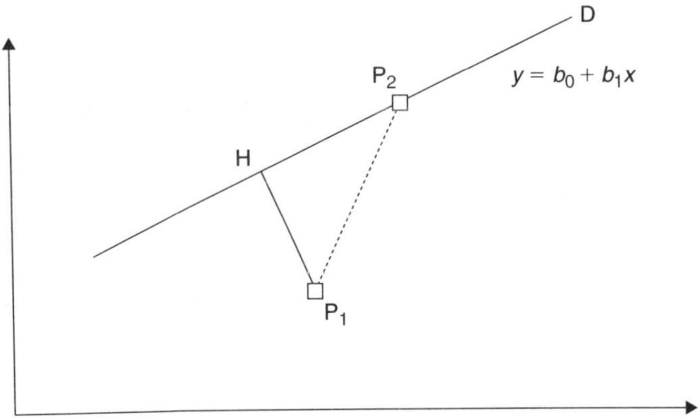

Figure 9.14 Computing the Euclidian distance.

The square of the distance between these points P_1 and P_2 points, when P_2 is on the line D is given by:

$$d_{1,2}^2 = (x_1 - x_2)^2 + (y_1 - b_0 - b_1 x_2)^2$$

This distance will be minimum for a particular position of point P_2 corresponding to the abscissa of point H, defined by the fact that P_1H is perpendicular to D.

For this value we will have:

$$\frac{\partial \delta_{1,2}^2}{\partial x_2} = 0$$

The coordinates of point H will therefore given by:

$$x_H = \frac{x_1 + b_1 y_1 - b_0 b_1}{b_1^2 + 1}$$
$$y_H = \frac{b_1}{b_1^2 + 1}(x_1 + b_1 y_1 - b_0 b_1) + b_0$$

Now it is possible to compute the Euclidian distance δ_1 (equal to P_1H) from P_1 to line D. We compute:

$$\delta_1 = \frac{1}{\sqrt{b_1^2 + 1}} \times |b_1 x_1 + b_0 - y_1|$$

Using now our set of data points, we look for the line D which goes as closer as possible from all these data points, "close" being quantified by the Euclidian distance. This means we want to compute the couple b_0, b_1 which minimizes the sum:

$$\text{Sum} = \sum_i \delta_i^2 = \frac{1}{b_1^2 + 1} \times \sum_i (b_1 x_i + b_0 - y_i)^2$$

Let us demonstrate that this line goes necessarily through the center $\langle \bar{y}, \bar{x} \rangle$ of the set: this Sum will be minimum when the two equalities are simultaneously satisfied:

$$\frac{\partial \text{Sum}}{\partial b_1} = 0$$
$$\frac{\partial \text{Sum}}{\partial b_0} = 0$$

Let us develop the second one:

$$\frac{\partial \text{Sum}}{\partial b_0} = \frac{2}{b_1^2 + 1} \sum_i (b_1 x_i + b_0 - y_i) = 0$$

which will be satisfied when:

$$\sum_i (b_1 x_i + b_0 - y_i) = 0$$
$$b_1 \sum x_n + I \cdot b_0 - \sum y_i = b_1 \cdot \bar{x} + b_0 - \bar{y} = 0$$

The center of the set, defined as the point of which coordinates are $(\bar{x}, \bar{y})$, satisfies the equation of the line: the line passes through this point.

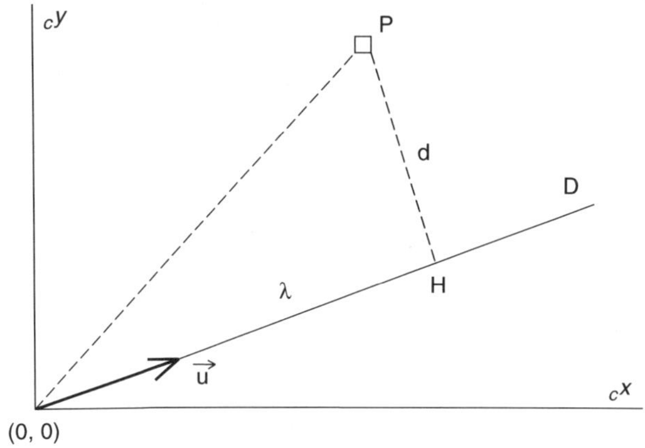

Figure 9.15 Solving the problem.

Minimizing Sum is difficult to do. But the problem may be solved by standard linear algebra: it suffices to note that Euclidian distances are invariant by a translation and/or a rotation of the axes. Therefore the center of the axes can be transferred to the point $(\bar{x}, \bar{y})$, which means that we can more easily work with centered values (Figure 9.15):

$$_c y = y - \bar{y} \quad \text{and} \quad _c x = x - \bar{x}$$

Now one can write, for each data point P (cf. Figures 9.14 and 9.15):

$$\delta^2 = \overline{_c\mathrm{P}_c\mathrm{H}}^2 = \overline{_c\mathrm{PO}}^2 - \overline{\mathrm{O}_c\mathrm{H}}^2 = \overline{_c\mathrm{PO}}^2 - \lambda^2$$

Consequently for all the data:

$$\sum_i \delta_i^2 = \sum_i \overline{_c\mathrm{P}_i\mathrm{O}}^2 - \sum_i \lambda_i^2$$

In this expression $\Sigma_{ic}\overline{\mathrm{P}_i\mathrm{O}}^2$ is a constant (in respect to b_1). To the minimum of $\Sigma_i\delta_i^2$ will therefore corresponds a maximum of $\Sigma_i\lambda_i^2$.

The problem is then to find b_1 which will maximize $\Sigma_i\lambda_i^2$. Have we made any progress? Yes, because this problem is already solved by theorems of linear algebra. It is the way the principal component analysis (PCA) works, as it is reminded in the following paragraph.

A Third Solution: Rotating the Axes The third solution can easily be explained in a figure (Figure 9.16.): we are looking for a straight line defined by the fact that the sum of the Euclidian distances is minimized. If we succeed to make a rotation of angle θ of the coordinates axes, transforming x in U_1 and y in U_2, in such a way that axis U_1 will be parallel to the line we are looking for, the Euclidian distances will be very easy to compute: they will be equal to the residuals!

The question is then: What should be the value of angle θ? The question is not so difficult to answer: we expect, once the axis is turned off, that the extension of the

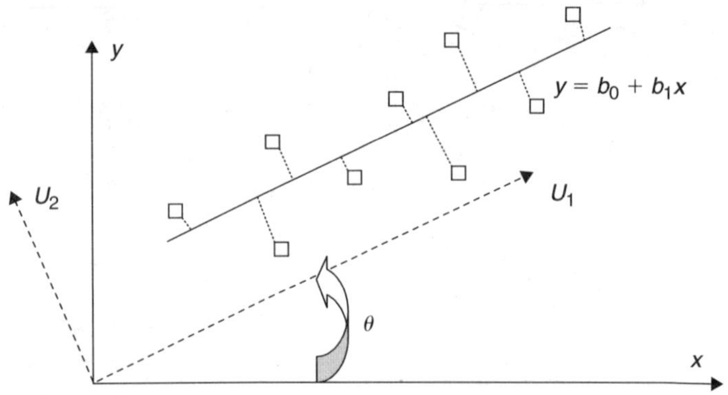

Figure 9.16 Rotation of the axis.

values on U_2 will be far less than the extensions on U_1. The problem is then to find axis U_1 on which data have the maximum extension.

This is exactly what the PCA does (see Chapter 6).

The solution is then to make a PCA. Immediately note that the PCA makes a computation such as axes U_1 and U_2 are orthogonal: this means that the correlation between the new values (computed on the axes U_1 and U_2) is 0.

The equation of the line will then obviously be, in the coordinates system defined by U_1 and U_2:

$$U_2 = 0$$

Let us apply it to our example. The PCA works on centered and scaled values defined by:

$$_{cs}x = \frac{x - \overline{x}}{s_x} = \frac{x - 3.299}{1.720}$$

$$_{cs}y = \frac{y - \overline{y}}{s_y} = \frac{y - 906.462}{192.979}$$

and the new variables are defined as:

$$U_1 = 0.707 \times {}_{cs}y + 0.707 \times {}_{cs}x$$
$$U_2 = 0.707 \times {}_{cs}y - 0.707 \times {}_{cs}x$$

Note that these formula are standard when working on two variables centered and scaled: it means that the axis have been rotated by 45°. This is obvious because, due to the scaling effect, both variables have the same normalized scale.

Writing $U_2 = 0$ and returning to the original variables gives the equation of the line:

$$\hat{y} = 536.324 + 112.197 \times x$$

which is very close to the equation we found by taking the average of the regressions of y on x and x on y.

No complex computation is therefore required when working with two variables only. Just compute the mean and standard deviation of both variables and write:

$$_{cs}y = {}_{cs}x$$

The equation of the line is computed in 30 s!

The residuals are easily computed: their arithmetic mean amounts to −0.026, very small indeed, and their standard deviation to 85.952.

Conclusion The three solutions, to be used when, let us say, the Bravais–Pearson correlation coefficient is less than 0.9 (unless you want to estimate far away from the center point $(\bar{x}, \bar{y})$ in which case start using it when this coefficient is smaller than 0.95) give very similar results, the third one being obviously the simplest one.

A Second Serious Problem: The "Weight" Given to High-Cost Values

As previously stated, the linear regression is based on the minimization of the squares of the deviations between the cost values y and the searched dynamic center. Let us remind the reader that absolutely no hypothesis is required for the computation; using a linear relationship and the metric based on the differences are *decisions*, not hypotheses.

This section tries to answer the following question: Are the data points considered, when preparing a formula with the linear regression algorithm, on an equal basis?

This is true in absolute terms: the algorithm tries to minimize the sum of the square of the deviations, and no difference is made between the data points.

Is that still true in relative terms? The question must be asked[4] because, in the cost domain, the accuracy of the cost values are always known in percentages: one can say that the costs are, for instance, known with an accuracy of 5%. This means that a cost of 10k€ is known with an absolute accuracy of 0.5k€, whereas a cost of 1000k€ is known with an absolute accuracy of 50k€.

It may happen that the cost figures you have in your database may represent a large range, for instance from 10k€ to 1000k€ corresponding to size in a range from about 1 to 100, whatever the unit.

The linear regression algorithm tries to minimize the absolute deviations – not the relative ones. This means that the residuals which will be discarded by the moving center once it is computed can very well represent an average of 2k€ for all costs. Consequently the dynamic center will pass much more closely – in relative terms – to the high-cost values than to the low ones.

This can be:

- Either interesting, if, for instance, you will have to estimate in the future figures close to the high costs: the relative deviation will be rather small, 0.2% in the given example. After all it can be appreciated to estimate the high costs with a very good – relative – accuracy.
- Or an inconvenience if, for instance, you have to produce 1000 items of low cost. In the given example the total cost will be 10M€, estimated with an accuracy of 2M€, which is rather poor!

[4] Stephen A. Book and Philp H. Youngs mention this point in their paper.

This problem has only to be solved if the cost range exceeds a ratio of, let us say, about 3; it becomes a serious question if this range exceeds a ratio of 10. The cost analyst should be aware of this problem: it helps explain why some analysts may decide to use other metrics, such as the dynamic median or the "multiplicative" residuals which are defined by the ratio $y_i/\hat{y}_i$. These metrics will be analyzed in the following section.

This discussion, once again, shows that several dynamic centers can be computed from the same data and that the linear is only one of them, maybe not the best one, depending on the circumstances.

A Related Problem: The Lack of Homoscedasticity in the Cost Domain

The term "homoscedasticity" of the residuals comes from the Greek and means "similar spread" which means that the dispersion of the residuals is supposed to be the same, whatever the value of x.

The lack of homoscedasticity (also called the presence of heteroscedasticity) is not a problem when computing a formula – except the one described in the previous section – because, as it was said, no hypothesis is required for making it. However, as we will see it in Part IV, such a hypothesis is required, when dealing, in the classical way, with the residuals.

In the domain of costs, the lack of homoscedasticity is obvious. For this reason some authors, such as Saporta [50] (p. 370) suggest to mitigate the problem by replacing the cost in the formula by the ratio of the cost to the causal variable (this ratio, when using the mass as the descriptor of the size, is the cost per kilogram).

The usual presentation:

$$y_i = b_0 + b_1 \times x_i + e_{+i}$$

can be written

$$\frac{y_i}{x_i} = b_1 + \frac{b_0}{x_i} + \frac{e_{+i}}{x_i}$$

which becomes if:

$$y_i' = \frac{y_i}{x_i},\ x_i' = \frac{1}{x_i} \quad \text{and} \quad e_{+i}' = \frac{e_{+i}}{x_i}$$

$$y_i' = b_1 + b_0 \times x_i' + e_{+i}'$$

This formula, if the e_{+i} are proportional to the size, satisfies the hypothesis of homoscedasticity.

It is solved as usual; returning to the original formula is simple, as the slope and the intercept have just to be switched.

9.2 Using Other Metrics

The first part of this chapter dealt with the most common metrics (the OLS or Euclidian metrics): the square of the differences of two values. We now investigate the use of the metrics defined in Chapter 8.

There is some correlation between the metric and the way the residuals are defined for the computations: it was mentioned at the beginning of Section 9.1 of this chapter that, when using the usual metric (the square of the differences), the residuals were defined as to be "**additive**" as we wrote:

$$y_i = \hat{y}_i + e_{+i} = b_0 + b_1 x_i + e_{+i}$$

where the symbol e_+ is used to remind the user that the residuals are defined in an additive way. We then decided that b_0 and b_1 should be such as to minimize the sum $\Sigma_i e_{+i}^2$ of the squares of the additive residuals.

In general terms, two solutions may be tested:

1. Choosing another definition.
2. Selecting another function to be minimized.

9.2.1 Choosing Another Definition of the Residuals

The residuals can also be defined as being "multiplicative", if, for instance, we define them from the formula:

$$y_i = \hat{y}_i \times e_{\times i}$$

as we used it for the second ratio. Dealing with such residuals is a bit special because these residuals should have a mean in the vicinity of 1.

For this reason, some authors (including Stephen A. Book and Philip H. Young) prefer to use a formula such as:

$$y_i = \hat{y}_i \times (1 - e_{\%i})$$

where the residuals $e_\%$ are now defined in order to get a mean of these residuals in the vicinity of 0, which is more familiar:

$$e_{\%i} = 1 - \frac{y_i}{\hat{y}_i} = \frac{\hat{y}_i - y_i}{\hat{y}_i}$$

They can also be defined in a completely different manner, as for instance:

$$e_{\bullet i} = \log \frac{y_i}{\hat{y}_i}$$

which has the advantage of having a mean in the vicinity of 0:

$$y_i = \hat{y}_i \times 10^{e_{\bullet i}}$$

9.2.2 Selecting Another Function to be Minimized

As explained in Chapter 8, although a lot of functions could be used, the function to be minimized is generally limited, whatever the way the residuals are defined, to either:

$$\sum_i |e_{+i}|^\alpha$$

or, when working with $e_\times$:

$$\prod_i e_{\times i} - 1$$

The exponent α can take, as illustrated in Chapter 8, a lot of values, but the most common one is 2, except when looking for the median (in which case it takes the value 1).

Standardization, for Comparison Purposes, of the Residuals

It would not be useful to use different metrics if we were not able to compare the results. This comparisons can be based on two points:

1. The way the dynamic center is computed. This type of comparison is mostly based on judgment or subjective criteria.
2. The distribution of the residuals (mean, standard deviation, and – why not? – skewness and kurtosis).

The residuals used in the minimization functions cannot be compared. In order to be able to compare the results given by different metrics, the residuals have to be "normalized". As we are primarily interested in the differences between the observed costs and the values computed for the dynamic center, the residuals, only for comparisons purposes, will be recomputed as:

$$e_{+i} = y_i - \hat{y}_i$$

whatever the metric which is used. So we do not compare the residuals used to build the formula giving the dynamic center of the distribution φ, but the absolute difference between the actual costs and the costs given by the dynamic center.

As the reader may be also interested in relative residuals, the residuals will also, be computed as:

$$e_{\times i} = \frac{y_i}{\hat{y}_i}$$

The residuals: $e_{\%i} = \dfrac{y_i - \hat{y}_i}{\hat{y}_i}$ could also be computed if necessary.

A Preliminary Comment

The linear regression uses the only metric for which analytic procedures are available. The other procedures can only be computed, not analytically, but by successive iterations. Consequently it is not possible to present, as we did for the linear regression, formulae from which practical examples could be dealt with. On the contrary each example has to be studied as a particular subject.

Therefore all the other metrics will be presented with the example which was introduced at the beginning of this chapter. For each metric the formula giving the dynamic center will be given, plus the four characteristics of the distribution ψ of the residuals: the arithmetic mean, the standard deviation, the skewness and the kurtosis.

9.2.3 Using the Metric Based on Differences with $\alpha = 2$: The Standard Regression

This is a reminder of the previous section.

This metric aims at minimizing $\Sigma_i |e_{+i}|^2$.

The Absolute Residuals

The characteristics of the residuals e_{+i} distribution are given by:

- Arithmetic mean: 0. This is a characteristic of the standard regression analysis.
- Standard deviation: 83.805.
- Skewness: 0.219.
- Kurtosis: 1.952.

The Relative Residuals

One finds for $e_{\times i}$:

- Arithmetic mean: 1.001. As a value different from 1 for the arithmetic mean of the relative residuals is sometimes considered as a "bias" of the formula, one can say that the formula is not biased.
- Standard deviation: 0.097.
- Skewness: 0.219.
- Kurtosis: 1.916.

9.2.4 Using the Metric Based on Differences with $\alpha = 1$: The Dynamic Median

The dynamic median is an interesting formula when the data are a bit scattered, for reasons explained in Chapter 8.

When using the median, the residuals are still defined to be additive. But we decide that b_0 and b_1 should be such as to minimize the sum $\Sigma_i |e_{+i}|$ of the absolute values of the residuals.

The formula is then given by:

$$\hat{y} = 521.563 + 114.880 \times x$$

which is different from the formula found by the linear regression and has the major advantage to be extremely robust.

The Absolute Residuals

The characteristics of the residuals e_{+i} distribution are given by:

- Arithmetic mean: 5.883. As a value different from 0 for the arithmetic mean of the absolute residuals is sometimes considered as a "bias" of the formula (do not confuse this bias with the bias of the coefficients which will be introduced in Chapter 8), one can say that the formula is (very) slightly biased; slightly compared to the intercept or to the average value of the cost (906.4).
- Standard deviation: 87.4.
- Skewness: 0.045.
- Kurtosis: 2.325.

The Relative Residuals

One finds for $e_{\times i}$:

- Arithmetic mean: 1.013. As a value different from 1 for the arithmetic mean of the relative residuals is sometimes considered as a "bias" of the formula (do not confuse this bias with the bias of the coefficients which will be introduced in Chapter 8), one can say that the formula is (very) slightly biased: 1.3%.
- Standard deviation: 0.103.
- Skewness: 0.399.
- Kurtosis: 2.327.

9.2.5 Using the Metric "Product" $\Pi_i e_{\times i} - 1$

This metric tries to directly minimize $\Pi_i e_{\times i} - 1$

The formula is then given by:

$$\hat{y} = 572.966 + 99.919 \times x$$

The Absolute Residuals

The characteristics of the residuals e_{+i} distribution are given by:

- Arithmetic mean: 3.84. One can say that the formula is slightly biased.
- Standard deviation: 83.829.
- Skewness: −0.049.
- Kurtosis: 1.833.

The Relative Residuals

One finds for $e_{\times i}$:

- arithmetic mean: 1.005,
- standard deviation: 0.097,

- skewness: 0.208,
- kurtosis: 1.89.

9.2.6 Using the Metric Based on the First Ratio

When using this metric, the residuals are defined to be multiplicative: we write:

$$\hat{y}_i = y_i \times e_{r1i} \quad \text{or} \quad e_{r1i} = \frac{\hat{y}_i}{y_i}$$

As we expect e_{r1i} to be in the vicinity of 1, we decide that b_0 and b_1 should be such as to minimize the sum:

$$\Sigma_i(e_{r1i} - 1)^2 = \Sigma_i((\hat{y}_i/y_i) - 1)^2$$

of the squares of these (multiplicative) residuals − 1.

The formula is then given by:

$$\hat{y} = 568.355 + 97.499 \times x$$

The Absolute Residuals

The characteristics of the residuals e_{+i} distribution are given by:

- Arithmetic mean: 16.435. As a value different from 0 for the arithmetic mean of the absolute residuals is sometimes considered as a "bias" of the formula (do not confuse this bias with the bias of the coefficients which will be introduced in Chapter 8), one can say that the formula is slightly biased.
- Standard deviation: 84.031.
- Skewness: 0.054.
- Kurtosis: 1.826.

The Relative Residuals

One finds for $e_{\times i}$

- arithmetic mean: 0.991,
- standard deviation: 0.095,
- skewness: 0.052,
- kurtosis: 1.893.

9.2.7 Using the Metric Based on the Second Ratio

When using this metric, the residuals are defined to be divider: it is the counterpart of the previous ratio. It may look strange (and it is very rarely used!) but we are interested here in its properties only:

$$\hat{y}_i = \frac{y_i}{e_{r2i}} \quad \text{or} \quad e_{r2i} = \frac{y_i}{\hat{y}_i}$$

As we expect e_{r2i} to be in the vicinity of 1, we decide that b_0 and b_1 should be such as to minimize the sum:

$$\Sigma_i(e_{r2i} - 1)^2 = \Sigma_i((y_i/\hat{y}_i) - 1)^2 = \Sigma_i e_{\%i}^2$$

of the squares of the (divider) residuals -1.

The formula is then given by:

$$\hat{y} = 592.765 + 97.392 \times x$$

The Absolute Residuals

The characteristics of the residuals e_{+i} distribution are given by:

- arithmetic mean: -7.622,
- standard deviation: 84.045,
- skewness: 0.054,
- kurtosis: 1.823.

The Relative Residuals

One finds for $e_{\times i}$:

- arithmetic mean: 0.991,
- standard deviation: 0.095,
- skewness: 0.154,
- kurtosis: 1.757.

9.2.8 Using the Metric Based on the Log of the Ratio

Distances between two values are now defined as the square of the log of their ratio. In other words the residuals are defined as:

$$e_{\bullet i} = \left(\log \frac{\hat{y}_i}{y_i}\right)^2$$

The coefficients which minimize the sum of these residuals are given by:

$$\hat{y} = 579.661 + 97.698 \times x$$

The Absolute Residuals

The characteristics of the residuals e_{+i} distribution are given by:

- arithmetic mean: 4.472,
- standard deviation: 84.007,
- skewness: −0.054,
- kurtosis: 1.833.

The Relative Residuals

One finds for $e_{\times i}$:

- arithmetic mean: 1.005,
- standard deviation: 0.097,
- skewness: 0.177,
- kurtosis: 1.813.

9.2.9 Using the Metric Based on the Biweight

As we did in Chapter 8 when introducing this metric, we will limit its description to its use combined with the linear regression.

The reader is reminded that a "weight" (between 1 and 0) is attached to each product and that this weight is used to decrease the influence of data points which are too far away from the bulk of the other data. It was said in the same part that the center, now the dynamic center, of the y distribution depends on the weights and the weights, at their turn, depends on the center. Therefore an iterative computation is necessary. These iterations allow to analyze this metric in a few steps.

The formulae necessary to compute the coefficients of the formula giving the dynamic center are described in the Section 9.1 of this chapter.

Initialization

The procedure has to start from a preliminary formula, called $\hat{y}^{(0)}$, for the dynamic center. This formula is computed with the classical regression analysis:

$$\hat{y}^{(0)} = 572.972 + 101.081 \times x$$

and from this formula, residuals are computed as $e_{+i}^{(0)} = y_i - \hat{y}_i^{(0)}$. The results are given in Figure 9.17.

Figure 9.18 displays the residuals values, according to their rank.

$$e\_0 = \begin{pmatrix} 14.644 \\ -69.328 \\ -148.08 \\ -112.891 \\ -21.328 \\ 90.747 \\ 18.98 \\ 7.99 \\ 130.272 \\ 113.212 \\ 83.217 \\ -56.112 \\ -51.323 \end{pmatrix}$$

Figure 9.17 Residuals values according to rank.

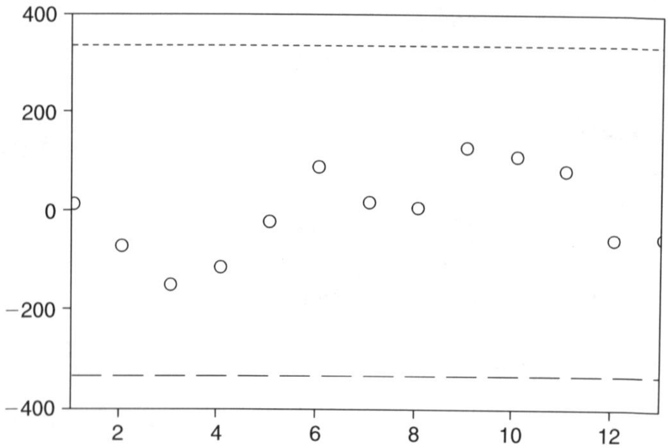

Figure 9.18 Residuals – according to their ranks – and cut-off values.

Step 1

First a cut-off value has to be chosen: we selected, although it is not the best one, here a cut-off value equal to four times the standard deviation of the residuals. This cut-off value amounts to 335.2; it is also displayed on Figure 9.17.

The residuals allow then to compute a preliminary value for the "weight" given to each product, according to the formula presented in Section 8.7.1. The list of these weights is displayed on Figure 9.19.

In the presentation situation, the residuals are not very large: nevertheless high values (corresponding to the ranks 3 and 9) receive a weight different from 1.

Given these "weights" a new formula for the dynamic center is computed:

$$\hat{y}^{(1)} = 564.721 + 103.678 \times x$$

which is different from the first one.

$$w\_0 = \begin{pmatrix} 0.996 \\ 0.916 \\ 0.648 \\ 0.786 \\ 0.992 \\ 0.859 \\ 0.994 \\ 0.999 \\ 0.721 \\ 0.785 \\ 0.881 \\ 0.945 \\ 0.954 \end{pmatrix}$$

Figure 9.19 First values of the "weights" given to the products.

Other Steps

From this new formula, new weights can be computed and the process goes on as many times it is needed to "stabilize" the formula.

After step 4 (generally speaking four to five steps only are required), the final formula is computed:

$$\hat{y}^{(4)} = 563.278 + 104.098 \times x$$

The result is not very different from the formula computed with the classical linear regression because the residuals are not too much scattered. But suppose that the cost value y_9 for product number 9 becomes 2800. Such an outlier should have been detected in the procedures developed in Chapter 5 or 6, but suppose that for any reason it was not. The linear regression then computes the formula:

$$\hat{y}^{(0)} = 832.668 + 65.244 \times x$$

which is very different from the previous one and the difference is due *only to one data*: it shows that the linear regression is not robust at all!

This formula produces the following residuals (Figure 9.20) where e_{+9} is now very close to the cut-off value. Also note that this cut off value is not robust also; using several times the median absolute deviation (MAD), defined in Chapter 4, would have given a more robust cut-off.

New "weights" can now be computed; product 9 receives now a weight of only 0.076. Then, after 4 steps, we get the following formula:

$$\hat{y}^{(4)} = 558.586 + 103.101 \times x$$

which is not far from the previous one: the problem has been corrected.

Conclusion

The biweight is an easy way to make the linear regression very robust. It computes, for the given example, a formula which is between the ordinary regression and the median.

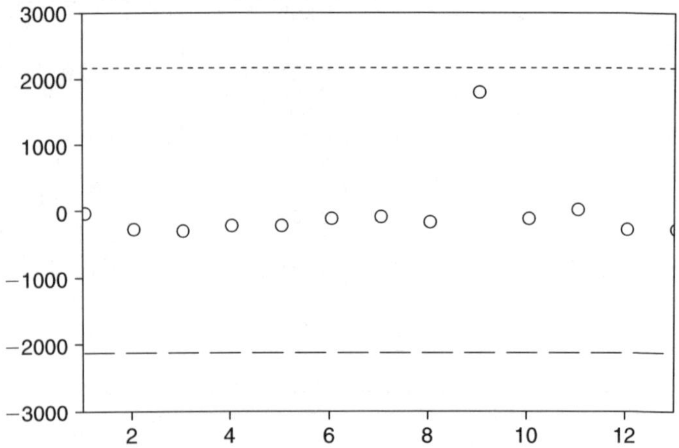

Figure 9.20 Formula showing residuals and cut-off values.

In case of rather scattered data, using either the median or the biweight is therefore the recommended procedure.

The distribution of the residuals was not computed at this stage, because the biweight is not done for this purpose, but for, by elimination of the outliers, computing a more robust formula for the dynamic center.

9.2.10 Comparison of the Distribution of the e_{+i} Based on the Various Metrics

An Algebraic Perspective

The following table summarizes the distribution of the residuals:

Use of	Mean	Standard deviation	Skewness	Kurtosis
Median	5.883	87.100	0.045	2.325
Least squares	0	83.805	−0.045	1.952
Product	3.84	83.829	−0.049	1.833
Ratio type 1	16.435	84.031	−0.054	1.826
Ratio type 2	−7.622	84.045	−0.054	1.823
Log	4.472	84.007	−0.054	1.823

A look at the values of this table shows that:

- All metrics, except the least squares, present a small (less than 2% of the average cost value) "bias", the ratio type 1 having the larger one.

- The least square method presents the smaller standard deviation of the residuals. This is an important point (often mentioned as a criteria for using this metric) but which must be tempered by the fact other metrics have just a very small increase of this standard deviation: compared to the standard deviation of the residuals, this point can be considered as negligible for practical purposes.
- The skewness are very small for all metrics.
- The kurtosis are about the same with a minor exception for the median for which the distribution ψ of the residuals is a bit more "normal" than the other ones.

The conclusion at this stage is that **the distribution of the residuals does not provide a criteria for choosing one or the other metric**. The choice must be based on other criteria, such as the robustness (which gives a preference for the median and the biweight) and/or the need to avoid the bias introduced by the linear regression (criteria which gives a definite advantage to the median).

A Geometric Perspective

The coefficients of the dynamic center are displayed on the following table:

Use of	Intercept	Slope
Median	521.563	114.880
Least squares	572.972	101.081
Product	572.966	99.919
Ratio type 1	568.355	97.499
Ratio type 2	592.765	97.392
Log	579.661	97.698
Biweight	563.278	104.098

The slopes appear rather similar, with the exception of the median which has the larger slope. It is interesting to see that this slope is very close to the one we found when averaging the linear regressions of y/x and x/y (112.828).

The same conclusions can be drawn when looking at the intercepts (the average of the linear regressions of y/x and x/y being equal to 534.216). One can add that the least squares metric (linear regression) presents an intercept in the middle of those of the two ratios, whereas the slopes of these two ratios do not change so much: a problem does appear.

Let us present now the results on two figures, the least squares, as it is the most frequently used metric, being used as the Ariane thread to compare the dynamic center:

- In Figure 9.21 presents the dynamic center for three different metrics: least squares as a thick full line, median as a thick dotted line and log as thin broken line. It appears that the log curve is very close to the least squares (and remember that the median is very close to the average least squares).
- In Figure 9.22 presents the dynamic center for three different metrics: least squares as a thick full line, ratio 1 as a thick dotted line and ratio 2 as a thin broken line.

Two points must be noticed:

1. As the three lines in Figure 9.22 have about the same slope, they suffer of the same "bias" as the linear regression,

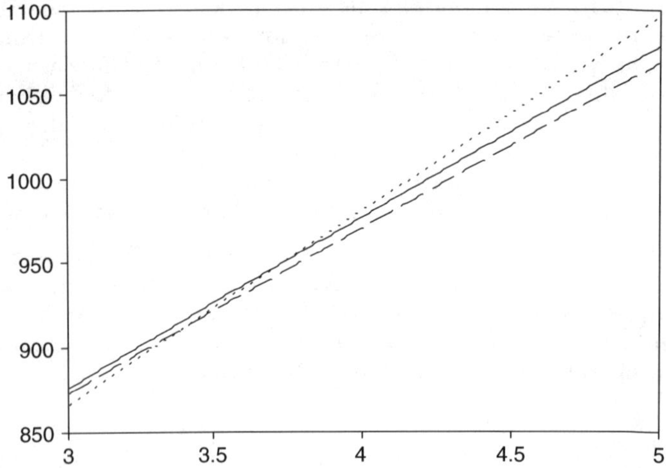

Figure 9.21 The dynamic center computed from the following metrics. Thick full line: least squares; thick dotted line: median; thin broken line: log.

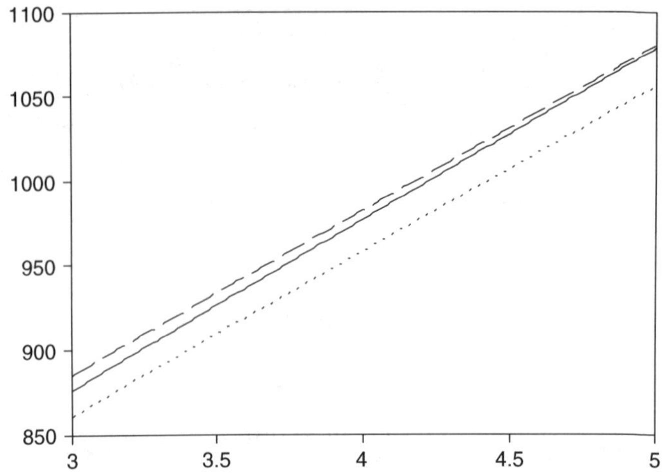

Figure 9.22 The dynamic center computed from the following metrics. Thick full line: least squares; thick dotted line: ratio 1; thin broken line: ratio 2.

2. The intercepts of both ratios are, compared to the least squares, biased, but in opposite directions. Ratio 1 metric gives a dynamic center lower than the least squares (for the average mass of 3.299, least squares computes a dynamic center equal to 906.4, whereas ratio 1 computes a value of 890.0 – a difference of 16.4, exactly equal to the difference of the mean of the residuals). This can easily be understood by looking at the figure: one can say that this metric is "attracted" by small values. The reverse is true for ratio 2 (for the same average mass, ratio 2 computes a value equal to 914.1 – a difference of 7.7).

9.2.11 Comparison of the Distribution of the $e_{\times i}$ Based on the Various Metrics

The following table summarizes the distribution of these residuals:

Use of	Mean	Standard deviation	Skewness	Kurtosis
Median	1.013	0.103	0.399	2.327
Least squares	1.001	0.097	0.219	1.916
Product	1.005	0.097	0.208	1.893
Ratio type 1	0.991	0.095	0.052	1.890
Ratio type 2	0.991	0.095	0.154	1.757
Log	1.005	0.097	0.177	1.813

A look at the values of this table shows that:

- The "bias" for these residuals is in the vicinity of 1%, which is very small.
- The standard deviations have the same order of magnitude.
- The skewness and kurtosis do not seem really abnormal. The reader will refer to Part IV for a more detailed discussion about these characteristics.

Therefore, at this stage, no solution clearly appears "better" than any other, from a pure mathematical point of view. Selection of a solution must often be based on more subjective criteria.

9.3 What Conclusion(s) at This Stage?

This section studied the consequences of choosing such or such metric on the results, the results being, as usual, the formula for the dynamic center on one hand, the distribution ψ of the residuals on the other hand.

Using different metrics produces, as expected, different results.

We noticed that the dynamic center is not the same, and we called "bias" in this section the differences with the linear regression. But it must be noted that these "biases" in the intercepts and the slopes are not dramatic, as the distribution of the residuals shows it: if they were dramatic we would have seen it on the distribution of these residuals. The biases are very well "contained " inside the range of the data; more precisely, they are much smaller than the standard deviations of the residuals: their impact on the accuracy of future cost estimates will therefore be very limited.

Is it possible at this stage to recommend a particular metric? Yes and no as it depends on the future use of the results.

First of all this discussion has an interest only *when the data are scattered.* If the data are not scattered, all metrics give identical responses and the easiest one to use should be selected. However, the more scattered the data, the more the cost analyst should be concerned by selecting the "right" metric; unfortunately this is very frequent in the domain of cost.

9.3.1 You Have to Estimate Within the Range of the Causal Variable

If the Range Is Small

By "small" we mean that the range goes between a ratio (between the maximum and the minimum of the values of the causal variable) of about 3 and certainly less than 10.

As a general comment, all metrics do their best: those which have the greatest slope "compensate" by smaller intercepts and vice versa: this is very logic. On the average the results are rather similar, the differences between the formulae being well inside the standard deviation of the residuals around the dynamic center. Maybe the ratio type one could be avoided as it is "attracted" by low values and therefore give lower estimates.

If the scattering of the data is small, choose the easiest procedure which is certainly the linear regression, available in all the statistical manuals. Do not worry too much about the procedure and spend the time available to improve the data, if it is possible.

If the scattering is large, prefer the median or the biweight, or – at least – the average between the regressions *y*/*x* and *x*/*y* which is so easy to compute.

Let us compute the cost estimates by the various formulae at both end of the range (1.25, 6.83 kg):

Use of	1.25 kg	6.83 kg
Median	665.2	1306
Least squares	699.3	1263
Product	697.9	1255
Ratio type 1	690.2	1234
Ratio type 2	714.5	1258
Log	701.8	1247

For the small mass the cost values go from 665.2 to 714.5, a difference of 49.3; for the high mass, from 1234 to 1306, a difference of 72. The differences are still inside one standard deviation of the residuals, but they start not to be negligible (about 6%). Do not forget these differences have nothing to do with the accuracy of the estimate: they all are "nominal" costs computed with a formula established by different algorithms which can all pretend they are right!

If the Range Is Large

In the previous example the range is limited, but you may expect something larger if the range becomes large. In such a case the important thing is not to give too much "weight" to the high values.

Unless you have to estimate in the future in the vicinity of the large values of the causal variable, prefer the metric given by the ratio 2 : all data will have a similar "weight" in the formula. But the median could also be used in order to avoid any bias.

9.3.2 You Have to Estimate Outside the Range of the Causal Variable

Now the important thing is to get the best slope as possible: if you have to estimate outside the range, the choice of the metric becomes very important.

The real problem is when you want to estimate at the limits of the range of the existing data and, *a fortiori*, outside this range even if you are confident in the stability of the technology outside this range (this is another problem). As the slopes are rather different (even if, in the range, the intercepts bring some compensation) cost estimates may differ in an important way as a function of the metric: the differences in the values of the intercept is unable to compensate outside this range. The important thing at this stage is to have confidence in the value of the slope.

Generally speaking our preference – when dealing with cost – goes to the use of the median, for the following reasons:

- The dynamic center computed by the median is very well in line with the average of the two linear regressions which can be made on y/x and on x/y. This can be very important if the data are very much scattered (we demonstrated that the "bias" of the linear regression is related to the poor correlation between the dependent variable and the causal variable: the less the correlation, the larger the bias).
- The median is a very robust metric: it is nearly completely insensitive to outliers.
- The median is insensitive on the accuracy of the cost: the dynamic center computed by the median does not care about the exact values of the costs, but about their rank. The fact that cost accuracy is relative and not absolute is irrelevant here.
- Consequently it automatically solves the "weight" given, in the standard linear regression, to large cost values and therefore, can be used even if the range of costs becomes very large.
- Intuitively, unless we do have other information, the median is, as explained in Section 8.7, the best choice.

The second choice, always when data are rather scattered, to be considered is the biweight used in conjunction with the linear regression.

Many cost analysts still prefer to use the linear regression, because Gauss demonstrated some characteristics which seem to be very interesting, completely forgetting that Gauss had to make hypotheses (mostly on the distribution of the residuals) to be able to carry out these demonstrations; we will see in Chapter 15 that we are never sure if these hypotheses are valid in the domain of cost.

Laplace used the linear regression for small deviations.

What we are interested in is the quality of the estimates we will make from our data: the median is certainly the first metric to be investigated for this purpose.

The characteristics of the dynamic center is probably not "optimal" in terms of pure mathematics. But the differences with the optimum is, as we saw it, so small that it cannot be an objection to the use of this metric.

In the domain of cost adding some constraints on the metric, such forcing the bias to be 0 when using the ratio 2, is, in our opinion, interesting from an academic perspective but irrelevant for practical applications: you better concentrate on the data (normalizations, corrections, checking the homogeneity of the product family, looking for outliers, etc.).

9.3.3 A Last Remark

No algorithm can improve poor data!

If your data are poor, which means here very scattered, it is clear that no algorithm can improve the level of confidence you will get in the validity of the dynamic center.

Therefore the solution, instead of looking for algorithms, should concentrate on the data:

1. Add variables (parameters): most often the scattering of the data comes from a poor "description" of the products. You cannot expect to get a reliable formula if you mix – and sometimes you have no other choice, due to the rareness of the data – inside the same product family inhomogeneous products (for instance differing in the material they are manufactured from, or using different production technologies, or developed – for advanced products – at different times, or composed of a widely different number of components or parts[5], etc.) described by only their size (the mass generally speaking).
2. Analyze carefully your data. This is the reason why a whole part was dedicated to this analysis: it is more important to spend a lot of time on this analysis, than to try to improve a solution by "playing" with mathematics.
3. And never forget that mathematics are there to quantify a solution, rarely, but this happens nevertheless sometimes especially when many data are available, something which is rather rare in the domain of cost, to find it!

[5] One of the first thing you learnt, when you follow a course on cost reduction in products manufacturing, is: "reduce the number of parts". This number is often a significant cost driver.

10 Using Several Quantitative Parameters: The Linear Cases

Summary

In the previous chapter, we studied the information φ in a sample containing two variables (the cost and one parameter) between which a bilinear relationship was looked for.

In this chapter we study information of a sample containing several variables, one of them being of course the dependent variable, the other ones being the causal, quantitative, variables.

We assume that the analysis of the sample data (according to the procedures described in Chapter 6) was carried out:

1. The potential problems related to possible outliers and/or multi-collinearities between the causal variables were discovered and solved.
2. An "interesting" *linear* correlation (measured by the Bravais–Pearson correlation coefficient) between the dependent variables and the (selected) causal variable(s) was found, which legitimates to devote some effort to go on with this sample. This chapter tries to use this correlation to carry out the analysis of the sample, the purpose being to get a better understanding of the cost distribution φ. The presence of the correlation between the cost and the other variables suggests that we can reduce the importance of the residuals by adopting a better value for the dynamic center.

The relationship we are looking for is not necessarily bilinear, as a formula such as:

$$\hat{y} = b_0 + b_1 x_1 + b_2 x_1^2$$

is perfectly acceptable, x_1^2 being considered as another variable.

The relationship we are looking for is a linear one between the cost and the coefficients $b_0, b_1, b_2, \ldots$. This means that, geometrically, we are looking for a hyper-plane (just an ordinary plane if only two parameters are involved) which passes, in the best possible way, through the data. In order to do that we will of course need one of the metric previously defined.

This chapter is limited to the basic points for finding this relationship:

- We discuss only the "additive" residuals defined by:

$$e_{+i} = y_i - \hat{y}_i$$

The reader can easily extrapolate the results to other residuals.

- Only the ordinary least squares (OLS) procedure is described, based on the metric of the squares of the differences. Other metrics can also be used with several parameters. Computations are then made by iterations, starting from the least squares procedures.

The purpose of this analysis must be clearly understood: we will compute a dynamic center of φ for the purpose of "improving" the distribution ψ of the residuals (which means here reducing its standard deviation).

This chapter presents the most important algorithms for finding the dynamic center and reminds the properties of the OLS.

It then introduces other procedures for computations. These procedures put some light on the OLS.

Eventually it describes on an example the "Ridge" regression.

10.1 Introduction

The same hypotheses as described at the beginning of Chapter 9 apply here, with the exception of the first one: several quantitative variables are going to be used.

Using several quantitative parameters is a very important concept in cost estimating for the following reason: we said earlier that a specific model is related to a product family and we added that the more homogeneous a product family is, the more efficient will the model be. It may happen that you have to work inside a homogeneous product family; in such a case using just one parameter – it is then the product size – may be sufficient. But most often, according to our experience, cost analysts manipulate much more frequently non-purely homogeneous product families; then they have to compensate these inhomogeneities by the introduction of other parameters. As we said, this is the only purpose for adding other parameters.

These added parameters can be quantitative or qualitative. The first ones are dealt with here; the second ones will be discussed in the following chapter.

About the Variables

Using several quantitative variables is a natural extension of the case studied in the previous section. These variables are represented by the symbols $V_1, V_2, \ldots, V_j, \ldots, V_J$: there are J such variables.

It is assumed that the variables are strictly non-collinear; the analysis developed in Chapter 6 is supposed to have been done. We do not expect that the variables are "completely" non-collinear, but that the amount of collinearity, which can be caused just by chance, is limited. This does not mean they are independent: for instance it quite acceptable to use $V_2 = V_1^2$ if we think that the link between the dependent variable and variable V_1 is quadratic.

Due to the number of variables involved, a concise notation is necessary, as well as the computations based on matrices.

About the Relationship

As the relationship we are searching in this chapter for the dynamic center is supposed to be linear (in terms of the coefficients), we write that the value of the dynamic center corresponding to product i (the word "dynamic" meaning that the center has a different value for each product) is given by:

$$\hat{y}_i = b_0 + b_1 x_{i,1} + b_2 x_{i,2} + \cdots + b_j x_{i,j} + \cdots + b_J x_{i,J} = b_0 + \sum_j b_j x_{i,j}$$

Let us remind for the reader who is not familiar with this notation that:

- the first index of $x_{i,j}$ represents the product number (it varies from 1 to I);
- the second index represents the variable number (it varies from 1 to J);
- consequently $x_{i,j}$ represents the value taken by variable V_j for product i:

$$x_{i,j} = x_{\text{product\_number,variable\_number}}$$

Why Adding Variables?

In the previous section we saw that replacing the static center by a dynamic one was a powerful tool to reduce the standard deviation of the residuals around the (dynamic) center of the distribution φ of the costs in the sample.

The logic is exactly the same here: other variables are introduced for the purpose of still reducing, if it is possible, this standard deviation. We do that because we believe that the product family we are working with is not homogeneous enough for using one parameter only; as other parameters are introduced for mitigating the inhomogeneities, we have to look at their influence on the cost.

If the scattering of the data is not due to this inhomogeneity, but for any other reason – such as working with price information instead of cost – we may have some doubt about the result of adding parameters. But we have to try this solution.

As we already said it, we try to replace the complex distribution φ of the cost by something simpler to handle: the distribution ψ of the residuals; adding parameters is generally a good way to do it. The first criteria for saying that ψ is easier to handle than φ is the reduction of its standard deviation; therefore it will be the first value to look at.

Example

The algorithms will be illustrated in Figure 10.1.

The first three columns are the same as the one we used in the previous section.

10.2 Computing the Solution

10.2.1 The Basic Computation

The solution can only be written with the matrix notation. It uses the matrix $||^{+}x||$ of the J (see four parameters in the example) causal variables – to which is added a

Name	Cost	Mass	Components	Connections	Boards
A	1278	6.83	1264	1274	10
B	724	2.18	1032	480	6
C	809	3.80	812	656	6
D	920	4.55	516	786	8
E	772	2.18	1032	480	6
F	877	2.11	1548	394	6
G	1064	4.67	2722	942	6
H	865	2.81	807	671	3
I	961	2.55	1598	872	6
J	856	1.68	737	450	5
K	1293	6.30	715	1400	19
L	717	1.98	186	430	7
M	648	1.25	228	257	6

Figure 10.1 Example with several parameters.

column of 1 for computing the intercept if it is needed – and the vector $\vec{y}$ of the cost. $\|{}^{+}x\| \in \Re^{I\times(J+1)}$ unless we force the intercept to be 0.

The solution, which is the vector $\vec{b} \in \Re^{(J+1)\times 1}$ of the coefficients of the causal variables in the formula, is given in all elementary books of statistics:

$$\vec{b} = \left(\|{}^{+}x\|^{t} \otimes \|{}^{+}x\|\right)^{-1} \otimes \|{}^{+}x\|^{t} \otimes \vec{y}$$

and the formula giving the value of the dynamic center as a function of the causal variables is then written as:

$$\hat{y} = \|{}^{+}x\| \otimes \vec{b}$$

The solution does not require making any hypothesis. If we replace in this last formula the vector $\vec{b}$ by its value, we have:

$$\hat{y} = \|{}^{+}x\| \otimes \left(\|{}^{+}x\|^{t} \otimes \|{}^{+}x\|\right)^{-1} \otimes \|{}^{+}x\|^{t} \otimes \vec{y} \;=\; \|h\| \otimes \vec{y}$$

where:

$$\|h\| = \|{}^{+}x\| \otimes \left(\|{}^{+}x\|^{t} \otimes \|{}^{+}x\|\right)^{-1} \otimes \|{}^{+}x\|^{t}$$

is the HAT matrix we met in Chapter 6.

As it appears, the solution has to invert the matrix $(\|{}^{+}x\|^{t} \otimes \|{}^{+}x\|)$. This solution will therefore exist only if this inversion is possible. It is a well-known theorem in the matrix algebra that a matrix which has the values of two columns proportional cannot be inverted. All the discussion we had in Chapter 6 about multi-collinearities problems refers to this question. This problem is here supposed to have been solved; nevertheless the solution of the "Ridge" regression, for the cost analyst who wants to keep collinear variables, is presented in Section 10.5.

Examples

Let us see on the example how much we win by adding parameters.

We saw in the previous section that, when only the first parameter is used, the following formula was found:

$$\hat{y} = 572.97 + 101.08 \times \text{mass}$$

The residuals having a standard deviation equal to 83.805.

Example with Two Parameters

With two parameters we get:

$$\hat{y} = 535.27 + 95.79 \times \text{mass} + 0.054 \times \text{components}$$

The residuals having a standard deviation equal to 76.67. Adding one parameter was a successful process: the standard deviation of the distribution ψ of the residuals was reduced.

Example with Three Parameters

With three parameters the formula becomes:

$$\hat{y} = 477.61 + 68.98 \times \text{mass} + 0.084 \times \text{components} + 16.01 \times \text{boards}$$

The residuals having a standard deviation equal to 68.025. Adding a third parameter was also a successful process: the standard deviation of the distribution ψ of the residuals was reduced.

Example with Four Parameters

With four parameters:

$$\hat{y} = 479.28 + 13.15 \times \text{mass} + 0.052 \times \text{components} + 0.393 \times \text{connections} + 7.55 \times \text{boards}$$

with this time a standard deviation of the residuals equal to 52.25.

In this example using up to four parameters significantly improved the result, as the standard deviation of the residuals went always down. This is not always the case, and we will see in Chapter 16 a method for selecting the most interesting parameters, as far the cost-estimating process is concerned.

You certainly have noticed that the coefficients do change when a new parameter is added; this is due:

1. to the fact that the influence of one parameter (for instance the mass, of which coefficient goes from 101.08 to 95.79, 68.58 and 13.15) is now replaced by other variables.
2. to the correlation between the variables, as previously explained. This correlation reduces the accuracy with which the coefficients are computed.

It shows that reducing the standard deviation of the residuals is not the only thing to consider when preparing a specific model: the precision with which the coefficients are determined also is an important subject, which will be discussed in Chapter 15.

10.2.2 How Does Each Observation Influence the Coefficients?

The formula giving the vector $\vec{b}$ (the coefficients of the linear regression) uses all the data of the sample. We expect that all the data contribute, about, equally to these coefficients. It may be therefore interesting to see how much each data point influences them, the idea being to check if one data point has not a too large influence.

In order to do that, we must define a "distance" D_i between the value $\vec{b}$ computed for the coefficients when all data points are present and the value $\vec{b}_{(i)}$ when data point i is removed. The procedure is similar to the one which was used to detect potential outliers (Chapter 6) and the objective is about the same, except we now directly compare the vectors.

R. D. Cook[1] proposed to use the distance defined by:

$$D_i = \frac{(\vec{b}-\vec{b}_{(i)})^t \otimes \left(\|^+x\|^t \otimes \|^+x\|\right) \otimes (\vec{b}-\vec{b}_{(i)})}{(J+1)\times \hat{S}^2}$$

which is, as the reader can easily check, a scalar. R. D. Cook indicates that a distance D_i larger than 1 generally shows an abnormal influence.

Draper and Smith (Ref. [20], p. 170) mention a formula that is easier to compute:

$$D_i = \frac{e_i^2}{\hat{S}^2(1-h_{i,i})} \times \frac{h_{i,i}}{1-h_{i,i}} \times \frac{1}{J+1}$$

where the HAT matrix appears once more. This formula attracts the reader's attention to the value of $h_{i,i}$: the closer these values are to 1, the larger will the distance be: the diagonal elements of the HAT matrix are therefore interesting to observe. Pay attention nevertheless to the fact that the HAT matrix only considers the values of the parameters: a data point can be far away from the other ones (its $h_{i,i}$ will be close to 1) without creating a problem if its residual is small.

Example

The procedure can be applied to the example. Using the four parameters leads to the following vector:

$$\vec{b} = \left\| \begin{array}{c} 479.275 \\ 13.654 \\ 0.052 \\ 0.393 \\ 7.547 \end{array} \right\|$$

The diagonal elements of the HAT matrix are given in Figure 10.2.

The distances between vectors $\vec{b}$ and $\vec{b}_{(i)}$ are now computed; the results appear in Figure 10.3. The average distance is equal to 0.119, with a standard deviation of 0.142.

[1] R. D. Cook. *Technometrics* Volume 19, 1977.

0.496
0.128
0.21
0.291
0.128
0.413
0.648
0.437
0.673
0.166
0.895
0.223
0.29

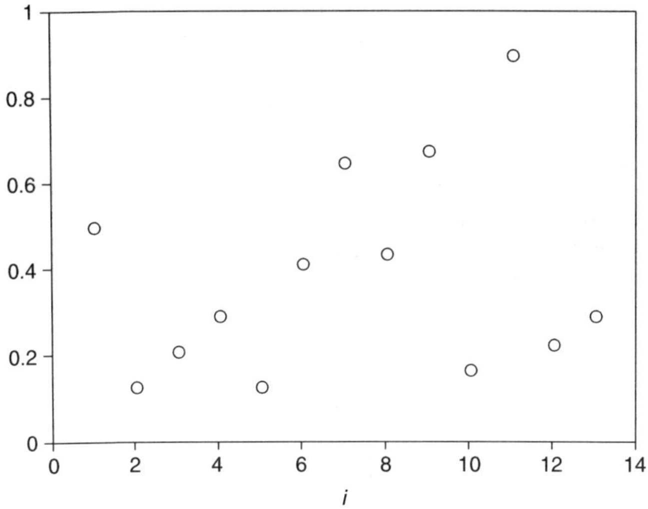

Figure 10.2 The diagonal elements of the HAT matrix.

0.357
0.04
0.069
8.002×10^{-3}
4.583×10^{-3}
0.424
0.302
0.023
0.164
0.109
0.036
7.206×10^{-3}
1.089×10^{-3}

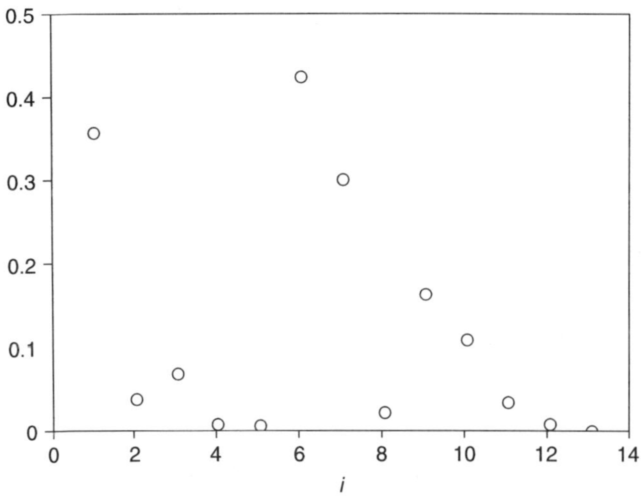

Figure 10.3 Change of $\vec{b}$ when each point is successively deleted.

This standard deviation is rather high compared to the average value, this effect being due to the data points A, F and G; these data points are not really "abnormal" but they should be verified.

Referring to Chapter 6, one can observe that these three data points were already discovered as potential outliers when looking at the dependent variable. Let us say that this new approach is another interesting way to confirm a first "impression".

10.2.3 The "Weighted" Least Squares

In a set of data, you sometimes think that some of them are less reliable than others. In such a case you would certainly like that the formula be less dependent on these data. The way you can deal with this problem is to weight each data according to its level of reliability.

We saw in Chapter 8 an automatic way to compute these weight. This section proposes a manual weighting procedure, based on your own analysis of the data.

The weight given to product i is called w_i. From all the weights, a "weight" (diagonal) matrix can be created:

$$\|W\| = \begin{Vmatrix} w_1 & 0 & \dots & 0 & \dots & 0 \\ 0 & w_2 & \dots & 0 & \dots & 0 \\ \dots & \dots & \dots & \dots & \dots & \dots \\ 0 & 0 & \dots & w_i & \dots & 0 \\ \dots & \dots & \dots & \dots & \dots & \dots \\ 0 & 0 & \dots & 0 & \dots & w_I \end{Vmatrix}$$

We have now to minimize:

$$\sum_i w_i e_i^2 \text{ or } \vec{e} \otimes \|W\| \otimes \vec{e}$$

The new vector $\vec{b}$ of the coefficients is then given (Ref. [20], p. 109) by:

$$\vec{b} = \left(\|^+x\|^t \otimes \|W\| \otimes \|^+x\|\right)^{-1} \otimes \|^+x\| \otimes \|W\| \otimes \vec{y}$$

and its variance by:

$$\mathrm{var}(\vec{b}) = S^2 \times \left(\|^+x\|^t \otimes \|^+x\|\right)^{-1} \otimes \|^+x\|^t \otimes \|W\| \otimes \|^+x\| \otimes \left(\|^+x\|^t \otimes \|^+x\|\right)^{-1}$$

10.3 The Properties of the Classical Solution

10.3.1 The Basic Properties

These properties are the same as the ones mentioned for the case of one parameter only:

1. The dynamic center, now defined as an hyper-plane, passes exactly through the center of the data, defined as $\bar{y}, \bar{x}_1, \bar{x}_2, \dots, \bar{x}_j, \dots, \bar{x}_J$.
2. The sum of the residuals is equal to 0 (when the intercept is not forced to 0).
3. The value of the coefficients are not symmetrical.
4. These values are strictly correlated.

10.3.2 The Difficulties with This Metric

The difficulties, in the domain of cost, are exactly the same as the ones we found with one parameter only (see Chapter 9):

- The linear regression is biased.
- Large costs receive too much "weight".

About the Bias

In Chapter 9 we proposed three solutions:

1. Averaging the regression of y on x and of x on y.
2. Using the euclidian distance.
3. Rotating the axis (using the results of the principal component analysis, PCA).

The first two solutions would be here either impossible (for the fist one) or rather complex. Consequently only the third one has therefore to be investigated.

The investigation was carried out in Chapter 9. The reader must not forget that, for solving the problem, the PCA must be done on all the variables, including cost.

The computations are not difficult and there is consequently no reason not to use them.

About the "Weight" Given to Large Costs

The problem is of course exactly the same as the one discussed about the use of only one parameter, and the same solution can be applied.

10.4 Introduction to the Other Forms

10.4.1 Introduction to the "Canonical" Form

The "canonical" form of the linear regression is not described here for solving the equations – it is not simpler than the standard matrix analysis described in Section 10.2 – but because it spreads a very interesting light on the problem in general and on the multi-collinearities in particular.

According to the singular values analysis, the matrix $\|{}_c x\|$ of the centered data (it is easier here to use the centered data, as the intercept disappears without changing anything about the multi-collinearities problem) can be written as the product of three matrices:

$$\|{}_c x\| = \|U\| \otimes \|K\| \otimes \|V\|^t$$

with $\|U\| \in \Re^{I \times J}$, $\|K\| \in \Re^{J \times J}$ and $\|V\| \in \Re^{J \times J}$, matrices $\|U\|$ and $\|V\|$ being orthogonal. $\|K\|$ is a diagonal matrix of which elements are the singular values d_i of matrix $\|{}_c x\|$.

The example given at the beginning of this section leads to the following matrices (the centered data do not need a column of 1 for computing the intercept):

$$_c\|x\| = \left\| \begin{matrix} 3.531 & 248.846 & 574.615 & 2.769 \\ -1.119 & 16.846 & -219.385 & -1.231 \\ 0.501 & -203.154 & -43.385 & -1.231 \\ 1.251 & -499.154 & 86.615 & 0.769 \\ -1.119 & 16.846 & -219.385 & -1.231 \\ -1.189 & 532.846 & -305.385 & -1.231 \\ 1.371 & 1.707\times 10^3 & 242.615 & -1.231 \\ -0.489 & -208.154 & -28.385 & -4.231 \\ -0.749 & 582.846 & 172.615 & -1.231 \\ -1.619 & -278.154 & -249.385 & -2.231 \\ 3.001 & -300.154 & 700.615 & 11.769 \\ -1.319 & -829.154 & -269.385 & -0.231 \\ -2.049 & -787.154 & -442.385 & -1.231 \end{matrix} \right\|$$

$$\|V\| = \left\| \begin{matrix} -9.136\times 10^{-4} & 4.816\times 10^{-3} & -1.913\times 10^{-3} & 1 \\ -0.978 & -0.209 & 2.27\times 10^{-3} & -1.185\times 10^{-4} \\ -0.209 & 0.978 & -9.481\times 10^{-3} & 4.918\times 10^{-3} \\ 2.341\times 10^{-4} & 9.755\times 10^{-3} & 1 & -1.866\times 10^{-3} \end{matrix} \right\|$$

$$\|K\| = \left\| \begin{matrix} 2.366\times 10^3 & 0 & 0 & 0 \\ 0 & 1.118\times 10^3 & 0 & 0 \\ 0 & 0 & 7.816 & 0 \\ 0 & 0 & 0 & 2.188 \end{matrix} \right\|$$

Starting from the previous relationship given in Section 10.1:

$$\vec{y} = \|_c x\| \otimes \vec{b} + \vec{e}_+$$

and using the properties of orthogonal matrices, one can write:

$$y_i = \|_c x\| \otimes \|V\| \otimes \|V\|^t \otimes \vec{b} + e_{+i} = \|Z\| \otimes \vec{c}$$

if we note $\|Z\| = \|_c x\| \otimes \|V\| \in \Re^{I\times J}$ and $\vec{c} = \|V\|^t \otimes \vec{b} \in \Re^{J\times 1}$.

The expression $\vec{y} = \|Z\| \otimes \vec{c} + \vec{e}_+$ is exactly the same as the one we started from in order to compute $\vec{b}$ in the previous section. A similar computation gives:

$$\vec{c} = \left(\|Z\|^T \otimes \|Z\|\right)^{-1} \otimes \|Z\|^T \otimes \vec{y}$$

This expression can be simplified if one notices[2] that:

1. $\|Z\|^t \otimes \|Z\| = \|V\|^t \otimes \|c^x\|^t \otimes \|c^x\| \otimes \|V\|$.
2. $\|_c x\|^t \otimes \|_c x\| = \|V\| \otimes \|S\| \otimes \|U\|^t \otimes \|U\| \otimes \|S\| \otimes \|V\|^t = \|V\| \otimes \|S\|^2 \otimes \|V\|^t$.
3. Consequently $\|Z\|^t \otimes \|Z\| = \|S\|^2$.

[2] You just have to remember that the transpose of a product of matrices is equal to the product of the transposes written in opposite order, plus the properties of the orthogonal matrices.

Then we can write:

$$\vec{c} = ||S||^{-2} \otimes ||Z||^{t} \otimes \vec{y}$$
$$\text{var}(\vec{c}) = s^{2} \times ||S||^{-2}$$
$$\vec{b} = ||V|| \otimes \vec{c}$$

The expression giving the variance of $\vec{c}$ is particularly interesting. As we have:

$$||S||^{-2} = \begin{Vmatrix} \frac{1}{d_1^2} & 0 & \dots & 0 \\ 0 & \frac{1}{d_2^2} & \dots & 0 \\ \dots & \dots & \dots & \dots \\ 0 & 0 & \dots & \frac{1}{d_J^2} \end{Vmatrix}$$

We see a direct correlation between the singular values of the data matrix $||_{c}x||$ and the variances of the components of $\vec{c}$: in the presence of collinearities between the variables, some singular values are going to be small and this will produce high variances for some components. These high variances will of course produce high variances for some components of $\vec{b}$.

This analysis shows that the singular values of the data matrix convey very useful information about this matrix. This explains why, in Chapter 6, the analysis of the variances based on the singular values was introduced for understanding the damage which can be caused by multi-collinearities.

10.4.2 Using the QR Decomposition

Previous computations of $\vec{b}$ uses the computations based on inverting the matrix $||^{+}x||^{T} \otimes ||^{+}x||$. Another type of computation, which does not involve matrix inversion, is possible. It is based on the QR decomposition of the matrix $||^{+}x||$. It is presented succinctly here because you may find it in other books, but it is not necessary for usual computations.

Any matrix (Ref. [31], p. 223) such as[3] $||^{+}x|| \in R^{I\times(J+1)}$ can be written as the product of two matrices $||Q|| \in R^{I\times I}$ and $||R|| \in R^{I\times(J+1)}$ where $||Q||$ is orthogonal and $||R||$ is upper triangular. This is the basis of the QR decomposition.

The procedure goes along with the following steps:

1. From matrix $||R||$ one extracts the upper triangle, which gives the matrix $||R_1|| \in R^{(J+1)\times(J+1)}$.
2. One then computes the vector $\vec{c} = ||Q||^{t} \otimes \vec{y}$ from which the $J + 1$ first rows generate a vector $\vec{c}_1 \in \Re^{(J+1)\times 1}$.
3. The coefficients of the formula are given by the vector $\vec{b} = ||R_1||^{-1} \otimes \vec{c}_1$.

This procedure provides a very elegant solution to the linear regression analysis. Obviously the most difficult part of it is the QR decomposition. If you ever try it, keep a lot of significant numbers in your computations: this decomposition uses a lot of iterations and lost figures in the first ones will entail a loss of precision.

[3] The $J + 1$ number of columns is of course due to the presence here of the intercept.

Example

Let us do a very simple example with 11 products and one parameter only (but the procedure can be used with any number of parameters). The data are the following ones:

$$y := \begin{bmatrix} 1200 \\ 1281 \\ 1700 \\ 1866 \\ 2250 \\ 3179 \\ 4032 \\ 4566 \\ 7752 \\ 9577 \\ 11971 \end{bmatrix} \quad ||^{+}x|| := \begin{bmatrix} 1 & 3.4 \\ 1 & 4.2 \\ 1 & 6.3 \\ 1 & 8.2 \\ 1 & 9.9 \\ 1 & 16.3 \\ 1 & 21.2 \\ 1 & 35.9 \\ 1 & 46.5 \\ 1 & 64.5 \\ 1 & 84.5 \end{bmatrix}$$

The matrix $||Q||$ is computed as:

$$Q := \begin{pmatrix} 0.302 & -0.277 & 0.912 & 0 & 0 & 0 & 0 & 0 & 0 & 0 & 0 \\ 0.302 & -0.268 & -0.181 & 0.897 & 0 & 0 & 0 & 0 & 0 & 0 & 0 \\ 0.302 & -0.244 & -0.174 & -0.209 & 0.881 & 0 & 0 & 0 & 0 & 0 & 0 \\ 0.302 & -0.222 & -0.167 & -0.201 & -0.245 & 0.855 & 0 & 0 & 0 & 0 & 0 \\ 0.302 & -0.202 & -0.161 & -0.194 & -0.237 & -0.304 & 0.81 & 0 & 0 & 0 & 0 \\ 0.302 & -0.128 & -0.139 & -0.168 & -0.206 & -0.265 & -0.371 & 0.772 & 0 & 0 & 0 \\ 0.302 & -0.071 & -0.121 & -0.147 & -0.182 & -0.235 & -0.331 & -0.472 & 0.669 & 0 & 0 \\ 0.302 & 0.099 & -0.07 & -0.086 & -0.11 & -0.146 & -0.209 & -0.313 & -0.562 & -0.63 & 0 \\ 0.302 & 0.222 & -0.032 & -0.042 & -0.058 & -0.082 & -0.121 & -0.198 & -0.371 & -0.686 & 0.43 \\ 0.302 & 0.43 & 0.031 & 0.033 & 0.03 & 0.027 & 0.028 & -3.015\times10^{-3} & -0.048 & -0.227 & -0.816 \\ 0.302 & 0.661 & 0.101 & 0.117 & 0.127 & 0.149 & 0.194 & 0.213 & 0.311 & 0.283 & 0.387 \end{pmatrix}$$

and the matrices $||R||$ and $||R_1||$ as:

$$R := \begin{pmatrix} 3.316625 & 90.724764 \\ 0 & 86.428278 \\ 0 & 0 \\ 0 & 0 \\ 0 & 0 \\ 0 & 0 \\ 0 & 0 \\ 0 & 0 \\ 0 & 0 \\ 0 & 0 \\ 0 & 0 \end{pmatrix} \quad R_1 := \begin{pmatrix} 3.316625 & 90.724764 \\ 0 & 86.428278 \end{pmatrix}$$

From the product $||Q||^t \otimes \vec{y}$ vectors $\vec{c}_1$ and $\vec{b}$ are computed:

$$c := \begin{pmatrix} 14886.8 \\ 11548.2 \end{pmatrix} \quad b := \begin{pmatrix} 833.534321 \\ 133.615991 \end{pmatrix}$$

The result is of course the same as the one computed by the ordinary procedure.

10.5 A Particular Case: The "Ridge" Regression

The Ridge regression was introduced by Hoerl and Kennard as a way for solving the problems caused by multi-collinearities.

Let us remind the origin of these problems. As we saw it in Section 10.2.1 of this chapter, finding the coefficients of the linear regression, and then the variances of these coefficients, involves inverting the matrix $(\|^{+}x\|^{t} \otimes \|^{+}x\|)$. Inverting a matrix implies to compute its determinant which will be used as a divisor. If this determinant is very small – the matrix is said to be "ill conditioned" – the result of the division will give very high values: the coefficients will then be very imprecise.

The idea of Hoerl and Kennard is simple: if the matrix $(\|^{+}x\|^{t} \otimes \|^{+}x\|)$ is ill conditioned, let us improve its conditioning by adding to it a small value; this small value will render its determinant clearly different from 0. Then its inverse will be properly defined.

There is obviously a "price to pay"; this price is a, slight, bias.

We will demonstrate this Ridge regression on an example.

Example

Let us an extreme example (Figure 10.4) for showing the power of the method:

$$\|x\| = \left\| \begin{matrix} 2.053 & 2.053 \\ -0.651 & -0.653 \\ 0.291 & 0.294 \\ 0.727 & 0.723 \\ -0.651 & -0.647 \\ -0.692 & -0.694 \\ 0.797 & 0.799 \\ -0.284 & -0.287 \\ -0.436 & -0.432 \\ -0.942 & -0.944 \\ 1.745 & 1.746 \\ -0.767 & -0.769 \\ -1.192 & -1.188 \end{matrix} \right\| \qquad \|y\| = \left\| \begin{matrix} 1.925 \\ -0.945 \\ -0.505 \\ 0.07 \\ -0.697 \\ -0.153 \\ 0.816 \\ -0.215 \\ 0.283 \\ -0.261 \\ 2.003 \\ -0.982 \\ -1.339 \end{matrix} \right\|$$

Figure 10.4 Example for demonstrating the "Ridge regression".

This example, built on two variables V_1 and V_2, plus the cost values, uses centered and scaled variables: the sum of each column values is equal to 0. The correlation between V_1 and V_2 is extremely high: 0.999996! And nevertheless the procedure works!

10.5.1 The Result of the Standard Regression Analysis

A standard linear regression on these values provides the following relationship for the dynamic center:

$$\hat{y} = -7.793 \times x_1 + 8.693 \times x_2$$

A simple look at this formula reveals that it is not acceptable: both variables being extremely well correlated, we expect them to have nearly the same coefficients in the formula. But these coefficients differ widely and the signs are opposite! Clearly the influence of both variables compensate each other by a very large extent.

What are the variances of these coefficients? Both values are the same and amount to 2433! This means that their standard error is 49.3, extremely high compared to the coefficients values; the "*t*"-values are of course very low: 0.158 for the first one, 0.176 for the second one. Clearly this formula, even if it has no bias, cannot be used for predictions.

10.5.2 Making the Matrix Better Conditioned

The matrix to be inverted is given by:

$$\|x\|^t \otimes \|x\| = \begin{Vmatrix} 13 & 12.99995 \\ 12.99995 & 13 \end{Vmatrix}$$

of which determinant is equal to 1.187×10^{-3}.

The singular values of matrix $\|x\|$ are: 5.099 and 6.758×10^{-3}. This means that this matrix is ill conditioned, the condition factor being given by 754!

Let us introduce a small correction to this matrix; as $(\|^+x\|^t \otimes \|^+x\|) \in \Re^{2\times 2}$ the correction will be the unit matrix belonging to $\Re^{2\times 2}$ multiplied by a constant k. The matrix becomes:

$$\|x\|^t \otimes \|x\| + k \otimes \begin{Vmatrix} 1 & 0 \\ 0 & 1 \end{Vmatrix} = \begin{Vmatrix} 13 & 12.99995 \\ 12.99995 & 13 \end{Vmatrix} = \begin{Vmatrix} 13+k & 12.99995 \\ 12.99995 & 13+k \end{Vmatrix}$$

which we have now to invert. Figure 10.5 shows how the determinant of this matrix depends on the value of k.

Let us compute it for $k = 1$. This computation gives the following result for the vector of the coefficients:

$$\vec{b} = \begin{Vmatrix} 0.433 \\ 0.434 \end{Vmatrix}$$

which is very good: now, as expected, both variables influence the cost nearly the same way.

About Their "t"

Does this operation improve the "*t*" of the coefficients? The variances–covariances matrix is computed the usual way; it gives a "*t*"-value of 6.85 for the first one, 6.862 for the second one, values which are quite acceptable.

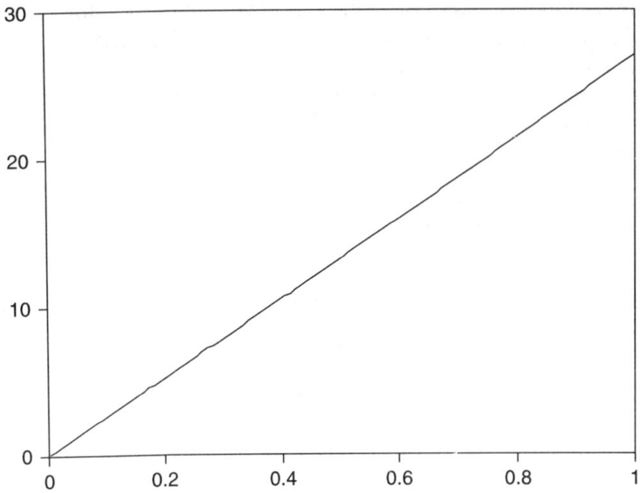

Figure 10.5 Value of the determinant according to k.

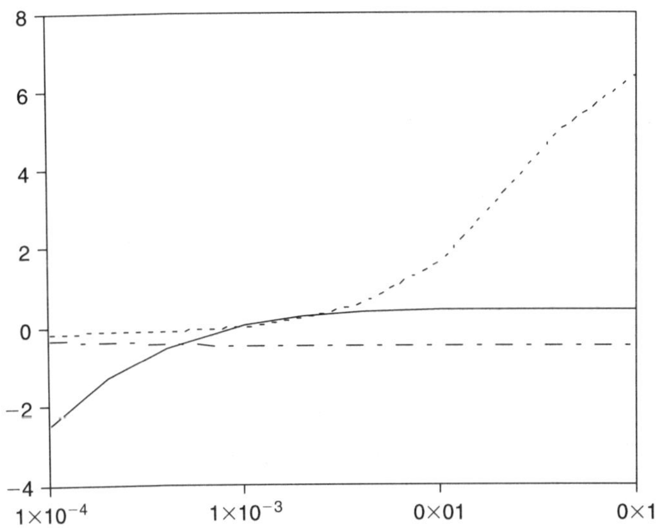

Figure 10.6 Results of the Ridge regression. Coefficient b_1 (full line); its "t"-value (dotted line); its bias (mixed line).

About Their Bias

We said earlier that the price to pay for improving the "t"-values was a bias in the value of these coefficients. What is the value of this bias? It is given by:

$$\text{bias} = -k\left(||x||^t \otimes ||x|| \oplus k \otimes ||1||\right)^{-1}$$

which gives here a value of 0.518 for both. Compared to their values this bias is here rather large. The reader must pay attention to the fact that the bias does not mean that the computed value is not correct: the bias is an asymptotic property.

Can we improve it? We should because the bias is directly proportional to the constant k we introduced. Figure 10.6 displays, as a function of k, the values of:

- the coefficient b_1: its value is nearly stabilized as soon as k is greater than about 3.10^{-3};
- its "t": this value becomes "reasonable as soon as k is greater that 0.02;
- its bias: which is also stabilized as soon as k is greater than about 3.10^{-3}.

An acceptable value of k for this example is 0.02. This does not really decrease the bias (which still is about 0.5) but the bias is not the most important characteristic, the standard error (measured here by its "t"-value) being more important: introducing the Ridge regression moves the "t"-value from -0.158 to 3.143.

11 Using Qualitative Variables

Discussion about qualitative variables was postponed up to this chapter, because it requires a preliminary understanding of the role of quantitative variables.

A model can very well include qualitative variables, the variables which are not quantifiable.

The first section of this chapter gives examples of such variables.

Then we insist on the major constraint when using this type of variable which is that the slope (for an additive formula) or the exponent (for a multiplicative formula) remains the same whatever the value of the qualitative variables. This is very important in order to avoid frequent mistakes: qualitative variables do not allow to work outside homogeneous product families; they are not there to escape this constraint, but to better "describe" similar products in the family. One can say that it slightly expands the accepted level of non-homogeneity in this family.

Before working with qualitative variables, the cost analyst should first check if they improve the quality of the model: it may happen that this quality is in fact degraded. A simple test is proposed to do that.

Then we briefly explain how to structure the data in order to be able to compute with the qualitative variables. This uses the concept of "dummy" variables.

The use of the quantitative variables is briefly presented: it uses algorithms already developed and the slight modifications they require is just mentioned.

11.1 Preparing the Qualitative Variables

11.1.1 What Are Qualitative Variables and Why Use Them?

Qualitative variables are very important in the way we think about products and consequently in cost estimating.

When you think about a product, it is probably the first variable which comes up to your mind. When you buy an airplane ticket for going from Paris to Chicago, the first question is: do you want first class, business class or economy. If you want to go to a restaurant, the first choice is: fast food, traditional, luxury, etc. All these variables are qualitative.

If it is not the first variable, it is very likely the second one. When you want to buy a house, the size of the floor is certainly the first (quantitative) variable you consider. Then come a lot of qualitative variables from the style (modern, traditional, etc.), the distance to the schools (it could be a quantitative variable, but you do not care about the exact distance and your choice could be: at a short walking distance, at a longer one, or far enough to impose driving the children to school), the distance to the shopping places or to the railway station (same comment), the type of neighborhood (which is purely qualitative), etc.

Qualitative variables are so important in cost estimating that we are very reluctant to use a model which does not include at least one qualitative variable. This comes from the fact we rarely believe that product families are homogeneous enough to avoid employing such variables.

Let us take examples in the industry:

- The quality level is a very common qualitative variable: we saw an example with the airplane tickets, but this can be applied to cars, houses, software, etc.
- The type of rock in which a tunnel has to be bored can be described by a qualitative variable (soft, hard, very hard, etc.)
- You are working with a set of products fulfilling the same function, but at different levels. Some of these products are not made from the same material as the other ones. A qualitative variable must be used to take into account this difference of material.
- In the product family you are working with, some products have slight changes in design that you want to take into account.
- Inside the same product family, you know that some products may have or not a particular function: the "Yes" or "No" is a qualitative variable.
- You want to work with the specific cost. You noticed that this specific cost does change with the product size, but you are still unwilling to use a quantitative variable to "describe" this size, or you know that, at certain thresholds, technology has to change due to the increased size (existing machines cannot cope with a largest size and/or another process must be used). In such a case, you can define different intervals, called for instance *A*, *B* and *C*. To each product is assigned an interval: this is a qualitative variable.
- When you buy the same types of products (that you consider belonging to the same product family) from different suppliers, it is sometimes very useful – and interesting – to use the supplier's name as a qualitative variable. This may help a lot the procurement officer.
- You work inside a project for modifying a railway and a road, for instance for installing a crossing. Should the traffic on the railway and/or the road to be maintained or not? This qualitative variable has obviously a major impact on the project cost.

Qualitative variables are therefore extremely useful to properly distinguish the products inside a family: they should be largely used. This explains our reluctance to use a model in which no qualitative variable appears.

Another Possible Application?

When dealing with cost (and sometimes with other data) we know quite well that the level of confidence we have about the data is not uniform: we generally consider that some information are more reliable than others.

The solution which is generally recommended for dealing with this question is to attribute a "weight" to each data: to a data we are very confident in, we attribute a weight of 1; to another one we are less confident in we attribute a weight of 0.7 and to another one we may attribute a weight of 0.5. It is possible to carry out the search of the dynamic center on these weighted variables. This solution is detailed in Chapter 10.

But this procedure presents two inconveniences:

1. The first one is the necessity to quantify our level of confidence. This may be difficult, as this level of confidence is more often an "opinion" than a "certainty": How can we say that we are 10 times more confident in this value than in another value?
2. The second one is that this quantification does not allow to indicate if we believe that such a value is probably too low, another one probably too high.

Consequently the use of qualitative variables seems sometimes more convenient for dealing the reliability of an information, as it solves both problems at the same time. For instance, using a qualitative variables with five modalities (but three can be sufficient) such as:

A: the value is probably much too low,
B: the value is probably too low,
C: the value can be considered as normal,
D: the value seems too high,
E: the value is probably much too high,

does not force the user to quantify the level of confidence and allows to indicate if the cost analyst believes that the value is too low or too high.

But what is going to be the result of the computations? As previously said, the algorithm will generate five straight lines (if an additive formula is used), each line corresponding to each modality. Therefore you will get formula for "too low cost", "low cost", etc., plus of course the "nominal" cost. This might be helpful, but requires further analysis and has to be interpreted for preparing a cost estimate. Nevertheless it gives very valuable information about the scattering of the cost.

This procedure might be sometimes recommended if your level of confidence is difficult to quantify: the equation giving the dynamic center will convey this level and will therefore be more reliable.

11.1.2 Definition and Constraints About the Use of Qualitative Variables

What Is a Qualitative Variable?

A qualitative variable is a variable which can take a *finite* number of non quantitative nature. Each qualitative variable is therefore described by a set of "attributes", for instance "good, medium, low", or "aluminium, steel", "*A*, *B*, *C*, *D*". Each attribute is called a "modality" of the qualitative variable.

There is no theoretical limit to the size of this set, but it must be finite. This distinguishes the qualitative variable from the quantitative one: quantitative variables are generally continuous, which means they can theoretically take an infinite number of values.

Breaking the Family into Sub-Families?

It is a question which is frequently asked: Why include all the products in the same family – which sometimes forces to use quantitative variable(s) – instead of creating as many "sub-families" as there are modalities, the sub-families being dealt with independently?

As usual there is no definite answer to this question and we can only give some hints:

- If you can built sub-families each having a number of products large enough to produce reliable models (the questions about reliability are dealt with in Volume 1), do so.
- If you have a limited number of data points – which is unfortunately often the case for cost estimating – prefer to use just one product family and use possibly qualitative variable(s). The reason for this choice is that, in this case, all products help *together* to build the rate of change of cost with the quantitative variable(s): the reliability of this rate of change will be much higher with this solution than working with small, sometimes very small, product families. Just make sure, before using this solution that it is the correct one.

The Major Constraint

The major constraint when using qualitative variables is that you assume that the way the dependent variable – the cost – changes with the quantitative variables is always the same, independent of the qualitative variables.

Geometrically it means that, if you use one quantitative variable only, the cost, when you change the qualitative variable, moves on parallel lines, or parallel planes – or parallel hyper-planes – if you use several quantitative variables. This constraint is illustrated in Figure 11.1: to each modality corresponds one straight line parallel to the other ones.

This constraint is extremely important: many difficulties when using qualitative variables come from forgetting about this constraint.

Example

Look at the following data set (Figure 11.2): it gives the price (the unit is irrelevant) and the power (in W) of electrical engines designed with 2 or 6 poles (the qualitative variable then takes the value 0 or 1).

This looks nice until we look at the graph displaying the price according to the power (Figure 11.3). This figure clearly shows that the change in cost with the power does not follow the same rate depending on the value of the qualitative variable: when the number of poles is equal to 6, the cost increases more severely when the power changes than when this number is equal to 2.

This is a very important rule when using a qualitative variable:

> *When you use one (or several) qualitative variable(s), you always* ***assume*** *that the rate of change with the quantitative variable(s) remains the same, irrespective of the qualitative variables.*

In other words, the graph should display here two parallel straight lines: the only change in the graph when going from one modality to the other one is the intercept.

Never forget about that! One may say that a qualitative variable is not there to say "this is a bicycle, this is a washing machine" (that will never work, unless you are lucky), but to say "this is a bicycle without a changing gear, this is a bicycle with one".

A Comment

If the scales of the previous graph are changed from linear to logarithmic, the graph becomes the one of Figure 11.4.

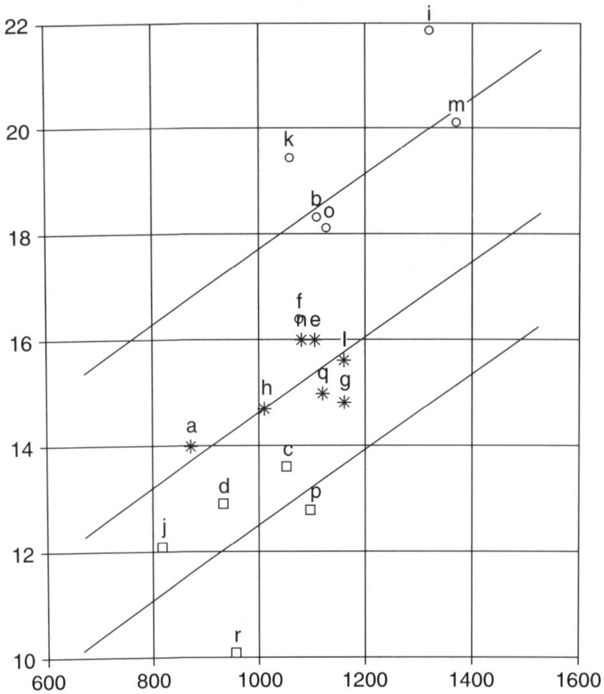

Figure 11.1 The role of the qualitative variable.

Name	Price	Power	Qualitative variable
a	819	250	0
b	934	370	0
q	934	120	1
r	983	180	1
c	1038	500	0
d	1212	750	0
e	1403	1000	0
g	1643	1500	0
h	2049	2200	0
i	2474	3000	0
s	2807	2200	1
j	2830	4000	0
t	3540	3000	1
k	3726	5500	0
u	4315	4000	1
l	4710	7500	0
v	5462	5500	1
m	6883	11 000	0
w	7211	7500	1
n	8742	15 000	0
x	9877	11 000	1
o	10 402	18 500	0
p	12 357	22 000	0
y	12 739	15 000	1

Figure 11.2 The set of price and power values.

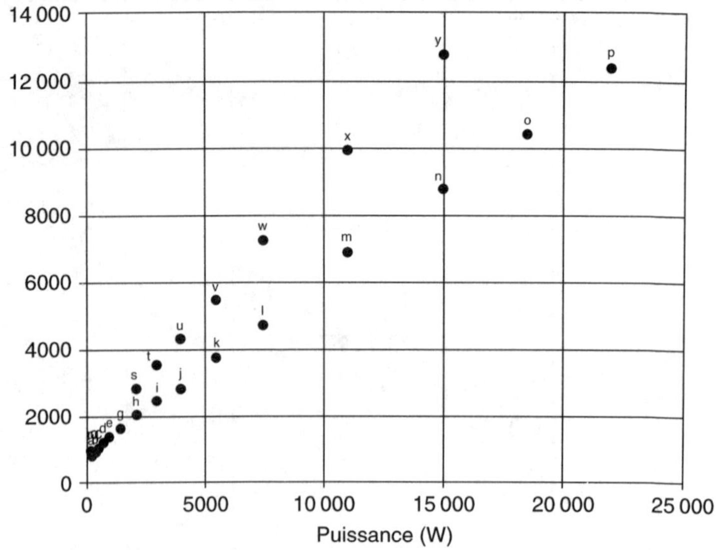

Figure 11.3 The linear scale representation of price and power values.

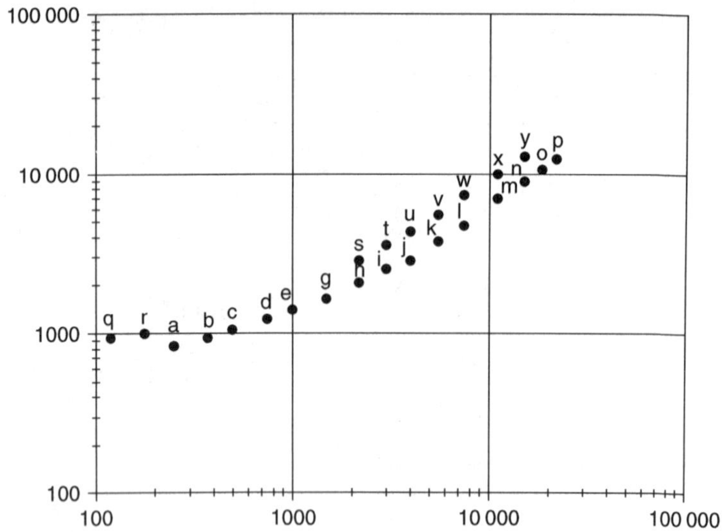

Figure 11.4 Using logarithmic scale.

Now we get two parallel curves, the relationship between non-linear between *log(price)* and *log(power)*. Therefore you have here the following choice:

- Either using two different straight lines:

$$\text{2 poles} \quad \hat{y} = b_0 + b_1 \times Wt$$
$$\text{6 poles} \quad \hat{y} = b_0' + b_1' \times Wt$$

- Or using one non-linear (establishing non-linear moving centers is described in Chapter 12.) relationship:

$$\log \hat{y} = b_0 + f(\log(power))$$

where the "intercept" changes with the attribute of the qualitative variable.

The Consequence of This Constraint

The constraint about the use of qualitative variables is described in the previous section: the relationship between the dependent variable (the cost) and the quantitative variable(s) must be the same (the slope for the additive model, the exponent for the multiplicative model, the constant for which the quantitative variable is the exponent for the exponential model).

The fact that this constraint is fulfilled by our data must therefore be checked: it is really the first thing to check when you want to use qualitative variables. A test has to be designed for this purpose.

The Logic of the Test

The logic of this test is based on the fact that the use of the qualitative variables is made in order to improve the quality of the equation giving the value of the dynamic center. This quality will be here based on the reduction of the residuals we would like to find when using these qualitative variables.

There are three major ways of using qualitative variables:

1. We can disregard the qualitative variables and use only the quantitative ones.
2. We can use them the standard way, by computing the formula giving the dynamic center, the qualitative variables being there just to change the value of the intercept (if we use linear relationships). This assumes that the constraint is fulfilled.
3. We may decide that the constraint does not seem to be fulfilled: in such a case, we may consider that the qualitative variables allow to distinguish several "sub-families" inside the family. Then each family will be dealt with independently of the other ones: each one will have its own dynamic center.

How can we compare the results? These three ways will produce three different vectors $\vec{e}_+$, each component e_{+i} of each vector being attached to a particular product. An easy way to compare both vectors is to compare their euclidian norm, simply given by $\Sigma_i e_{+i}^2$.

Going One Step Further

As we are dealing with this test, we can go one step further and answer the following question: Does the use of qualitative variable really help? Should we really bother with these variables?

Consequently three analysis are recommended:

1. In the first one, called α, we do not care about the qualitative variables. We will look for the dynamic center. The euclidian norm of the residuals is named (because it follows a χ^2 distribution) χ_α^2.

2. In the second one, called β, we consider that the constraints are fulfilled and we dealt the normal way (explained in Chapter 7) with the qualitative variables. The euclidian norm of the residuals is named χ^2_β.
3. In the third one, called γ, we consider that the constraints are not fulfilled (the relationship between the cost and the quantitative variable(s) may depend on the qualitative variables) and we consider we have different sub-families depending on the qualitative variables; each sub-family is dealt with independently. The euclidian norm of the residuals is named χ^2_γ.

Of course, if several qualitative variables are present, each one with its own set of modalities, such an analysis must be carried out on each variable.

Illustration

In order to illustrate the procedure, let us consider the example given in Figure 11.5. This example uses one quantitative variable (the mass of the product) and one qualitative one with three modalities (the quality level).

Name	Cost	Mass	Quality
A	106.49	10	A
B	127.10	20	A
C	146.58	30	A
D	197.51	40	A
E	182.06	50	A
F	220.08	60	A
G	231.97	70	A
H	271.63	80	A
I	273.77	90	A
J	295.99	100	A
K	130.14	13	B
L	148.78	23	B
M	188.86	33	B
N	199.15	43	B
O	224.72	53	B
P	222.73	63	B
Q	256.28	73	B
R	270.14	83	B
S	278.75	93	B
T	326.04	103	B
U	145.30	17	C
V	180.34	27	C
W	183.21	37	C
X	222.51	47	C
Y	222.57	57	C
Z	260.86	67	C
A1	279.35	77	C
B1	308.01	87	C
C1	305.99	97	C
D1	342.98	107	C

Figure 11.5 The set of values for the example.

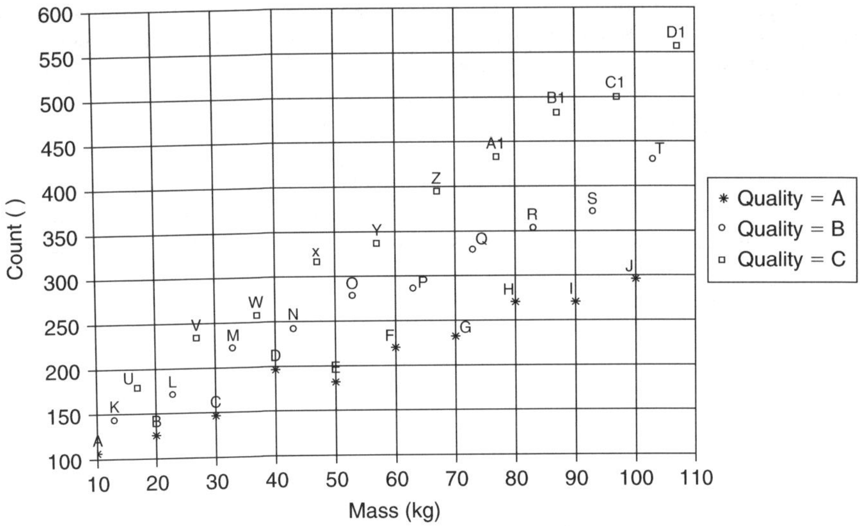

Figure 11.6 Visual representation of the data.

As we are using just one quantitative variable, it is easy to display these data on a chart: this is done in Figure 11.6.

On this figure, it clearly appears that the slopes of the dynamic center depend on the modalities of the qualitative variable and no computation – except for quantifying the phenomenon – is really required. But the situation is sometimes more complex, especially if several quantitative variables are involved: in such a case a computational procedure is required.

For this example this procedure gives the following values:

$$\chi^2_\alpha = 119\,904.39$$

$$\chi^2_\beta = 19\,268.89$$

$$\chi^2_\gamma = 2719.64$$

It is clear on this example that solution β is much better than solution α: the residuals are considerably reduced. But in fact qualitative modalities should be dealt with independently: solution γ which considers three sub-families is much better than solution β.

If the graphical representation can be considered as sufficient when using just one quantitative variable, it cannot be used when there are several. In such a case you have to rely on the algorithms.

A Comment

The previous discussion is just a part of the story. Another point of view has to be considered: it is the precision with which the coefficients are known when we transfer the results found in the sample to the population for which the specific model is built; we will see in Chapter 15 that this precision depends on the number of data points on which the formula is built (this is logic). Consequently we should also have a look on what happens to this precision when the family is split into sub-families.

For the time being, let us consider how the "deviations" for the whole population behave.

Let us, for that, return to the example of the electrical engines in order to illustrate another test. In the test we have just described, we are only interested in the sample: the procedure only uses the $\Sigma_i e^2_{+i}$. This solution can be considered as satisfactory from the sample point of view, but is it still valid from the population point of view? We will see in Chapter 15, dedicated to the population, that an estimate of the euclidian norm of the deviations for the whole population is given by:

$$\hat{S}^2 = \frac{\sum_i e^2_{+i}}{I - J - K}$$

where I is the number of data points in the sample, J the number of quantitative parameters and K the number of modalities of the qualitative parameters (which include the intercept).

Does working on the sample or on the population changes the conclusion?

If we work from the sample point of view, we get the following values:

$$\chi^2_\alpha = 5\,400\,000$$

$$\chi^2_\beta = 4\,636\,000$$

$$\chi^2_\gamma = 3\,574\,000$$

The third solution γ (considering two independent families) should be preferred.

If we work from the population point of view, then the following values are found:

$$\chi^2_\alpha = 234\,800$$

$$\chi^2_\beta = 210\,700$$

$$\chi^2_\gamma = 362\,800$$

Solution β (working normally with the qualitative parameters) is now the most interesting.

So the point of view changes completely the situation. As we are mainly concerned by the population (the sample is there just to help us), the second point of view should be preferred.

11.1.3 From Qualitative to "Dummy" Variables

It is quite possible to work with a qualitative variable defined by its attributes: you saw examples in Chapter 7 when, for instance, you wanted to quantify the possible correlation between a quantitative variable and a qualitative one.

When it goes to cost prediction, it is generally much more convenient to quantify the qualitative variables: that will allow us to use existing algorithms. Such a quantification produces what is generally called "dummy variables"; a dummy variable is a variable which can only take a value of 0 or 1.

The logic for going from a qualitative variable to a set of dummy variables is simple, but requires a minimum of precaution.

Quantification of One Qualitative Variable with Only Two Attributes

Let us start with a qualitative variable having only two attributes *A* and *B*. The first idea could be to create two variables, named x_1 and x_2 and to say: if the attribute of a product is *A*, then $x_1 = 1$, otherwise 0; if the attribute is *B*, then $x_2 = 1$ otherwise 0. You can do that if you force the intercept to be 0, which means that you work with the "pure" $\|x\|$ matrix, which could have the form (for five products only):

$$\|x\| = \begin{Vmatrix} 1 & 0 \\ 0 & 1 \\ 1 & 0 \\ 1 & 0 \\ 0 & 1 \end{Vmatrix}$$

As most cost-estimating software work with the $\|{}^{+}x\|$ matrix, we would have to work with the following matrix:

$$\|{}^{+}x\| = \begin{Vmatrix} 1 & 1 & 0 \\ 1 & 0 & 1 \\ 1 & 1 & 0 \\ 1 & 1 & 0 \\ 1 & 0 & 1 \end{Vmatrix}$$

This *must* be avoided due to a simple mathematical theorem, the matrix:

$$\|{}^{+}x\|^{t} \otimes \|{}^{+}x\| = \begin{Vmatrix} 6 & 3 & 3 \\ 3 & 3 & 0 \\ 3 & 0 & 3 \end{Vmatrix}$$

is such that the first column is a linear combination of columns 2 and 3. In such a case (see the introduction to this volume on "what you need to know about matrices") this matrix is singular: its determinant is equal to 0, and therefore the inverse:

$$\left(\|{}^{+}x\|^{t} \otimes \|{}^{+}x\|\right)^{-1} \text{ does not exist.}$$

Therefore the logic is to work with just one dummy variable; which takes the 1 if the attribute is *A*, the value 0 if it is *B*. Note this is possible because we have an intercept: the value of the intercept then corresponds to the value $x_1 = 0$. The conversion table is given by:

Attribute	x_1	
A	1	
B	0	Corresponds to the intercept

Quantification of One Qualitative Variable with Several Attributes

Let us now look at a qualitative variable with three attributes: A, B and C. It is not possible to use just one quantitative variable taking the values 0, 1 and 2, because this will force the mathematical routine to consider that attribute C represents two times attribute B, which does not really make sense.

Force is to use two quantitative variables in order to describe these three attributes x_1 and x_2; x_1 will be used as previously indicated, whereas x_2 will be used if the attribute is C: it takes the value 0 for attribute A or B, and 1 for attribute C, in which case x_1 must be at 0. The table of conversion is now as follows:

Attribute	x_1	x_2	
A	1	0	
B	0	0	Corresponds to the intercept
C	0	1	

The conclusion is, when dealing with three attributes, that we use three combinations between two variables which can take only the values 0 and 1: two each set of the couple (x_1, x_2) corresponds one and only one attribute.

What about four attributes? It is impossible to add a new line to the previous table: this will give the new table.

Attribute	x_1	x_2	
A	1	0	
B	0	0	Corresponds to the intercept
C	0	1	
D	1	1	

and then D is a linear combination of A and C: the matrix of the data will be singular.

As we already used all the possibilities available with two variables x_1 and x_2 we have to add a third variable, giving the conversion table.

Attribute	x_1	x_2	x_3	
A	1	0	0	
B	0	0	0	Corresponds to the intercept
C	0	1	0	
D	0	0	1	

The conclusion is obvious: if the number of modalities of one qualitative variable is equal to k, we need $k - 1$ quantitative variables – taking only the values 0 or 1 – to represent them.

Quantification of Several Qualitative Variables Each One Having Several Attributes

Suppose we have two qualitative variables $K1$ and $K2$ (for example the quality level and the supplier): $K1$ may have two attributes and $K2$ three of them.

Note now that we have necessarily one attribute for $K1$ and one attribute for $K2$: we cannot have $K1$ or $K2$ alone. Therefore the value of the intercept corresponds to a couple $(K1, K2)$. For quantifying the five remaining attributes, we need four variables.

Consequently for quantifying two independent qualitative variables with each k_1 and k_2 attributes, we need only $k_1 + k_2 - 2$ variables.

The conversion table can take the following form:

Modality	x_1	x_2	x_3	
$A1$	1	0	0	
$B1$	0	0	0	} Corresponds to the intercept
$A2$	0	0	0	
$B2$	0	1	0	
$C2$	0	0	1	

Interactions

The basic idea in the previous section is that the influences of variables $K1$ and $K2$ are independent: if $B1$ and $C2$ are simultaneously present, then the influence on the dynamic center will be the sum of the response to $B1$ on one hand, to $C2$ on the other hand.

But other situations are possible: suppose, to take an example, that we are interested in the prices of refrigerators and that $K1$ is the variable for the presence or not some feature, $K2$ (with three modalities) being the quality level. Using only three dummy variables assumes that the price of the refrigerators is the sum of the price of the quality + the price of the feature. But it may happen that the price of the feature for a high-quality refrigerator is higher than the price of a standard feature: there is what is called an "interaction" between the qualitative variables.

It is quite possible to take into account these interactions of qualitative variables. In the case where we think that only the presence of $B1$ and $C2$ may interact (the other variables do not and the response of a couple is expected to be the sum of the individual responses), we have to add another variable x_4 which will be defined as the product (Ref. [34] p. 182) of x_1 and x_2.

Q1	Q2	x_1	x_2	x_3	x_4	
$A1$	$A2$	1	0	0	0	
$A1$	$B2$	1	1	0	1	
$A1$	$C2$	1	0	1	0	
$B1$	$A2$	0	0	0	0	Corresponds to the intercept
$B1$	$B2$	0	1	0	0	
$B1$	$C2$	0	0	1	0	

This is rather rare in the domain of cost, but may happen.

11.1.4 The Matrix of the Data

What is the form of the matrix data now?

If we keep the intercept (which is the value when all dummy variables are 0) the matrix has the following form (for illustration purposes only):

$$\begin{Vmatrix} 1 & 0 & 0 & x_{1,1} & x_{1,2} & \cdots & x_{1,J} \\ 1 & 1 & 0 & x_{2,1} & x_{2,2} & \cdots & x_{2,J} \\ 1 & 0 & 1 & x_{3,1} & x_{3,2} & \cdots & x_{3,J} \\ \cdots & \cdots & \cdots & \cdots & \cdots & \cdots & \cdots \\ 1 & 1 & 0 & x_{I,1} & x_{I,2} & \cdots & x_{I,J} \end{Vmatrix}$$

intercept — dummy variables — quantitative variables

This matrix is still called $||^{+}x||$ because dummy variables are true variables.

11.2 Defining the Variables

11.2.1 Working with Dummy Variables Only

You define the number of dummy variables you need; suppose there is only two, taking the values 0 (corresponding to attribute A of the qualitative variable) or 1 (corresponding to attribute B of the qualitative variable). Applying the results given in the previous section will provide a dynamic center such as:

$$\hat{y} = b_0 + b_1 x_1$$

But generally the results – due to the fact that x_1 can only take the value 0 or 1 – are not given this way but according to the following way:

$$\begin{aligned} &\text{if attribute } A\text{, then } \hat{y} = b_0 \\ &\text{if attribute } B\text{, then } \hat{y} = b_0 + b_1 \end{aligned}$$

This illustrates the idea that, when working with qualitative variables only, the dynamic center takes only discrete values.

For this reason, the use of only qualitative variables is not taken into account by this procedure: in such a case, it is easier to compute the center of the distribution of the dependent variable corresponding to attribute A, then the center of the distribution of the dependent variable for attribute B, etc.

The real interest of the qualitative variable is when they are used with at least one quantitative variable. This is the focus of the rest of this chapter.

11.2.2 Using a Quantitative or a Qualitative Variable?

It may happen that you have a choice, when you work with two causal variables of which one, which describes a change in design, could be defined as a quantitative or a qualitative variable.

Let us take an example. You are working with a product family: electrical engines, the mass *Wt* being used for quantifying the engines sizes. You have also three slightly different designs: some of the engines have 2 poles, some have 4 poles and the remaining 6 poles; you do not expect to have a different number of poles. Then you have a choice for taking the number of poles into account:

- You may use a quantitative variable of which the value is the number of poles: 2, 4 or 6.
- Or you may use a qualitative variable (two as a matter of fact because you have three possible modalities).

What solution should you choose?

Instinctively we would prefer to select the second option: after all the number of poles is a design alternative and there is a limited – discrete – choice of them. It so happens here that this design change is described by a number, but it does not change its true nature. If you select this option, you may find the following dynamic centers (the slope is the same for all the centers):

$$\begin{array}{ll} \text{2 poles} & \hat{y} = 262 + 32 \times Wt \\ \text{4 poles} & \hat{y} = 274 + 32 \times Wt \\ \text{6 poles} & \hat{y} = 282 + 32 \times Wt \end{array}$$

If you select the first option (the use of several quantitative variables is described in Chapter 10), you may find the following dynamic center:

$$\hat{y} = 254 + 32 \times Wt + 5 \times \text{poles}$$

Let us compare both solutions:

- In the first option, when you go from 2 poles to 4 poles, you increase the cost by 12; then going to 6 poles increases the cost by 8. This gives an average of 10 when you increase the number of poles by 2 units.
- In the second option, increasing the number of poles by 2 units always gives an increase of the cost of 10.

On an average basis, both solutions give the same result, but in the second option you *force* the change in cost to be always the same, whereas in the first solution, the change is much more precise.

The conclusion is to be very careful to use a quantitative variable when there is only a limited set of discrete options: experience shows that it is generally better to use a qualitative variable in this respect.

11.2.3 Solving the Problem

Once the set of quantitative and qualitative variables has been converted in the new set of "quantitative + intercept + dummy variables", the data matrix looks, for the example of the electrical engines, as indicated earlier.

$$\left\| {}^{+}x \right\| = \left\| \begin{matrix} 1 & 1 & 4.5 \\ 1 & 1 & 5.5 \\ 1 & 1 & 6.5 \\ 1 & 1 & 9 \\ 1 & 1 & 10 \\ 1 & 1 & 13 \\ 1 & 1 & 16 \\ 1 & 1 & 21 \\ 1 & 1 & 25 \\ 1 & 1 & 37 \\ 1 & 1 & 42 \\ 1 & 1 & 76 \\ 1 & 1 & 85 \\ 1 & 1 & 95 \\ 1 & 1 & 120 \\ 1 & 0 & 5 \\ 1 & 0 & 5.5 \\ 1 & 0 & 27 \\ 1 & 0 & 39 \\ 1 & 0 & 46 \\ 1 & 0 & 54 \\ 1 & 0 & 79 \\ 1 & 0 & 93 \\ 1 & 0 & 145 \end{matrix} \right\| \qquad \left\| y \right\| = \left\| \begin{matrix} 819 \\ 934 \\ 1038 \\ 1212 \\ 1403 \\ 1643 \\ 2049 \\ 2474 \\ 2830 \\ 3726 \\ 4710 \\ 6883 \\ 8742 \\ 10402 \\ 12357 \\ 934 \\ 983 \\ 2807 \\ 3540 \\ 4315 \\ 5462 \\ 7211 \\ 9877 \\ 12739 \end{matrix} \right\|$$

Figure 11.7 The data including a qualitative variable with two modalities.

Looking for an Additive Formula

The solutions presented in Chapter 10 for multiple variables can be directly applied. For instance the conventional or ordinary least squares (OLS) gives:

$$\vec{b} = \left(\|{}^{+}x\|^{t} \otimes \|{}^{+}x\| \right)^{-1} \otimes \|{}^{+}x\|^{t} \otimes \vec{y}$$

For example we have:

$$\vec{b} = \left\| \begin{matrix} 152.669 \\ 376.978 \\ 94.213 \end{matrix} \right\|$$

and consequently, returning to the quantitative variable:

- for 2 poles cost $= 152.669 + 94.213 \times \text{mass}_{\text{kg}}$
- for 6 poles cost $= 529.647 + 94.213 \times \text{mass}_{\text{kg}}$

because, for the second case, we must add the value found for the second column of the matrix (the dummy variable) to the intercept.

As previously mentioned, the reader will note that the slope is the same (94.213 per kg), both formulae differing only by the intercept.

The problems with the OLS are here the same as the problems previously mentioned, including the bias. The solutions are the same.

Looking for a Multiplicative Formula

Anticipating on the results of the next chapter, we can also look for a multiplicative formula. In such a case, the data matrices $||^{+}x||$ and $\vec{y}$ must be transformed by the logarithms of the quantitative variables only (not for the dummy variable), including $\vec{y}$ (Figure 11.7).

The algorithm computes:

$$\vec{b} = \begin{Vmatrix} 2.342 \\ 9.794 \times 10^{-3} \\ 0.808 \end{Vmatrix}$$

Due to the fact that $10^{2.342} = 219.7$ and $10^{0.00979} = 1.023$, we get the following formulae:

- for 2 poles $\quad \text{cost} = 219.7 \times \text{mass}_{\text{kg}}^{0.808}$
- for 6 poles $\quad \text{cost} = 224.7 \times \text{mass}_{\text{kg}}^{0.808}$

Looking for Other Formulae

The same procedure can be applied to find out an exponential formula in the presence of qualitative variables.

It is not however usable for non-linearizable formulae, which are found by iterations. A special program should therefore be built for these types of formulae.

12 Non-Linear Relationships

Summary

It has been said (introduction to this volume) that the shape of the relationship between the dependent variable and the cost drivers (the parameters) results from a choice from the cost analyst.

A large proportion of these persons selects a linear relationship – and very often a bilinear relationship – linear meaning here that the dependent variable depends linearly of the variables – and of the coefficients in the bilinear case. However, as it is sometimes said,"nature is not linear", or more exactly the assumed linearity is an approximation of the true relationship. It is therefore important to study the non-linear relationships.

There is potentially an infinite number of non-linear relationships and we do not expect the cost analyst to test an infinite number of such relationships. Therefore, we will limit our investigation to the most standard ones. These standard forms can be classified into two sets:

1. The relationships that can be linearized by a change of the variable(s). These relationships can be – once the change of the variable has been accomplished – dealt with as linear relationships, with the procedures which were developed in the previous chapters. This chapter, for clarity purposes, divides these relationships into two subsets: relationships based on one variable only (this is the most frequent use of non-linear relationships) and relationships based on several variables. For the first subset several relationships are presented, starting by the most frequently used: the "multiplicative" formula and the "exponential" formula.
2. The relationships that cannot be linearized: these ones are the truly non-linear relationships. Studying these relationships is much more difficult; examples using one variable only are presented. A general solution is discussed for computing the value of the coefficients. Its application to the most important relationship of this type – called the "correction-by-constant" formula – will be developed.

In addition we show that using any type of relationship – linear or not linear – when a non-additive metric is selected has to be solved with the same tools as a purely non-linear relationship. A general solution is presented.

12.1 Linearizable Relationships

12.1.1 The "Multiplicative" Formula

The multiplicative formula is much in favor for cost analysis purposes, as soon as the size becomes large enough. In many industries this formula is so much used that it can really be called the "standard" formula; a few examples are given in Chapter 9 of Volume 1.

Generally speaking – and unless the size is defined by a length or a surface – one finds values of the exponent b_1 lower than 1. This is called the "economy of scale": the larger an object, the more expensive it costs, but the less expensive it is per unit of size. This is a general law of nature which is true for all mechanical objects and for building but is not true for software. Some exceptions are known, the most important being the polishing of mirrors: in such a case the cost grows faster than the size.

The order of magnitude of the exponent varies from about 0.6 to 0.9 depending on the technology. For most purely mechanical items used on the ground it is in the vicinity of 0.7, up to the point that this "standard" formula is known as the law in 2/3 power of the equipment mass. The interesting feature is that this value corresponds to the ratio "surface/volume". What is that so? A basic example illustrates the fundamental reason:

- Suppose you manufacture an object of which mass is 1 kg (it would be exactly the same if you prefer 1 ton) and that the breakdown of the cost is given by:
  - Raw material: €5
  - Machining: €20
  - Giving a total cost of €25.
- Suppose now you enlarge this object, doubling all its dimensions: the mass becomes $2^3 = 8$ kg. What about the cost?
  - The cost of the slug becomes €40.
  - And the cost of machining $2^2 \times 20 =$ €80 (because the machining effort is related to the surface of the thing, not its mass).
  - Giving a total cost of €120.
  - Therefore when the mass goes from 1 to 8 kg, the cost goes from €25 to €120, which is an increase given by $8^{0.75}$ (this power is higher than 2/3 because fix charges were not taken into account).

Consequently the origin of this power 2/3 comes from the fact that cost depends for one part on the volume, for another part on the surface of the object (it is the surface which is machined, not the volume): the power has to be less than 1. This is also true – but to a lesser extent – for electronic equipments: if you double the sizes of an electronic board, its weight is multiplied by 4 (this is a first approximation, not taking into account the fact that you may have to increase slightly the number of layers): the most important part of the cost – the electronic components – is also doubled and so is the cost of integration; therefore, one can expect a power in the vicinity of 1. But it has to be less than 1, due to the fix charges.

A more in depth explanation of this phenomenon is given in Chapter 13 of Volume 1.

An interesting paper was published by Donald McKenzie on the subject; we will return to it in the next chapter dealing with the residuals.

Using One Causal Variable Only

The multiplicative formula gets its name from the general formula in which all variables appear in a multiplicative form.

When dealing with one causal variable only, the formula giving the dynamic center is written as:

$$\hat{y}_i = b_0 x_i^{b_1}$$

where x is, as usual, the causal variable, b_0 and b_1 the coefficient and the exponent, respectively.

The shape of this relationship is illustrated in Figure 12.1 for three values of b_1: for $b_1 = 1$, the curve is a straight line, for $b_1 < 1$, the curve grows slower than the straight line and it is the contrary for $b_1 > 1$.

How Are the Residuals Defined?

As usual, the residuals can be defined in different form, such as:

- *Additive*, if one writes $y_i = \hat{y}_i + e_{+i} = b_0 x_i^{b_1} + e_{+i}$.
- *Multiplicative*, if one writes $y_i = \hat{y}_i \times e_{\times i} = b_0 x_i^{b_1} \times e_{\times i}$.
- A *ratio*, if one writes $e_{\%i} = \dfrac{y_i}{\hat{y}_i} - 1$.
- Or *any other.*

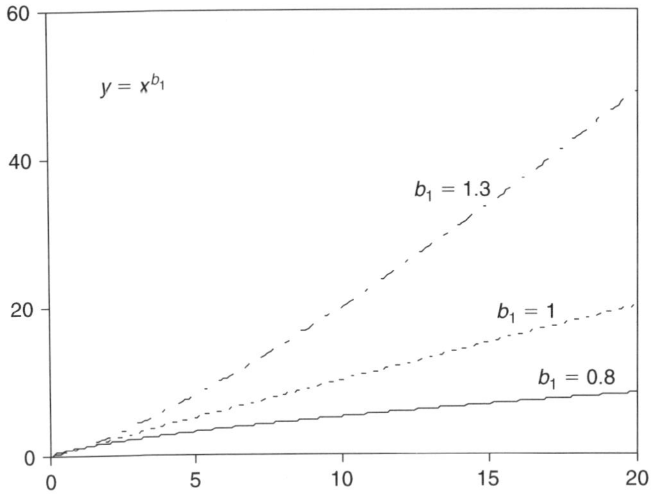

Figure 12.1 The shape of the multiplicative formula for three different values of b_1.

The most common definition, when using a multiplicative formula, is to the second one. This simplifies a lot the computations. Other definitions must be dealt with the procedures dedicated to the non-linearizable formulae.

Therefore in this section, the residuals are defined as:

$$e_{\times i} = \frac{y_i}{\hat{y}_i}$$

where the $e_{\times i}$ having of course an average value of 1.

Linearization

In order to use the algorithms defined for the linear case, it is convenient to redefined the variables and the residuals by using the logarithms:

$$e^*_{+i} = \log e_{\times i} = \log \frac{y_i}{\hat{y}_i} = \log y_i - \log \hat{y}_i = y^*_i - \hat{y}^*_i$$

$$\hat{y}^*_i = \log b_0 + b_1 \log x_i = b^*_0 + b_1 x^*_i$$

The e^*_{+i} are precisely the ones which was studied in Chapter 10.

The Metric

Working with the e^*_i, y^*_i and $\hat{y}^*_i$ (the asterisks reminding the reader that the values are in fact logarithms of the true values) allows now to use the procedures already established: the problem has been linearized. Using these procedures means that we try to minimize:

$$\sum_i (e^*_{+i})^2 = \sum_i (\log e_{\times i})^2$$

These procedures computes the coefficients:

$$b_1 = \frac{\sum_i (x^*_i - \bar{x}^*)(y^*_i - \bar{y}^*)}{\sum_i (x^*_i - \bar{x}^*)^2}$$

$$b^*_0 = \bar{y}^* - b^*_1 \bar{x}^*$$

Returning to the True Values

The formula is now written as:

$$\hat{y}_i = 10^{b^*_0} x_i^{b_1}$$

where "10" being replaced by e if natural logarithms were used.

Note that for the multiplicative formula b_1 is not affected by the linearization process.

The General Multiplicative Formula

This procedure can easily be extended to any number of variables, the dynamic center becoming:

$$\hat{y} = b_0 x_1^{b_1} x_2^{b_2} \ldots x_j^{b_j} \ldots x_J^{b_J}$$

The results can easily be extrapolated from the bilinear relationship with several causal variables: taking the log of both sides gives:

$$y_i^* = b_0^* + b_1 x_1^* + b_2 x_2^* + \cdots + b_i x_i^* + \cdots + b_I x_I^*$$

which is now a linear relationship.

Using Qualitative Parameters

Qualitative causal variables can – and should – very well be used with the multiplicative formula. It was seen that the use of qualitative parameters inside an additive formula consisted in adding columns with 0 or 1 in the data matrix $||^+x||$.

The same idea can be used here, these columns being of course added *once the change of variables is done*, which means that the new matrix about the data is now defined as $||^+x^*||$. The result of the computation will provide values of the constant b_0 which depends on the qualitative variable.

Using the Level of Confidence

The level of confidence is not a causal variable. It can therefore used here for solving the equations giving the coefficients.

Here again the "weight" matrix $||W||$ must be introduced when $||^+x^*||$ have been computed.

12.1.2 The "Exponential" Formula

This formula can help solve some problems: compared to the multiplicative one it grows less rapidly for low values of the causal variable, and then faster.

Of course it grows very fast when this variable becomes large and should therefore be used with caution.

The formula giving the dynamic center is given, in the case of one causal variable only, by:

$$\hat{y}_i = b_0 b_1^{x_i}$$

of which graph is illustrated in Figure 12.2 for two values of b_1, keeping constant $b_0 = 1$.

This formula can easily be linearized by taking the logarithms:

$$\log \hat{y}_i = \log b_0 + x_i \log b_1$$

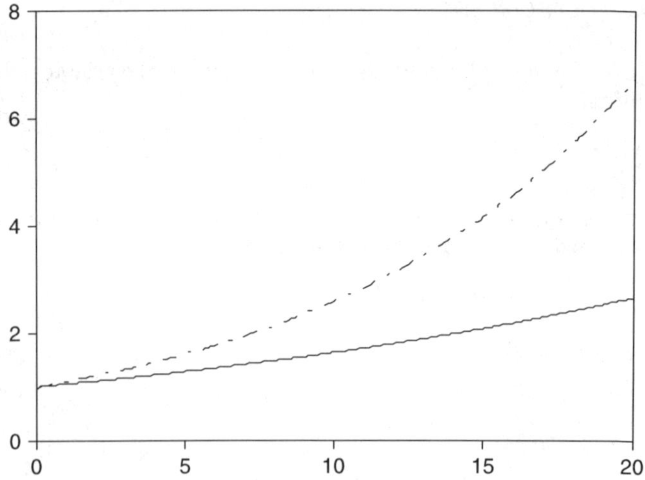

Figure 12.2 The shape of the exponential formula for two different values of b_1.

and using the same kind of change of variables, and the multiplicative error:

$$e^*_{+i} = \log e_{\times i} = \log \frac{y_i}{\hat{y}_i} = \log y_i - \log \hat{y}_i = y^*_i - \hat{y}^*_i$$

$$\hat{y}^*_i = \log b_0 + x_i \log b_1 = b^*_0 + b^*_1 x_i$$

Using the same metric as in the previous section, one gets the formula for the dynamic center:

$$\hat{y} = 10^{b^*_0} \times 10^{b^*_1 \times x}$$

This formula can easily be extended to any number of causal variables and to the use of qualitative variables, and of the level of confidence.

12.1.3 Mono-Variable: Other Relationships

Linearizable relationships are relationships which, by a change of the variable(s), can be transformed to linear relationships. There is obviously an infinite set of linearizable relationships, even with one parameter only; the presentation is here limited to one causal variable, but the idea can be extended to several such variables.

Choosing the type of relationship can be done in three ways:

1. There might be theoretical reasons – if the relationship between the dependent variable and the cost driver is well understood – for selecting a particular relationship.
2. Looking at the distribution of the data points on a graph may suggest such or such type of relationship. It has already been said that the eye is a powerful investigator for discovering "hidden" links in a set of not too much scattered data.
3. Several different relationships may be tested, one after the other, until the "best one" can be selected. This procedure is generally used when the data are too

much scattered for suggesting a particular relationship. This solution is rather artificial and should be avoided, especially if the relationship will be used for cost-estimating new products of which the cost driver value is outside the range of the present parameter: in the presence of scattered data, it is often quite possible to improve the "quality" (measured by the R^2 for instance) of the relationship, but – unless the shape of the relationship can be explained – there is absolutely no guarantee that the relationship remains true outside the range of the data points.

Using One Quantitative Causal Variable

For the cost analyst who has no preconceived idea about the relationship (point 1), the second solution is generally used. For this reason, a few useful relationships are displayed below, with their behavior (in order to help the cost analyst for selecting an appropriate one) and the change of variable which can be used for linearization. The new variables will be called y^* for the dependent variable and x^* for the causal variable (Figure 12.3).

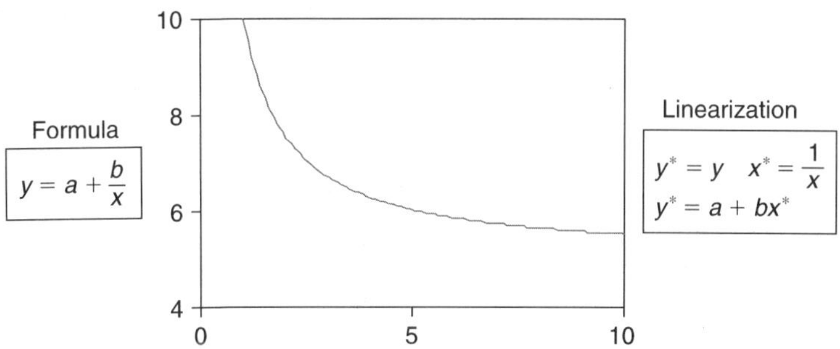

Figure 12.3 $a = 5$ and $b = 5$.

When dealing with cost, decreasing curves with x may be useful when working with the specific cost, whereas increasing curves with x are useful when working with the cost, at least in a limited interval (Figures 12.4 and 12.5).

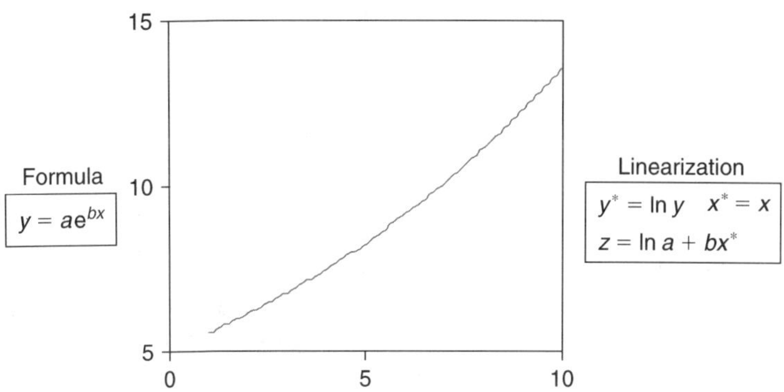

Figure 12.4 $a = 5$ and $b = 0.1$.

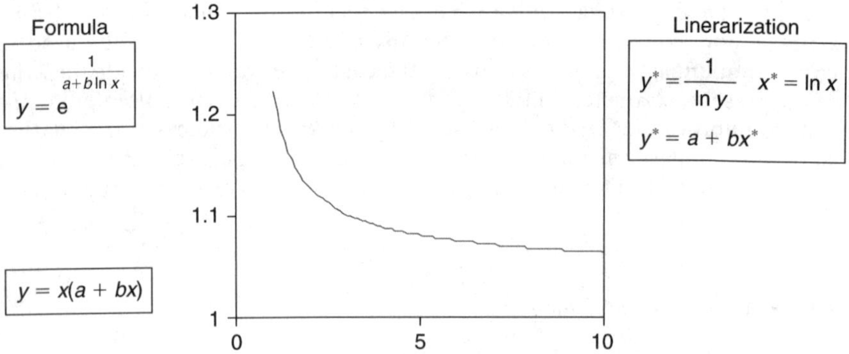

Figure 12.5 $a = 5$ and $b = 5$.

$$y^* = \frac{y}{x}(x^* - x)$$
$$y^* = a + bx^*$$

This formula can also be considered as a particular case of the more general formula:

$$y = a + bx + cx^2$$

dealt with as a formula containing two causal variables $x_1 = x$ and $x_2 = x^2$. Both new variables, although strictly correlated are not linearly correlated; therefore, the formula is perfectly usable (Figures 12.6–12.8).

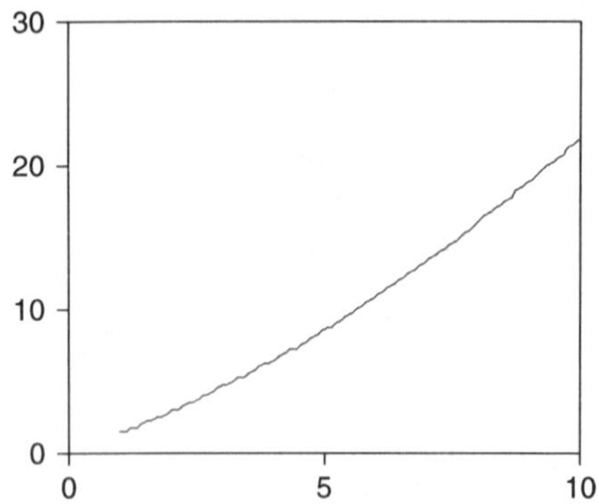

Figure 12.6 $a = 1.2$ and $b = 0.1$.

A lot of formulae are therefore possible and they cannot be here all studied in detail: one could add power, division, etc. The reader will note that sometimes the coefficients are also modified; for instance in the second example, the coefficients which appear in the linearized formula are $\ln a$ and b.

Once the change of the variables is done, finding the values of the modified coefficients is simple as it uses the results of Chapter 9.

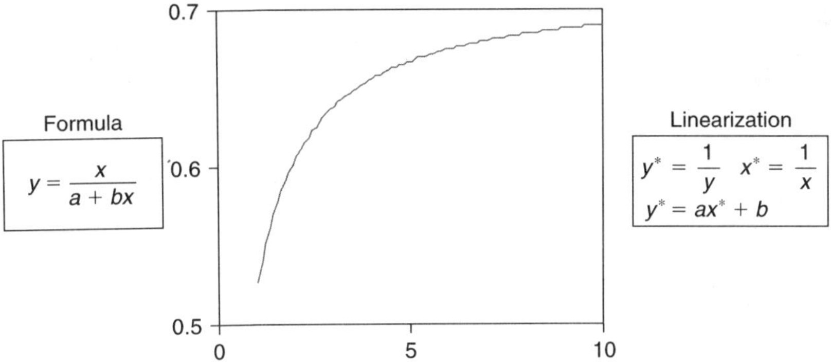

Figure 12.7 $a = 0.5$ and $b = 1.4$.

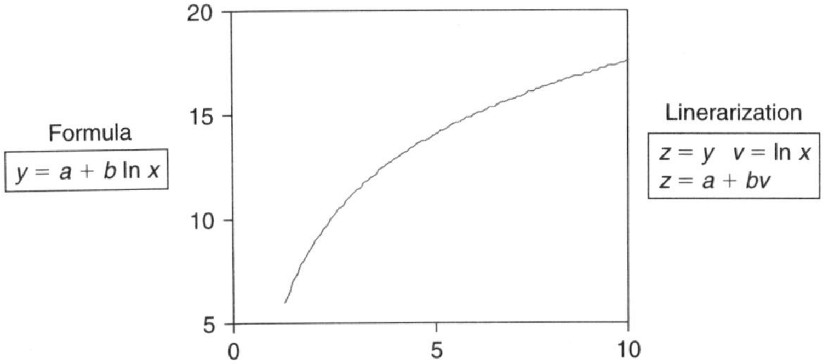

Figure 12.8 $a = 6$ and $b = 5$.

Adding Qualitative Variables

Linearizable formulae are always solved by looking for a linearized formula, which means are solved by linear algebra. Consequently the use of qualitative variables is always possible and will be used for filling added columns *once the variables transformation has been made.* Of course the final result may not be a simple change of the intercept (and there may be no intercept anymore).

Let us take an example with the desired formula:

$$y = e^{\frac{1}{a+b\ln x}}$$

which is found by solving the linear formula $y^* = a + bx^*$ where:

$$y^* = \frac{1}{\ln y} \quad \text{and} \quad x^* = \ln x.$$

If we add quantitative variables, the result will be that two – to limit the example, but it can be more than two – values of a, such as 1 and 2. The graph displaying this formula for $b = 1$ is given in Figure 12.9.

It is clear that no intercept is involved; the curve can be called "parallel", but not in the Euclidian meaning.

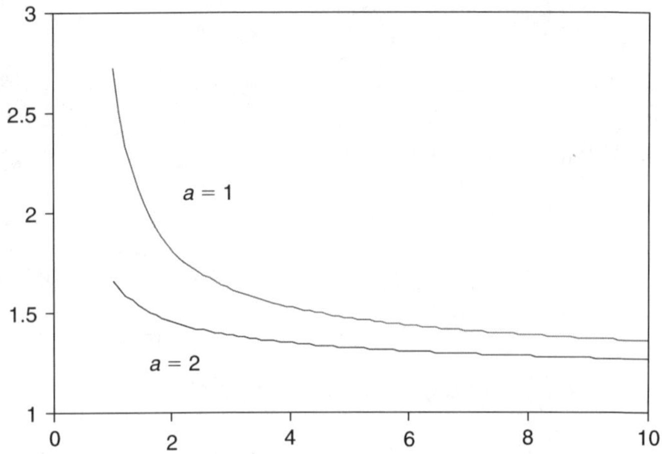

Figure 12.9 The graph of a non-linear formula using quantitative variable.

12.2 Strictly Non-Linear Cases

The algorithms for solving strictly non-linear cases are important in two domains:

- First of all for studying "pure" non-linear formulae, or formulae that cannot be linearized.
- Second for studying linear (or non-linear) relationships when you decide to use a metric which cannot be solved algebraically. An example is given by the metric:

$$\left(\frac{y}{\hat{y}} - 1\right)^2$$

First of all, several strictly non-linear formulae are presented. Then the way these relationships can be computed is described. Eventually the use of another procedure for solving nearly any type of relationship when a metric different from the simple one $(y - \hat{y})^2$ is described.

12.2.1 Examples of Strictly Non-Linear Formulae

In this section we introduce the case of the formulae which cannot be linearized. This introduction is limited to the use of one causal variable only (as such formulae are the only ones which are practically used in the domain of cost): x is, as usual, the causal variable, y being the dependent variable.

There is obviously an infinite set of non-linear relationships, even if we limit ourselves to the use of one variable only. This section presents a few of them which are sometimes used to describe the cost behavior or, more exactly, to represent the dynamic center of a cost distribution $\hat{y}$.

Around this dynamic center, the distribution of the residuals can be, as usual, studied.

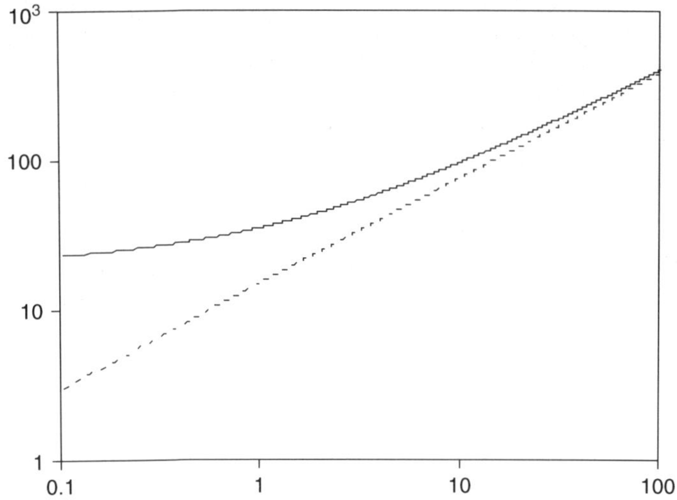

Figure 12.10 Example of a formula including a "correction-by-constant".

The "Correction-By-Constant" Relationship

This is the most frequently used strictly non-linear relationship. Its general shape is given by:

$$\hat{y} = b_1 + b_2 \times x^{b_3}$$

where b_1, b_2 and b_3 are the coefficients. An example of this relationship is illustrated on Figure 12.10.

On this figure the relationship $\hat{y} = 20 + 15x^{0.7}$ is displayed with a full line, the dotted line representing the function $15x^{0.7}$.

When using such a formula, x has to be the product size, whatever the way it is defined. It can be applied to any domain, because it describes properly the way, we, as humans, work.

Such a formula is very interesting for describing the cost behavior for the following reasons:

- This formula is the sum of two terms which can easily be interpreted as:
  - A constant term b_1 giving the set-up cost; the preparation of the work to be done (for instance tuning the machine, setting the raw material, etc.) is relatively independent of the product size.
  - The other term represents the machining cost. It obviously depends on the product size.
- The upper side of the curve – the "multiplicative" formula – is very often used for describing the cost behavior of products as soon as the size becomes large enough, as it was seen at the beginning of this chapter.
- When the product size becomes large, the "fix" cost becomes negligible compared to the "variable" cost. This explains why for large enough products, the second term of the formula is then only used.

This formula describes then the cost behavior in a very large range of sizes; obviously it should grow up on the left part of the graph, when the size becomes very small, but such products should be dealt with independently: there is no reason, at this stage, to search for a relationship which could cover inside the same product family product sizes from 1 μg to 100 ton!

The "Sigmoidal" Model

The sigmoidal model is quite often used by people studying the growth of vegetables. In our domain it can conveniently describe some phenomena such as the productivity: it can for instance be a substitute to the learning curve introduced in Chapter 8 of Volume 1.

Several models were described by different authors. One of the simplest is known as the "logistic" model:

$$\hat{y} = \frac{b_1}{1 + e^{b_2 - b_3 x}}$$

of which one example ($b_1 = 100$, $b_2 = 1$ and $b_3 = 0.5$) is given on Figure 12.11.

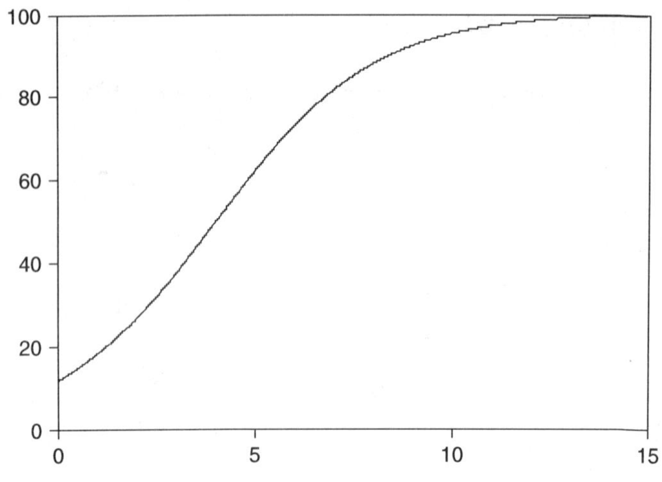

Figure 12.11 Example of the sigmoidal curve.

This curve presents a stable maximum (b_1) for large values of x. This maximum is reached, from a start (for $x = 0$), more or less rapidly depending on the value of b_3.

The Full Cost Production Model

Chapter 9 of Volume 1 introduced the modelization of the full production cost depending on the manufacturing load for a given production capacity. The model can conveniently be represented by the following relationship:

$$\hat{y} = b_1 + x^{b_2} \times e^{b_3 \times x^{b_4}}$$

This relationship requires four coefficients: this high number just reflects the complexity of the curve we try to modelize:

- b_1 corresponds to the true "fix" cost associated with no production at all ($x = 0$);
- b_2 is used to describe the first part of the curve;
- the last term of the relationship takes into account the second part of the curve.

An example is given in Figure 12.12 with the following coefficients: $b_1 = 2, b_2 = 0.4$, $b_3 = 0.001$ and $b_4 = 1.47$.

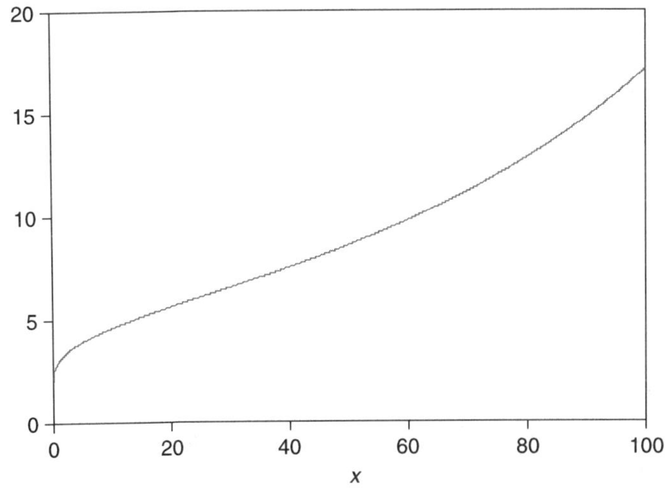

Figure 12.12 The full production cost.

12.2.2 Computation of the Coefficients

The Theory

The theory is known as the Newton, or Newton–Raphson, method (see for instance Ref. [32], p. 619). However, this method had to be improved (Ref. [47], p. 39) in order to take into account the values corresponding to several inputs.

Let us call $\vec{y} \in \Re^{I \times 1}$ the set of the values of the dependent variable $y_1, y_2, \ldots, y_i, \ldots, y_I$ corresponding to the set of the causal variable $x_1, x_2, \ldots, x_i, \ldots, x_I$.

Let us now consider a purely non-linear formula for the dynamic center of the y-distribution; this formula can generally be defined as:

$$\hat{y} = f(b_1, b_2, \ldots, x)$$

where $b_1, b_2, \ldots, b_p, \ldots, b_P$ are the unknown coefficients of the formula; this set of coefficients can also be written as a vector $\vec{b} \in \Re^{P \times 1}$. The values computed for the different x using this formula can also be defined as a vector $\vec{\hat{y}} \in \Re^{I \times 1}$.

In order to find out the values of the coefficients, we will have to minimize some function related to the "residuals" between the dynamic center of the cost distribution and the costs.

In this section we chose the additive residuals defined as:

$$e_{i+} = y_i - \hat{y}_i$$

but any other residual could of course be selected. The function to be minimized for simplicity purposes will be the sum of the squares of the deviations:

$$\sum_i e_{i+}^2 = \sum_i (y_i - \hat{y}_i)^2$$

Let us call Σ this sum. It can be conveniently written using the vector notation as:

$$\Sigma = (\vec{y} - \vec{\hat{y}})^t \otimes (\vec{y} - \vec{\hat{y}})$$

How can we find out the values of the $b_1, b_2, \ldots, b_p, \ldots, b_P$? The idea is the following one:

- Let us start from a set of initial values, selected as close as possible from the true values (graphs are extremely useful for estimating these values). Let us call $b_1^{(1)}, b_2^{(1)}, \ldots$ or $\vec{b}^{(1)}$ this set; this set cannot – except in rare circumstances (as the one illustrated in section "Finding initial values for the 'correction-by-constant' formula" of this chapter) be found algebraically. The value of Σ corresponding to this set is called $\Sigma^{(1)}$; it is the starting point.
- Now we add increments to these values, called $\vec{\delta}^{(1)} \in \Re^{P \times 1}$, carefully selected in order to decrease $\Sigma^{(1)}$.
- The new values of the coefficients are called $\vec{b}^{(2)} = \vec{b}^{(1)} + \vec{\delta}^{(1)}$ and the new value of the sum of the squares of the deviations $\Sigma^{(2)}$.
- Then we add another increments to the $\vec{\delta}^{(2)}$, etc., until we cannot decrease anymore Σ.

Finding the Increments $\vec{\delta}^{(1)}$

How can the $\vec{\delta}^{(1)}$ be selected? Let us suppose they are small enough to use a linear approximation of $\hat{y} = f(b_1, b_2, \ldots, x)$ around the set $\vec{b}^{(1)}$. This linear approximation is given by the development in Taylor's polynomial (Ref. [17], p. 609). For a particular value x_i one can write:

$$f(b_1, b_2, \ldots, x_i) = f(b_1^{(1)}, b_2^{(1)}, \ldots, x_i) + \sum_p \frac{\partial f(b_1^{(1)}, b_2^{(1)}, \ldots, x_i)}{\partial b_p} \times \delta_p^{(1)}$$

We have, for the different values of x, I such relationships. We can use the "matrix stenography" for representing all these relationships. If we create the matrix $\|J\| \in \Re^{I \times P}$, called the Jacobian matrix, as (the set of $b_1, b_2, \ldots$ has not been written in the f parenthesis for the sake of clarity):

$$\|J(b_1,b_2,\ldots)\| = \left\| \begin{matrix} \frac{\partial f(x_1)}{\partial b_1} & \frac{\partial f(x_1)}{\partial b_2} & \cdots & \frac{\partial f(x_1)}{\partial b_p} & \cdots & \frac{\partial f(x_1)}{\partial b_P} \\ \frac{\partial f(x_2)}{\partial b_1} & \frac{\partial f(x_2)}{\partial b_2} & \cdots & \frac{\partial f(x_2)}{\partial b_p} & \cdots & \frac{\partial f(x_2)}{\partial b_P} \\ \cdots & \cdots & \cdots & \cdots & \cdots & \cdots \\ \frac{\partial f(x_i)}{\partial b_1} & \frac{\partial f(x_i)}{\partial b_2} & \cdots & \frac{\partial f(x_i)}{\partial b_p} & \cdots & \frac{\partial f(x_i)}{\partial b_P} \\ \cdots & \cdots & \cdots & \cdots & \cdots & \cdots \\ \frac{\partial f(x_I)}{\partial b_1} & \frac{\partial f(x_I)}{\partial b_2} & \cdots & \frac{\partial f(x_I)}{\partial b_p} & \cdots & \frac{\partial f(x_I)}{\partial b_P} \end{matrix} \right\|$$

where the $(\partial f/\partial b_1), (\partial f/\partial b_2), \ldots$ are the partial derivatives of the function f, we can write:

$$\vec{f}(\vec{b}^{(1)} \oplus \vec{\delta}^{(1)}) = \vec{f}(\vec{b}^{(1)}) + \|J^{(1)}\| \otimes \vec{\delta}^{(1)}$$

where $\|J^{(1)}\|$ is the value of the Jacobien $\|J\|$ for $\vec{b} = \vec{b}^{(1)}$.

The value of $\Sigma^{(2)}$ is then given by:

$$\Sigma^{(2)} = [\vec{y} - \vec{f}(\vec{b}^{(1)} \oplus \vec{\delta}^{(1)})]^t \otimes [\vec{y} - \vec{f}(\vec{b}^{(1)} \oplus \vec{\delta}^{(1)})]$$

It is possible to analyze this expression:

$$\begin{aligned} \Sigma^{(2)} &= (\vec{y} - \vec{f}^{(1)})^t \otimes (\vec{y} - \vec{f}^{(1)}) - 2(\vec{y} - \vec{f}^{(1)})^t \otimes \|J^{(1)}\| \otimes \vec{\delta}^{(1)} + \vec{\delta}^{(1)t} \otimes \|J^{(1)}\|^t \otimes \|J^{(1)}\| \otimes \vec{\delta}^{(1)} \\ &= \Sigma^{(1)} - 2(\vec{y} - \vec{f}^{(1)})^t \otimes \|J^{(1)}\| \otimes \vec{\delta}^{(1)} + \vec{\delta}^{(1)t} \otimes \|J^{(1)}\|^t \otimes \|J^{(1)}\| \otimes \vec{\delta}^{(1)} \end{aligned}$$

from which the vector $\vec{\delta}(1)$ can be found by writing:

$$\frac{\partial \Sigma^{(2)}}{\partial \vec{\delta}^{(1)}} = 0$$

$$\vec{\delta}^{(1)} = [\|J^{(1)}\|^t \otimes \|J^{(1)}\|]^{-1} \otimes \|J^{(1)}\|^t \otimes [\vec{y} - \vec{f}^{(1)}]$$

This allows to get new values $b_1^{(2)}, b_2^{(2)}, \ldots$ with which the process will go on.

The major problem with this procedure is the ability to find "good" starting values $\vec{b}^{(1)}$ for the coefficients. Otherwise the procedure may diverge very quickly; this comes from the fact that the Taylor's series, based on a linear approximation of the function, suppose that the increments are small: if it is not the case the series gives a non-valid approximation which can lead to anything.

Finding Initial Values for the "Correction-By-Constant" Formula

The correction-by-constant formula is a formula which is especially interesting in the cost domain. It is always possible to find satisfactory initial values by looking at the graph: this is a major advantage when working with one causal variable only.

Another solution can be found by a simple computation. Such a solution was proposed[1] by Winklehaus and Michel. Their procedure uses four steps:

1. Using $x^* = \ln x$, a linear regression allows to find the coefficients α, β and γ of the following formula:

$$\hat{y}^* = \alpha + \beta \times x^* + \gamma \times x^{*2}$$

where the asterisk on $\hat{y}^*$ just reminds the reader that the regression works on the log of x.

2. Now we try to get a different expression of this relationship by developing the Taylor's series of $z = x^{b_3} = \exp(b_3 \times x^*)$ in the vicinity of:

$$\bar{x}^* = \frac{1}{I}\sum_i \ln x_i$$

This gives another development of $\hat{y}^*$. Equating both developments produces the following equations:

$$b_2 b_3 \times \exp(b_3 \times \bar{x}^*) - b_2 b_3^2 \times \exp(b_3 \times \bar{x}^*) \times \bar{x}^* = \beta$$
$$b_2 \frac{b_3}{2} \times \exp(b_3 \times \bar{x}^*) = \gamma$$

of which ratio gives a starting value of b_3:

$$b_3^{(1)} = \frac{1}{\dfrac{\beta}{2\gamma} + \bar{x}^*}$$

3. From this value the procedure described in the previous section could be used. In order to avoid – in this particular case – its instability, it is better to use a linear regression in order to find out the values $b_1^{(1)}$ and $b_2^{(1)}$ which, without changing $b_3^{(1)}$, minimize the sum:

$$\sum_i e_{+i}^2 = \sum_i \left(y_i - b_1^{(1)} - b_2^{(1)} \times x_i^{b_3^{(1)}} \right)^2$$

4. Now we can try by iterations, by slightly changing the value of b_3, to reduce this sum until no improvement is possible.

[1] See Ref. [7]. However, their article was full of typing mistakes. It had to be completely recomputed and demonstrated.

An Example

The data:

Name	Cost	Mass (kg)
A	161.89	0.1
B	300.00	1.0
C	1168.69	8.0
D	1949.75	15.0
E	3515.39	30.0
F	5484.32	50.0
G	7373.78	70.0
H	10 111.13	100.0

A quick drawing using a log–log scale reveals that the correction-by-constant could be used (Figure 12.13):

$$\hat{y} = f(b_1, b_2, b_3) = b_1 + b_2 \times x^{b_3}$$

the problem being to find out values for b_1, b_2 and b_3. A look at the graph suggests to start with $b_1 = 100$, $b_2 = 110$, $b_3 = 1$, which gives $\Sigma^{(1)} = 1.272 \times 10^6$. The first iteration gives $b_1 = 153.174$, $b_2 = 147.525$, $b_3 = 0.905$ and $\Sigma^{(2)} = 3.696 \times 10^5$; the second one $b_1 = 142.033$, $b_2 = 157.966$, $b_3 = 0.900$ and $\Sigma^{(3)} = 313.901$; the third one $b_1 = 142.001$, $b_2 = 158.000$, $b_3 = 0.900$ and $\Sigma^{(3)} = 4.869 \times 10^{-5}$. The iterations may stop here.

The procedure, of which convergence is very fast, eventually gives the following relationship:

$$\hat{y} = 142 + 158 \times x^{0.9}$$

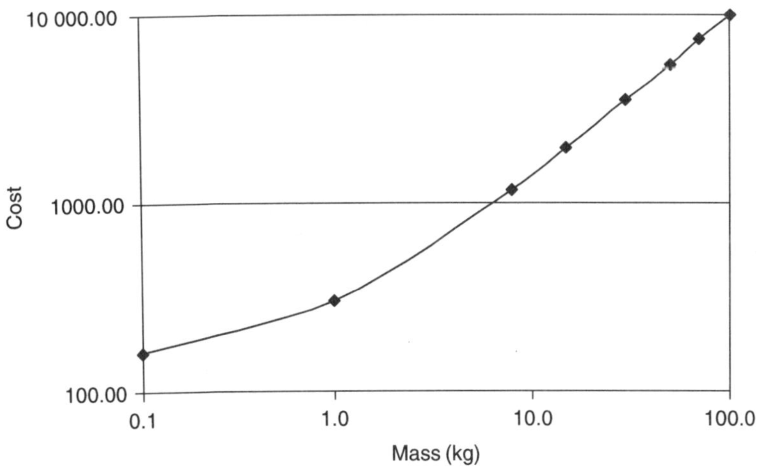

Figure 12.13 The data.

12.2.3 Using Different Metrics

You have seen in the preceding chapter that different metrics can be used. It is obvious that the use of these metrics cannot be done algebraically. The only solution is to "guess" an initial set of values for the coefficients and then proceed by iterations; in the present situation, the conventional regression analysis (ordinary least squares, OLS) can provide this initial set.

How can we proceed afterward? The Newton–Raphson method as described in the previous section cannot be used because it is based on the minimization of:

$$\Sigma = (\vec{y} - \hat{\vec{y}})^t \otimes (\vec{y} - \hat{\vec{y}})$$

which means that it uses the results of the linear algebra.

Minimizing the sum:

$$S = \sum_i \left(\frac{y_i}{\hat{y}_i} - 1 \right)^2$$

where $\hat{y}_i$ is defined by a simple linear relationship: $\hat{y}_i = a + bx_i$ (so written in order to simplify the notations) is strictly non-linear: another solution has to be found.

A Geometric Perspective

In the expression of S, all the y_i and the x_i are known (they are our data); therefore, S is only a function of a and b, and we write $S(a, b)$. Geometrically $S(a, b)$ defines a surface in the three-dimensional space,[2] as illustrated on Figure 12.14. The values of the initial set (a_0, b_0) defines a value $S(a_0, b_0)$ from which we want to find the set (a, b) which will minimize this function.

Starting from (a_0, b_0), the idea is then to find out the *direction* $\vec{d}$ in which $S(a, b)$ decreases the more rapidly: this will be approximately the direction of (a, b). This direction is characterized by two increments $(\delta a, \delta b)$; in order to find the direction, we will force these increments to satisfy the relationship:

$$\delta a^2 + \delta b^2 = k^2$$

which means that the point $(a_0 + \delta a, b_0 + \delta b)$ will be on the circle of radius k drawn around the point (a_0, b_0). The value of k must be chosen in order to go not too far, if, by chance, we start in the vicinity of (a, b).

The point:

$$\begin{aligned} a_1 &= a_0 + \delta a \\ b_1 &= b_0 + \delta b \end{aligned}$$

[2] If the expression giving $\hat{y}$ includes more than two coefficients, this can be generalized to more dimensions.

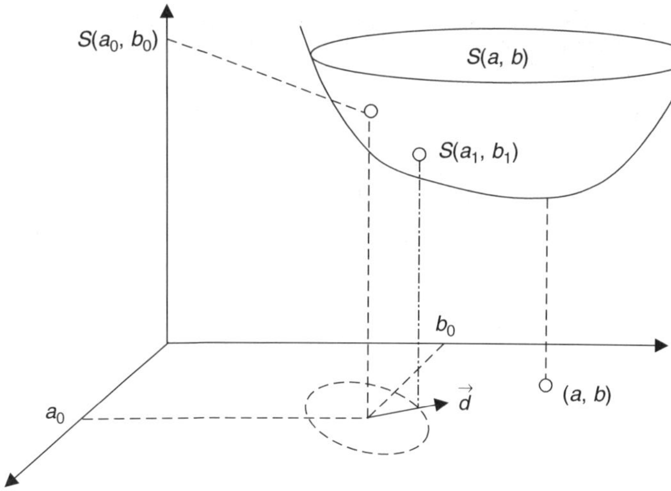

Figure 12.14 A geometric perspective.

will give a new value $S(a_1, b_1)$ from which the procedure can restart until we get a minimum value.

Algebraically

In the vicinity of (a_0, b_0) the value of $S(a, b)$ can be given by the Taylor's polynomial:

$$S(a,b) = S(a_0,b_0) + \frac{\partial S(a_0,b_0)}{\partial a}\delta a + \frac{\partial S(a_0,b_0)}{\partial b}\delta b + \text{second order terms} + \cdots$$

If the values $(\delta a, \delta b)$ are small enough (this depends on the value of k) the first order terms gives a sufficient approximation.

Writing

$$A = \frac{\partial S(a_0,b_0)}{\partial a}, \; B = \frac{\partial S(a_0,b_0)}{\partial b}$$

the problem is to compute $(\delta a, \delta b)$ which maximizes (with the right sign!) the sum $A\delta a + B\delta b$ with the constraint $\delta a^2 + \delta b^2 = k^2$. The solution of this problem is well known and uses the Lagrange's multiplier λ. We define a function:

$$\Phi(\delta a, \delta b, \lambda) = A\delta a + B\delta b - \lambda(\delta a^2 + \delta b^2 - k^2)$$

that has to be maximized. Notice that A, B and k are here constants. The solution is therefore given by writing the three partial derivatives are null:

$$\frac{\partial \Phi}{\partial \delta a} = A - 2\lambda \delta a = 0$$
$$\frac{\partial \Phi}{\partial \delta b} = B - 2\lambda \delta b = 0$$
$$\frac{\partial \Phi}{\partial \lambda} = \delta a^2 + \delta b^2 - k^2 = 0$$

Eliminating δa and δb allows to compute λ:

$$\lambda = \frac{\pm\sqrt{A^2 + B^2}}{2k}$$

the sign having to be chosen in order to decrease $S(a, b)$. Once λ is known, δa and δb are immediately given:

$$\delta a = -\frac{A}{2\lambda}, \ \delta b = -\frac{B}{2\lambda}$$

and the process can go on starting now from$(a_0 + \delta a, b_0 + \delta b)$, gradually reducing the value of k as soon as one gets closer from (a, b).

Practically

The computations are not very complex. However, finding A and B may seem to be a formidable task! However it is not: we do not need an algebraic value of these terms, but only their numerical values. $S(a, b)$ is a function of all y_i, x_i, which are constant values, and of (a, b). Returning to the definition of the derivative, we can get an excellent approximation of A and B by giving small increments Δa to a and Δb to b and write:

$$A = \frac{S(a_0 + \Delta a, \ b_0) - S(a_0, \ b_0)}{\Delta a}$$
$$B = \frac{S(a_0, \ b_0 + \Delta b) - S(a_0, \ b_0)}{\Delta b}$$

The rest of the computations is very straightforward.

12.2.4 Using a Metric Including a Constraint

We already saw such a metric with the biweight in Chapter 8. We also mentioned in the same chapter that some cost analysts sometimes imposed a constraint they called "with zero bias".

For instance the metric we saw in the previous section defined by:

$$\left(\frac{y_i - \hat{y}_i}{\hat{y}_i}\right)^2$$

or minimum percentage error (MPE) can also be used with the constraint:

$$\sum_i \frac{y_i - \hat{y}_i}{\hat{y}_i} = 0$$

called "with zero bias". This metric is then called minimum unbiased percentage error (MUPE).

How can this constraint be taken into account?

As we saw it in the section dealing with the biweight, the general procedure is also to work by iterations, but using a slightly different approach: the usual iteration, at step n, tries to minimize, for the example given, the sum:

$$\sum_i \left(\frac{y_i - \hat{y}_{i,n}}{\hat{y}_{i,n}} \right)^2$$

In the presence of a constraint we try to minimize the sum:

$$\sum_i \left(\frac{y_i - \hat{y}_{i,n}}{\hat{y}_{i,n-1}} \right)^2$$

where the previous iteration is used at the denominator.[3]

As mentioned by Book and Young, "the percentage error of a MUPE cost-estimating relationship will naturally be larger (than with the MPE), but its bias will be less, exactly 0 in the case of a linear functional form and apparently asymptotically 0 in other cases".

As we mentioned it upwards in this volume, this is an interesting academic exercise but it is not going to improve the relationship we are looking for. Once again we recommend to concentrate instead on the data, their normalizations, the solutions for potential problems, etc.

[3] See for instance Stephen A. Book and Philip H. Young "General-error regression for deriving cost estimating relationships". *DoD Cost Analysis Symposium*. Leesburg 1994.

Part IV

Studying the Residuals Is as Important as Finding the Formula

Part Contents

This chapter deals with the residuals in the classical way.

However, the cost is a very special subject and "laws" found for other subjects do not necessarily apply to cost. It must not be forgotten for instance that the linear regression was created by Gauss for finding the best parameters of the ellipsis that a planet should follow when several astronomical observations were made; these observations were subjected to "errors" (which was a good term) and Gauss was looking for the ellipsis which would go as well as possible through the observations. The hypotheses on which he based his computations were realistic for solving this problem.

Can we build a "theory" which would be more realistic for the cost domain? The modern approach, based on the Bootstrap – or the Jackknife when the number of available data points is large enough – need no hypothesis to be used. Consequently we are not forced at all to look for a normal distribution of the residuals (or their log) as it will be done in this chapter.

Practically considering that the distribution of the deviations from the dynamic center should be symmetrical is not realistic in our domain for the following reason: if it is always possible to add costs (and therefore an infinite right side of the distribution can be kept), the infinite left side of this distribution is not realistic at all: in order to make something there is certainly an absolute minimum cost under which it is impossible to go. Of course this minimum cost, as the 0 K, cannot be achieved: but it is an "absolute" barrier.

We therefore look for a non-symmetrical distribution with one infinite side on the right and a minimum value on the left. There are several candidates among the "available" distributions, such as the log normal one, or the χ^2.

This is still a domain of research, the problem being that the data are not too numerous. One interesting consequence of this approach is that, once the distribution has been found, the "absolute" minimum cost for doing something would be computable. And we strongly believe that there is such a minimum.

The Bootstrap – and sometimes the Jackknife – should be the normal way to deal with the residuals in the domain of cost. They will be investigated in some details in Chapter 15.

The Bootstrap can also easily be used for using other metrics (for which the Gauss' hypotheses are certainly not valid) for instance the median, which remains, as said upwards, one of the most interesting metric when no other – meaning beyond the sample – information is available.

But let us start with the conventional approach, which must be known by the cost analyst.

Definition

The residuals are "What is left when the values of the dynamic center are removed from each cost value?" Residuals should not be called "errors". The point of view the cost analyst should adopt when studying the residuals is that residuals are caused, in this order, by:

1. A lack of homogeneity in the products aggregated in the product family. As previously indicated, this lack of homogeneity should be compensated by the addition of parameters,

but we recognize it is not always possible theoretically (when the number of data points is limited)[1] or practically (when the information is missing). Experience nevertheless shows that many cost analysts use a very limited number of parameters – most often one (it has to be the size) – and they are strongly advised to improve the definition of their products instead of trying to improve the quality of the model by beautiful but counterproductive mathematical procedures.

2. A lack of knowledge about the production environment. Many things may have happened during the production which can make the costs more or less erratic: change in materials, change in material procurement cost, change in the manufacturing process, modifications, etc. All these things are very difficult to grasp; at least if the cost analyst works with internal production costs, he/she should try to know something about them.
3. A lack of proper normalization: refer to Part II of Volume 1 for a detailed investigation of the normalization. The normalization process should also check if the cost figures include the same things (part of the development cost, tooling, transport, packing, guarantee, etc.). This is especially true if price figures are used instead of cost figures, but it must be recognized that this situation is difficult to remedy to, as the pricing policy of the companies is difficult to perceive. In such a case the cost analyst should make a preliminary correction to the costs from what he/she knows about the market, or use the quantitative information called the "confidence level" he/she may have in the figures.
4. And, eventually, the measurement process or the human behavior which adds random fluctuations to the cost.

All these considerations are well known to most cost analysts and it must be recognized that it is often difficult to improve the situation. As previously mentioned, the quality of the model could be improved – which does not necessarily mean by reduction of the residuals – by finding the most realistic dynamic center through the data.

There are several ways to define the residuals; these ways are theoretically independent on the metric which was used to build the formula, even if the definition of the residuals is generally associated with this metric:

- A frequent definition of the residuals is the "**additive**" form:

$$e_{+i} = y_i - \hat{y}_i$$

This form is generally used with the metric defined by the difference, as its purpose is to minimize the sum $\sum |y_i - \hat{y}_i|^{\alpha}$ whatever the value of α. It is called additive because the observed cost of product is given by:

$$y_i = \hat{y}_i + e_{+i}$$

- Another definition is defined as "**multiplicative**" by the ratio:

$$e_{\times i} = \frac{y_i}{\hat{y}_i}$$

which is very often used with the "multiplicative formula" $\hat{y} = b_0 \prod_j x_{i,j}^{b_j}$ with:

$$y_i = \hat{y}_i \times e_{\times i}$$

[1] We investigate in Chapter 16 of Volume 1 what can be done when this number is limited.

- still another definition, a "**mixed**" form or percentage form, is given by:

$$e_{\%i} = \frac{y_i}{\hat{y}_i} - 1 = \frac{y_i - \hat{y}_i}{\hat{y}_i}$$

with the cost given as:

$$y_i = \hat{y}_i \times (1 + e_{\%i})$$

- and a log form could also be used:

$$e_{\bullet i} = \log \frac{y_i}{\hat{y}_i}$$

with

$$y_i = \hat{y}_i \times \exp e_{\bullet i}$$

In these formulae
- y_i represents the value of the dependent variable associated with product A_i of the sample,
- $\hat{y}_i$ represents the value of the dynamic center of the cost distribution in the sample, corresponding to product A_i.

The formula giving $\hat{y}_i$ is irrelevant here: we are just studying the residuals. The purpose of the analysis is nevertheless to make a judgment, based on these residuals, on the interest of using the dynamic center for cost-estimating purposes.

The distribution of the residuals in the sample is represented by the letter ψ. Studying this distribution ψ is as important as the search we made for finding an interesting dynamic center of the cost. The reader must never forget that the distribution of the cost in the sample is replaced by:

- the dynamic center $\hat{y}$ on one hand,
- the distribution ψ on the other hand.

Up to now the search we made was mainly an attempt to reduce the standard deviation of this distribution ψ, not taking too much in consideration the other characteristics of it. In this chapter we will study only this distribution in order to improve, if it is possible, the quality of our future cost-estimating model.

A Recommendation

Whatever the metric, it is recommended, once the computation of the formula is made, to recompute the residuals as *additive* (or any other form, the question being to get always the residuals under the same form). Working this way will allow you to make comparisons between the different solutions and use many tools developed for this kind of residuals. Otherwise you cannot really compare a multiplicative formula (which is generally associated with multiplicative residuals) with, for instance, an additive formula (for which the additive residuals are nearly always used).

Example

We will use in this chapter, the same example as the one which was introduced in Chapter 2. The values are repeated here for the sake of convenience:

Cost	V_1	V_2	V_3	V_4
1278	6.83	1264	1274	10
724	2.18	1032	480	6
809	3.8	812	656	6
920	4.55	516	786	8
772	2.18	1032	480	6
877	2.11	1548	394	6
1064	4.67	2722	942	6
865	2.81	807	671	3
961	2.55	1598	872	6
856	1.68	737	450	5
1293	6.3	715	1400	19
717	1.98	186	430	7
648	1.25	228	257	6

← *One product is described, inside the product family, by a row giving the value of its four variables, plus its cost.*

V_1: represents the mass
V_2: the number of connections
V_3: the number of components
V_4: the number of boards

13 Studying the Additive Residuals

Summary

This chapter is an introduction to the analysis of the residuals, the residuals being what is left over, in the sample, when the value of the dynamic center is removed from all data points.

This analysis is important because:

1. the idea for selecting a formula type, for adding parameters, etc. was based on a tentative for reducing these residuals;
2. studying the residuals may reveal problems about the data;
3. studying the residuals may suggest new formula, to look for other cost drivers, in order to improve the formula;
4. all we can do to estimate the quality of a specific cost model is based on the values of the residuals;
5. the confidence we may have about an estimate is highly correlated to these values.

Studying the residuals is therefore an important step in building a specific model.

This chapter, after redefining what we call "residuals", investigates first the classical approach with the most important – meaning those you are probably to use often – residuals: the additive ones. It shows that the simple display of these residuals is important as it can reveal interesting features, such as new (undetected up to now) outliers, a bad choice for the formula, or a trend in their values.

It then mentions the question of homoscedasticity and the sign test, to be used in particular circumstances.

The study of the residuals in the bilinear case, so often presented in most manuals, is developed. We mention the autocorrelation problem, even if it is rather rare in the cost domain, except if a wrong formula type was selected.

An interesting comment, mentioned by Mosteller and Tukey [43], allows, when a trend – even modest – in the residuals is apparent, to considerably improve the predictive capacity of the formula.

Multiplicative residuals are afterwards studied.

Eventually the modern approach based on the Bootstrap – and sometimes on the Jackknife – is briefly presented. It allows for using a more realistic, in the cost domain, distribution of the residuals and for computing the absolute minimum cost for doing something.

It also allows for using different metrics, among which the median is especially interesting.

13.1 Introduction

This chapter, which is as important as the chapters dedicated to the search of the dynamic center, deals with the additive residuals. These residuals are important, not so much due to their natural interest, but because they are mainly used by cost analysts (when they study the residuals).

Additive residuals are not at all specific to additive formulae, even if they are generally used in such circumstances. They can be, and should be, applied to any relationship.

Definition

Additive residuals are what is left, whatever the nature of the formula which was used for building the dynamic center, from the values y_i of the dependent variable (the cost) when the value $\hat{y}_i$ of the center (the dynamic center when causal variables are involved, whatever the way it is computed) is removed from them:

$$e_{+i} = y_i - \hat{y}_i$$

It is common practice to consider the set of the residuals as a vector $\vec{e}_+$ defined by its components, these components being the values for each product belonging to the sample.

The values of the example considered in the previous part (Chapter 10) with an additive formula using the four parameters give the following vector for the residuals:

$$\vec{e}_+ = \left\| \begin{array}{r} 63.642 \\ -72.522 \\ -67.422 \\ -17.514 \\ -24.522 \\ 88.487 \\ -35.759 \\ 19.099 \\ -24.019 \\ 100.94 \\ -3.114 \\ -20.82 \\ -6.476 \end{array} \right\|$$

13.2 Studying the Additive Residuals in General

13.2.1 The Distribution of the e_{+i}

Residuals are values which, in principle, do not depend on any parameter: they are supposed to be random.

The basic study of these residuals can therefore be dealt with according to the data analysis described in Chapter 4 for "one variable only", the variable being of course the value of the residuals:

- *Looking for outliers*: if the search for outliers has been studied for the data sample, there should not be any outlier in the residuals. But it is always a good idea to check it.
- *Visualization of the distribution*: this visualization is very important for discovering several potential problems. When several causal variables are involved it is recommended to visualize, on a two dimensions graph, the behavior of the residuals with each quantitative variable, one at its turn. The things you must look at are:
  - Is there any trend?
  - Are there abnormal values?
  - Do the residuals appear correlated?

Basic Values

There are three basic values for quickly getting an idea of the residuals importance:

1. The *arithmetic mean*. If the ordinary least squares (OLS) method is used, then this mean is exactly equal to 0. Otherwise it can be slightly different from 0.
2. The *average of the absolute values* of relative residuals (multiplied by 100 if you prefer to work with percentages), defined either as "locally" by:

$$\frac{1}{I}\sum_i \left|\frac{e_{+i}}{y_i}\right|$$

   or "globally" (this is preferred by some cost analysts) by:

$$\frac{\sum_i |e_{+i}|}{\sum_i y_i}$$

   of which results are about the same, except in rare circumstances: for our example, the values are 4.8% and 4.6%. The main interest of this value is that it is so easy to understand, even if it is not a generally considered statistical property: knowing that your data points are, on the average, at 5% of the dynamic center is easier to interpret than many statistical tests! The second reason for computing it is that, in the domain of cost, precisions are always given in percentages, not in absolute values: this value is therefore easy to compare with other information.
3. The *spread of the residuals* around their average value, spread classically given by their standard deviation. For the example, this standard deviation equals 52.25. The reader will note that the ratio of this standard deviation to the average cost

equals 5.76% has the same order of magnitude that the previous ratios: both convey the same idea.

Visual Examination of the Residuals

This is the first thing to examine: residuals can be displayed on graphs, the first concern being the choice of the variable to be used for the abscissas.

This choice is based on the response to the question: What are we looking for? As a matter of fact we are looking for several things:

- Is there any trend with the cost value? The utility of this search comes from the fact that one of the most common assumption about the residuals is what is called the homoscedasticity – from the Greek όμοσ, equal, and σκεδαννυμι, to spread: it means that the spread of the residuals is the same whatever the cost value (see Figure 13.1).

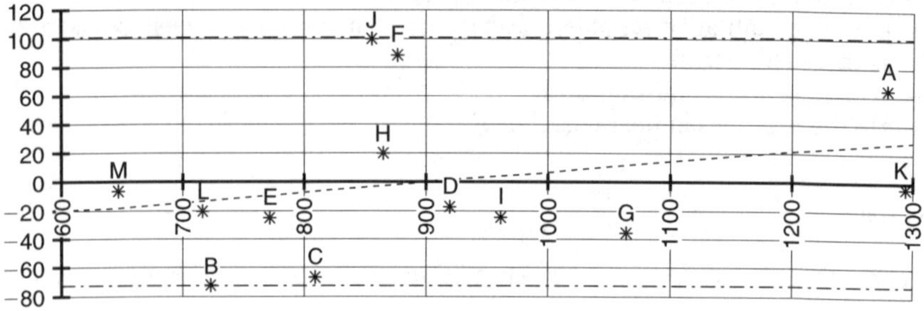

Figure 13.1 Residuals according to the cost values.

In Figure 13.1 the dynamic center was computed as a function of two variables only: the number of components and the number of connections. On this graph the trend line of the residuals was computed (it is represented by the dotted line): it shows a slight lack of homoscedasticity, as the values of the residuals increase with the cost values. This is not due to higher cost values because it remains if the residuals are expressed in relative values (Figure 13.2).

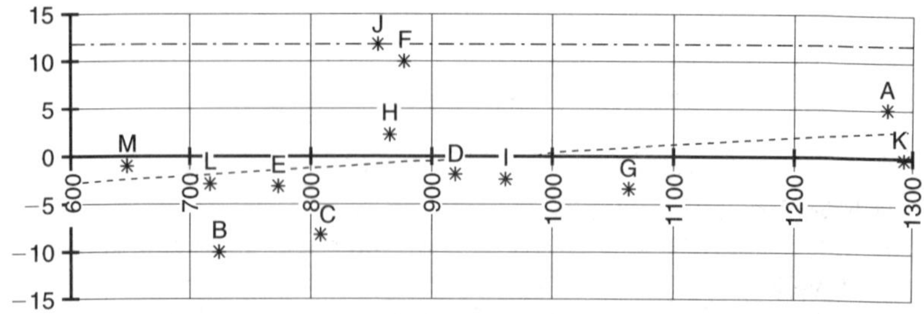

Figure 13.2 Residuals according to the relative cost values.

- Is there a trend with any variable used for establishing the formula giving the dynamic center? In the presence of several variables, each variable should be successively tested. On the following figure the number of components is used for abscissas: no trend does appear (the trend line, the dotted line, is exactly on the x-axis) (Figure 13.3).

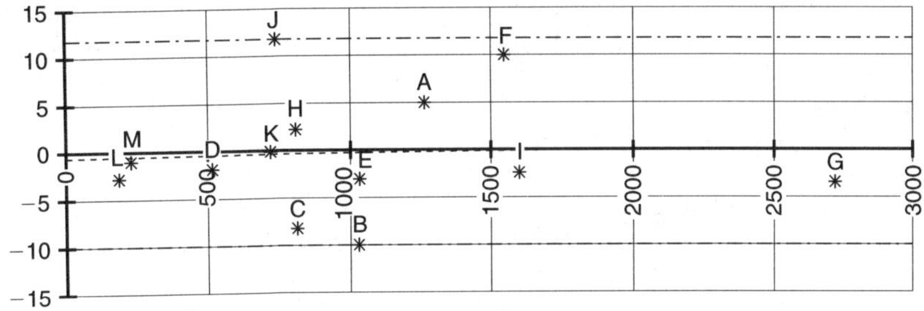

Figure 13.3 Residuals according to the number of components.

- Are the residuals autocorrelated? Residuals are said to be autocorrelated if the e_{+i} value depends on the e_{+i-k} value whatever the value of k. The discussion about autocorrelation is made in Section 13.3.5. There are mathematical procedures for determining the level of autocorrelation, but, for the time being, the eye can reveal, on any graph, a problem of this type.

  In the previous figure, no autocorrelation clearly appears. But let us take another set of data and use a linear formula for the dynamic center of the distribution of the sample values. The plot of the residual values is given in Figure 13.4.

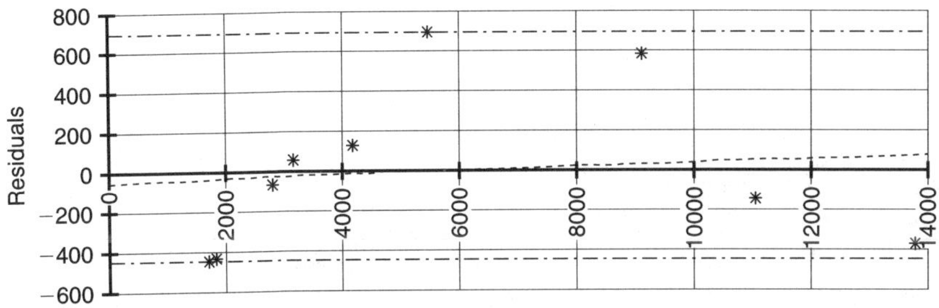

Figure 13.4 Another set of residuals based on a linear formula.

In this figure, one can see that the residuals are correlated: their shape is given by some inverted parabola. What is the origin of this correlation? In order to understand it, let us use, instead of a linear formula, the formula called "correction-by-constant", always defining the residuals as additive (Figure 13.5).

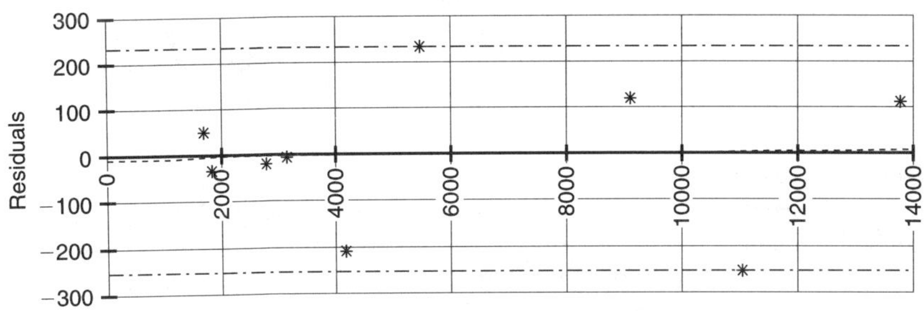

Figure 13.5 The same set of residuals based on a corrected-by-constant formula.

Two things can be observed on the second curve: first of all the values of the residuals are smaller (although the R^2 is about the same: it goes from 0.993 to 0.999); second the correlation between the residuals seems to have vanished.

In conclusion, and from our experience, when dealing with cost, **the autocorrelation generally comes from a wrong choice of the formula.** This is a good motivation to look at these graphs because not so may tests disclose this information about the choice of the formula type: the R^2 is about the same, the "t" values for the second formula are worse than those for the first formula. The average of the absolute values is an indicator, as it goes down from 8.9% to 2.1%, but it does not reveal the real cause of the improvement.

Another cause of such correlated residuals may come from the fact that an important variable is missing. If the cost analyst is convinced that the linear relationship is the right one, he/she should re-study the data for discovering, what is the variable which is missing.

- Are some residuals "far away" from the other ones? Such residuals may reveal outliers which should have been discovered earlier (see Chapter 6 for the search of outliers). If such data points were given a low confidence level, or if the biweight algorithm was used, these points did not cause any damage to the formula giving the dynamic center. Otherwise they might have.

Statistical Analysis of the Distribution ψ

The basic results are, are always, given by:

- the center,
- the standard deviation,
- the skewness,
- the kurtosis.

These values will be used in the next chapter in a test about the "normality" of the deviations in the whole population.

The Center of This Distribution

The center can be computed, as we saw it in Chapter 8 in different ways, but we concentrate here on the arithmetic mean.

As the dynamic center can also be computed in different ways, this mean is not necessarily equal to 0 (this is the case when the least squares, also called the linear regression, is used). The mean of the residuals distribution takes, generally speaking, a small value, sometimes called the "bias" although this word is not correct here.

The Standard Deviation Around the Center

For the present example, the standard deviation of the residuals is 52.25 which is small but not negligible as it amounts to about 5% of the cost (do not forget that the standard deviation of a distribution is not the full spread of its values: here this full spread is 173.5).

The Skewness
The level of asymmetry is equal to 0.652, which is reasonable.

The Kurtosis
Its level is equal to 2.401.

The distribution of the residuals is not, from these observations, too far away from a normal distribution. This does not prove the homoscedasticity of these residuals, as the parameters are global information, whereas homoscedasticity is a local observation.

13.2.2 Testing the Homoscedasticity

Many authors[1] have studied this property and the influence of a lack of it on the other tests.

The tests are generally built in creating two groups of data and comparing the sum of the squares of their residuals (this is a way to get a "local" information): Goldfeld and Quandt make two simple regressions on the first $I/2$ data points and the last $I/2$ data points: then the ratio $S2/S1$ of the squares of their deviations has an F-distribution and can therefore be used as a test.

However, it is very difficult to estimate the damages caused by heteroscedastic data, and in the cost domain, the number of data is generally too small to make a realistic test on this subject.

What can easily be done is a simple test in order to see if the residuals are about equally distributed. This test divides the range of costs in four intervals and looks at the residuals (their mean and the average of their absolute value):

Interval	Average of the residuals	Average of their absolute value
600–775	−31.1	31.1
775–950	+35.3	69.0
950–1125	−29.9	30.0
1125–1300	+30.3	33.4

Nothing looks abnormal in these values: the average of the absolute values is rather regular. There is a "peak" in the second interval, which seems mainly due to the fact that this interval contains more data points than any other interval.

It is also interesting to look at the relative values of the residuals, as we did in Figure 13.2. On this figure it seems that these relative values decrease from going from left to right; it is not generally the case when dealing with cost (one generally observes that the relative values remains about constant) and it seems to be mainly due to data points F and J, which, by the way, were found (with data point A) as potential outliers by the algorithms based on the variances–covariances matrix.

13.2.3 The Sign Test

We expect the residuals to be distributed randomly around their center. Consequently we should have about as many positive as we have negative values; it

[1] See for instance Malinvaud [38], p. 292 or Theil [56], p. 196.

should be detrimental to have, for instance, a few large positive values and a lot of small negative values.

The sign test is there to check. This information completes the one given by the test of normality and may help explain some discrepancies from normality.

In order to use the sign test, the number of positive values, noted I_+, and the number of negative values, noted I_-, are counted, the total being of course equal to the number I of products. As we expect to have as many positive and negative values (the probability P to get a positive value is therefore $P = 0.5$), the number of positive values has a binomial distribution with $P = 0.5$.

The binomial distribution is defined the following way (the general definition is here simplified because the probabilities to get a positive and a negative value are equal to 0.5): the probability to get exactly I_+ from a set of I residuals is given by:

$$\text{prob}(I_+/I) = C^I_{I_+} \times 0.5^I$$

where

$$C^I_{I_+} = \frac{I!}{I_+!(I - I_+)!}$$

with $I! = 1 \times 2 \times 3 \times \cdots \times I$.

For instance, if $I = 13$, the probabilities to find different values of I_+ are given by Figure 13.6.

As expected the highest probability is given for I_+ equal to 6 or 7, but the probability distribution is rather flat. With the data of our example, we get four positive values and nine negative values; the probability to get such a result, assuming that the signs are randomly distributed is 0.087. This is not bad and this value is consistent with a random distribution of the signs (this type of conclusion is studied in Part V).

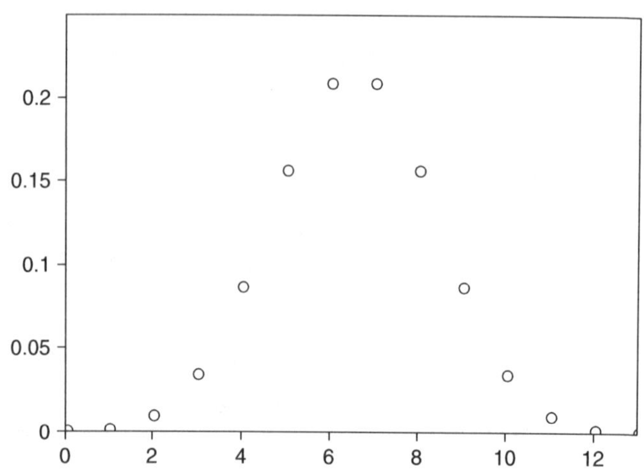

Figure 13.6 Probability to get I_+ values in I residuals.

13.3 Studying the Residuals in the Bilinear Case

As the bilinear formula is the most used by cost analysts, studying these residuals has to be developed more in details.

13.3.1 Computing the Residuals in the Bilinear Case

The additive residuals for the linear formula based on the OLS or "linear regression", are computed as:

$$\vec{e}_{+} = \vec{y} - ||^{+}x|| \otimes \vec{b} = ||M|| \otimes \vec{b}$$

where

$$||M|| = ||1|| - ||^{+}x|| \otimes \left(||^{+}x||^{t} \otimes ||^{+}x||\right)^{-1} \otimes ||^{+}x||^{t} = ||1|| - ||hat||$$

The "HAT" matrix is defined in Chapter 6 as:

$$||hat|| = ||^{+}x|| \otimes \left(||^{+}x||^{t} \otimes ||^{+}x||\right)^{-1} \otimes ||^{+}x||^{t}$$

It is a square, symmetrical, $\Re^{I \times I}$ matrix entirely defined by the causal variables only (I is the number of products in the sample). Its element are called $h_{\text{row,column}}$.

This matrix related to the example is computed and displayed in Chapter 6.

The important thing to remember at this stage is that the sum of its diagonal elements is equal to the number of variables $I + 1$ when the intercept is not forced to 0. From this expression, it is clear that, when there is only one causal variable, $\sum_{i} h_{i,i} = 2$: the sum of diagonal elements of this matrix is equal to 2.

13.3.2 Statistical Analysis of the Distribution ψ

The Center of This Distribution

In the bilinear case, the arithmetic mean of the residuals is equal to 0. This comes from the fact that using the standard regression analysis, as we wrote it in Chapter 9:

$$\sum_{i} e_{+i} = 0$$

The Variance of This Distribution

It can be noted that the variance of the residuals depends only, whatever the number of variables, on the variance of the cost and the correlation between the cost and the causal variables. This is easily demonstrated ([50], p. 366):

$$\text{var}(e_{+i}) = (1 - r^{2}) \times \text{var}(y_{i})$$

with

$$\text{var}(y_i) = \frac{1}{I}\sum_i (y_i - \bar{y})^2$$

As a rule of thumb, it can be said that the variance of the residuals is equal to the variance of the y_i, reduced by the difference between the square of the correlation coefficient and 1. This is an interesting result to remember (and it is quite logic): the closer the correlation coefficient is to 1, the less the residuals are scattered.

It is an interesting relationship because **it immediately reveals**, without effectively computing the formula, **what can be expected** once the distribution of the y_i and the r^2 have been computed. Both computations are required anyway when analysing the data. It also shows that, if several causal variables are available and if a formula with just one variable is considered, the choice of the variable to be used depends only on its r^2.

13.3.3 Other Measures Related to the Residuals

Some authors, instead of the variance for indicating the dispersion of the residuals, prefer to use other measures.

The "standard error[2]" is defined as:

$$\text{SE} = \sqrt{\frac{\sum_i e_{+i}^2}{\text{dof}}}$$

where dof stands for "degrees of freedom", equal to the number of data points I less the number of coefficients in the formula, including the intercept. When the standard linear regression is used, as $\bar{e} = 0$, this SE is closely related to the variance of the residuals; more precisely it is exactly how is estimated the standard deviation S for the population. This "standard error" is then an estimate of the standard deviation of what we call the deviations[3]. This will be explained in the next part.

The "standard percent error" is defined as:

$$\text{SPE} = \sqrt{\frac{\sum_i \left(100 \times \frac{e_{+i}}{\hat{y}_i}\right)^2}{\text{dof}}}$$

where the percentage residuals replace the residuals.

13.3.4 Normalization of the Residuals

Sometimes it is useful to compare the distribution of the e_{+i} for two different product families. In order to make this comparison easier, it is convenient to normalize

[2] We keep here the term "error", because it is generally used, but the term "residual" should be preferred.
[3] Deviation is the term used for the population, residual being reserved for the sample.

their e_{+i}. This normalization consists in dividing each residual by its standard error: it therefore requires to compute all the standard errors.

It has been shown ([56], p. 195) that given the Gauss' hypotheses (these hypotheses will be discussed in Chapter 15), the variances–covariances matrix of the residuals is given by the matrix:

$$S^2\left(||^+x||^t \otimes ||^+x||\right)^{-1} = S^2\left(||1|| - ||hat||\right)$$

where

- S^2 is the variance of the deviations for the whole population (this characteristic will be introduced in Chapter 15),
- $||1||$ a diagonal matrix containing the value 1 in all its main diagonal,
- $||hat||$ the "HAT" matrix.

The variances–covariances matrix contains, as usual, in its main diagonal the variances $h_{i,i}$ of the e_{+i}, all other non-diagonal elements quantifying the covariances between the e_{+i} and the e_{+k}.

As S^2 is unknown at the time the data analysis is performed, an estimate of its value is needed. This will be discussed in the next part. For the time being we use:

$$\hat{S}^2 = \frac{1}{I - J - 1}\sum_i e_{+i}^2$$

where I is the number of products and J the number of causal variables.

This normalization process replaces each e_{+i} by e_{+i}^* defined by:

$$e_{+i}^* = \frac{e_{+i}}{\hat{S}\sqrt{1 - h_{i,i}}}$$

Example

As an example, Figure 13.7 gives the normalized residuals computed for the example, according to the cost values: the graph must be compared to Figure 13.1.

The shape of the distribution is about the same, but the range is completely different.

The Use of the Normalized Residuals

When I is large, the normalized residuals should remain between -2 and $+2$.

The purpose of the normalization is not to "improve" the residuals, but only to be able to compare them, if different computations are carried out on the same product family, for instance. The normalized residuals are very convenient for comparing data because they have equal variances: a high value for a data point suggests that this data point is certainly an outlier (which does not mean that all outliers have necessarily a large normalized residual).

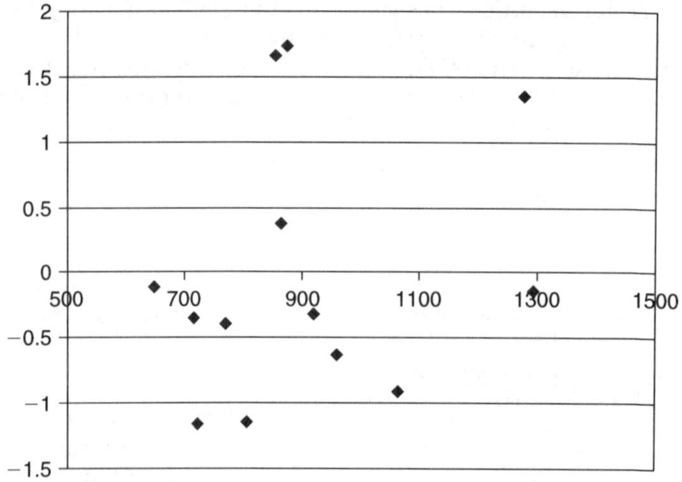

Figure 13.7 The normalized residuals.

13.3.5 The Autocorrelation

The autocorrelation problem has only been studied for the bilinear case. It should nevertheless be studied for all the formulae.

All the values encountered so far (the y_i, e_{+i}, ...) but not the x_i, are defined as realizations of random variables. So are the residuals. Consequently each one has its own distribution and residuals can be correlated. Correlation between residuals is a natural phenomenon: we saw in the section "The variance of this distribution" in Section 13.3.2 that the variances–covariances matrix of the residuals was given by:

$$\text{var}(e_{+i}) = S^2 \otimes \left(||1|| - ||hat|| \right)$$

The non-diagonal terms refer to the covariances between the residuals. The question is: Are these correlations not too high? Correlations between residuals are generally called "autocorrelation".

Autocorrelation was studied by several econometrists: these persons often study the change of data (for instance the consumption or the savings) according to time and correlation between the residuals can be expected. However, it is rarely a serious problem in cost analysis, especially for cost analysts who have to deal with costs or prices coming from various sources; however, the concept must be known to the cost analyst as it may happen in special circumstances. Therefore we will limit the discussion here to some basic points.

Autocorrelation between the residuals means that the residuals observed for different products are not independent: the value of e_{+i} depend on e_{+k}. Auto correlation is said to be the first order if $k = i \pm 1$, whatever may be i.

The general consequences of the autocorrelation are:

- It increases the variance of the coefficients. However, in the presence of autocorrelation, these variances are underestimated by the usual formulae of the standard regression analysis, which means we do not know exactly where we are anymore (do not trust too much these formulae!).

- And this, of course, produces larger imprecision in the cost estimates that can be done using the formula.

Testing for Autocorrelation

As the off-diagonal elements of the variances–covariances matrix cannot be expected to be 0 (as example shows it), the e_{i+} are "naturally" correlated: the correlation between e_{+i} and e_{+k} is given by:

$$\rho = \frac{\text{cov}(e_{+i}, e_{+k})}{\sqrt{\text{var}(e_{+i}) \times \text{var}(e_{+k})}}$$

It is interesting to note that the correlation coefficients depends mainly on the $\|{}^{+}x\|$ matrix, the cost effect appearing only through the term S as a multiplication factor: the correlation depend on the relative position of the data parameters.

Several authors have studied the question, limiting their investigation to the first-order autocorrelation, which they call "serial correlation". The best-known tests are:

- A test based on the sign test: it is clear that if the sign of the residuals is negative for low values of the residuals according to one parameter and positive for large values, a strong correlation does exist. Consequently the number of sign changes is the important point to consider.[4]
- The test elaborated by Durbin and Watson. It is based on hypothesis testing, the H_0 hypothesis being that the correlation is 0. In order to test this hypothesis, the following value is computed:

$$d = \frac{\sum_{i=2}^{I} (e_{+i} - e_{+i-1})^2}{\sum_i e_{+i}^2}$$

Tables for using this value for these tests were published by the authors; they appear in most books of statistics (For instance Draper and Smith [20], p. 164).

Practically the best way for a first approach of checking the possible autocorrelations is the graphs. This is why it is important to be able to display the distribution of the residuals according to all the parameters, as it is indicated upwards.

Assessing the Damages

J. Johnston ([34], p. 247), studying the first-order autocorrelation (writing that $e_{+i+1} = \rho \times e_{+i} + \varepsilon$ where ρ is the correlation coefficient), established that the variance of the coefficients should be multiplied by:

$$I - \frac{1+\rho^2}{1-\rho^2}$$

[4] See Draper and Smith [20], p. 158 for a detailed explanation, plus a table based on this number.

which produces a correction of 4% for $I = 20$ and $\rho = 0.5$, which is a quite high correlation. The consequence is that, in most situations, the cost analysts should not worry too much about the phenomenon.

Correction

Draper and Smith suggest ([34], p. 156) to use the "weight matrix" defined in Chapter 10, the question being of course to establish this matrix: they give some hints for doing so.

As this problem of autocorrelation is marginal in the vast majority of the problems of cost estimating, we will not discuss this point any further.

13.3.6 Analysis of Variance

The standard analysis of variance (ANOVA) has a limited interest because it is valid only in the bilinear case and does not reveal many things about the formula. It should nevertheless be known to the cost analyst.

This analysis starts with a quantity proportional to the variance of the data; this quantity is generally called in the literature as "sum of squares" or SS. The starting point is:

$$SS_0 = \sum_i (y_i - \bar{y})^2$$

If one considers that the average value $\bar{y}$ does not convey any information (if all y_i are equal to $\bar{y}$ we cannot progress in the analysis of our data and one could say that our data does not convey any useful information), then SS_0 can be called[5] the level of "*usable information*" contained in the sample (it has $I - 1$ degrees of freedom) whereas $\sum_i y_i^2$ can be called the level of "available information".

It can be demonstrated that, **only in the bilinear case solved by the conventional regression analysis:**

$$SS_0 = \sum_i (y_i - \hat{y}_i)^2 + \sum_i (\hat{y}_i - \bar{y})^2 = SS_1 + SS_2$$

where SS_2 can be considered as the information "captured" in the formula, whereas SS_1 is the information remaining in the residuals, information which is lost when the sample values are replaced by the formula. The fact that some information is lost advocates the fact of considering that the formula as such is NOT the model and that the true model is the formula plus the distribution of the residuals around it.

[5] We use here the word "information" in a very general meaning. It has nothing to do with the information as defined by Shannon and is used just to make the ANOVA easy to understand.

13.4 Improving the Forecasting Capabilities by Studying the Residuals

Is there anything we can do for improving the forecasting capabilities of our formula by studying the residuals? One can think about using them in the following circumstances: suppose you believe that two causal variables should be used for "explaining" the behavior of the cost inside a product family. However, you have a limited number of data points and consider that using these two variables immediately will not give you a reliable information about the quality of the model. What can be done is to start with one variable only and then to try "explain" the residuals with the other variable. This is very close to what is called in Chapter 6 the step-by-step analysis; it will be easy to use if the variables are not correlated.

Mosteller and Tukey made an interesting and different approach in Ref. [43]. Their idea is first to find out if there is any trend in the residuals, second to use this trend to improve the relationship. It will be introduced on our example.

13.4.1 Preparing the Data

The question is to find out some trend somewhere. As the residuals are rather scattered, Mosteller and Tukey use a smoothing process in order to discover some possible interesting phenomena.

First of all they decide to use the estimates as the "causal variable" (they explain why using the observed values can be misleading).

Second they smooth the residuals: the smoothing process consists in several steps:

- Data are sorted according to the values of the causal variable.
- Each residual value is compared with the preceding and the following values; its value is changed by the median of the three values. For instance, referring to Figure 13.8, the second residual (−18.88) is replaced by the median of the set {2.77, −18.88, 115.33) which is 2.77.
- This process is continued until there is no change (experience shows that no more than two smoothing actions are necessary).

Name	Estimates	Residuals	Smoothing process		New estimates
			1st	2nd	
M	645.23	2.77	2.775	2.775	648.00
L	735.88	−18.88	2.775	2.775	738.65
F	761.67	115.33	91.302	91.302	852.97
J	764.70	91.30	91.302	91.302	856.00
B	790.39	−66.39	−18.393	−18.393	772.00
E	790.39	−18.39	−66.393	−18.393	772.00
C	876.77	−67.77	−18.393	−19.580	857.19
H	884.58	−19.58	−19.580	−19.580	865.00
D	936.16	−16.16	−19.580	−19.580	916.58
I	1017.62	−56.62	−27.947	−27.947	989.67
G	1091.95	−27.95	−27.947	−27.947	1064.00
A	1220.46	57.54	23.628	23.628	1244.09
K	1269.37	23.63	23.628	23.628	1293.00

Figure 13.8 Computing "new estimates".

Third "new estimates" are computed by adding to the estimates the "smoothed residuals".

Let us illustrate on our example, using only two variables: the number of connections and the number of components. Figure in Part IV introduction gives the result of the computations.

13.4.2 Finding a Trend

The easiest way to find a trend between the new estimates and the first estimates is now to display the values on a graph (Figure 13.9).

The graph shows that a slight trend does occur: the shape of the data shows an upwards curvature, except for data points F and J which appear as outliers. It is therefore logic to find out the formula of this trend. Mosteller and Tukey use an exponential formula (this choice appears logic when the graph is considered), of which result is here, after eliminating data points F and J:

$$\text{New estimate} = 327.11697 \times 1.00109^{\text{first estimate}}$$

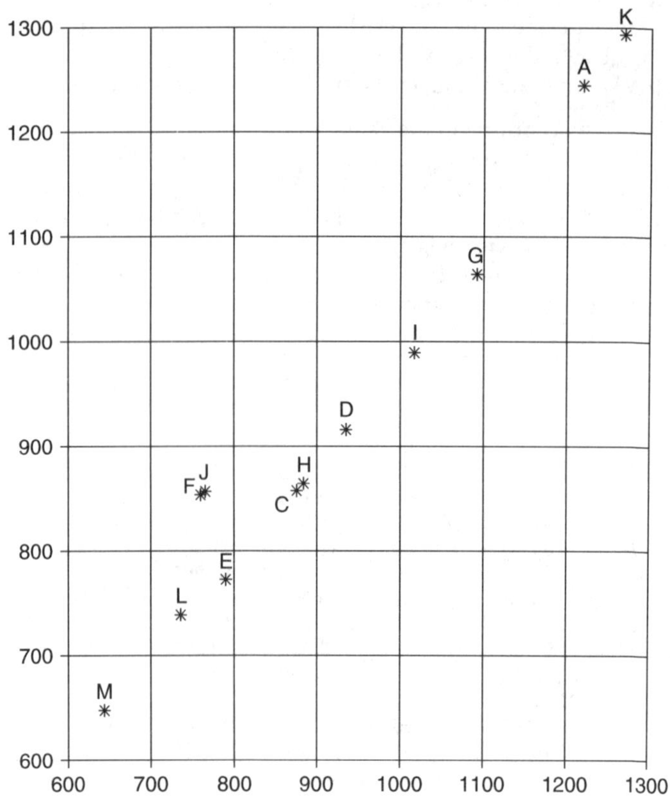

Figure 13.9 Plotting "new estimates" against first estimates.

The graph of this function appears in Figure 13.10 on a log–log scale.

Does the process really help? Figure 13.11 presents the cost figures as they appear in the database, the first estimates as they were computed by the regression analysis and the residuals as they appear after the correction: the average absolute value in the first case is 44.8, which becomes 23.7. It shows that the procedure really helps, as this absolute value is about divided by 2! Note that even the residuals for F and J have been considerably improved, although they were not considered in the process.

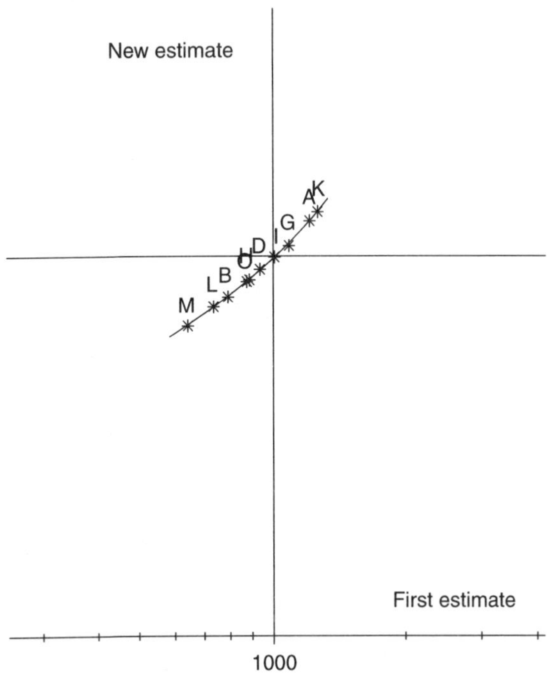

Figure 13.10 New estimates as a function of first estimates.

Name	Cost	First estimates	First residuals	Final estimates	Final residuals
M	648	645.2	2.8	662.7	−14.7
L	717	735.9	−18.9	731.4	−14.4
F	877	761.7	115.3	828.4	48.6
J	856	764.7	91.3	831.2	24.8
B	724	790.4	−66.4	758.5	−34.5
E	772	790.4	−18.4	758.5	13.5
C	809	876.8	−67.8	832.3	−23.3
H	865	884.6	−19.6	839.4	25.6
D	920	936.2	−16.2	887.9	32.1
I	961	1017.6	−56.6	961.5	−0.5
G	1064	1091.9	−27.9	1042.6	21.4
A	1278	1220.5	57.5	1268.6	9.4
K	1293	1269.4	23.6	1338.0	−45.0

Figure 13.11 Does the procedure really help?

The price to pay is, as it was guessed, that the average value of the residuals was 0 in the first case and 3.3 in the second case: this is a very low price indeed.

This example shows that the formula we use for building all models are, generally speaking, a "first-order approximation" of the true formula: it can be easily improved, most of the time. But obviously the formula we eventually get is a little bit more complex than the first one.

Conclusion: If there is a trend in the residuals, it is a good practice to exploit it in order to improve the process. This is the reason why the trend is always displayed on the graphs (see Figures 13.1–13.5).

14 The Other Residuals

Summary

This chapter presents a brief introduction to the use of residuals computed in a different way.

The discussion is however limited to the use of multiplicative residuals, because their use, in conjunction with the multiplicative formula, offers some potential interest to the cost analyst.

The reader is reminded on the fact that, whatever, the type of residuals used, comparison between formulae can only be done if the residuals are expressed the same way.

The previous chapter investigated the residuals defined as additive.

Let us remind the reader that the way residuals are defined has nothing to do with the choice of the metric, even if, most often, cost analysts select the additive residuals with the additive formula, the multiplicative residuals with the multiplicative or the exponential formulae, etc.

This chapter investigates other residuals, but limit the discussion to the multiplicative ones, because they present an interesting properties in the cost domain when they are used with the multiplicative formula.

14.1 Definition

Multiplicative residuals are defined as:

$$e_{\times i} = \frac{y_i}{\hat{y}_i}$$

which means that $y_i = \hat{y}_i \times e_{\times i}$, formula from which these residuals take their name.

These residuals can be used for any type of formula, but they are mainly considered for the multiplicative and the exponential formulae:

$$\hat{y} = b_0 \times x_1^{b_1} \times x_2^{b_2} \times \cdots$$
$$\hat{y} = b_0 \times b_1^{x_1} \times b_2^{x_2} \times \cdots$$

for the obvious reason that the logarithms allow for linearizing both the formulae and their residuals.

14.2 Returning to the Additive Formula

In the previous section, we have looked for an additive – or bilinear – formula and computed the residuals defined as $\ddot{e}_{+}$. But the same formula can be used for computing multiplicative residuals.

Let us use the simple example given in Chapter 9 (Figure 9.1). Trying to minimize the sum $\sum_i e_{+i}^2$ (which is the ordinary least squares or OLS) produces the formula:

$$\hat{y} = 572.97 + 101.08 \times x$$

and the residuals have a mean equal to 0 and standard deviation equal to 193.

One could also try, with an additive formula, to minimize $\prod_i e_{\times i} - 1$ with $e_{\times i} = (y_i)/(\hat{y}_i)$. The result is then given by:

$$\hat{y} = 572 + 100.2 \times x$$

very close to the previous one.

But let us keep the previous formula which minimizes the sum of the squares of the residuals and express now these residuals as multiplicative. The values are displayed in Figure 14.1; their average value is 1.00063, very close to 1, as expected. These values could very well be used instead of the e_{+i}.

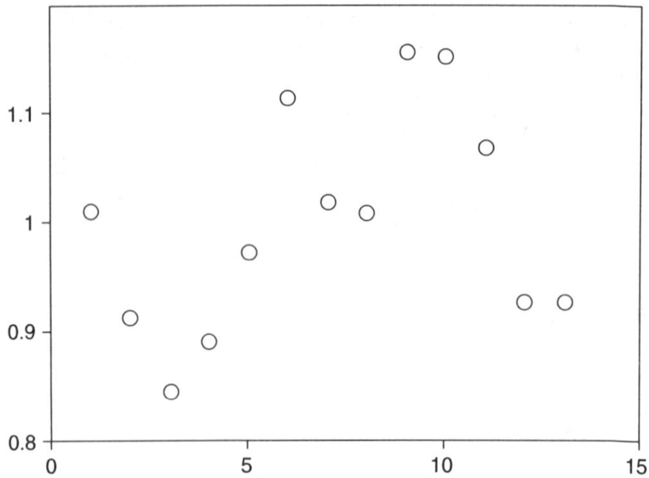

Figure 14.1 The residuals defined as multiplicative.

14.3 The Multiplicative Formula

The multiplicative residuals are however practically only used with the multiplicative formula defined as:

$$y_i = b_0 x_i^{b_1} \times e_{\times i}$$

because taking the log of both parts produces a linear relationship:

$$y_i^* = b_0^* + b_1 x_i^* + e_{\times i}^*$$

where $y^* = \ln y$, $b_0^* = \ln b_0$, etc.

14.3.1 The Distribution of the Multiplicative Residuals

Figure 14.2 presents the multiplicative residuals $e_{\times i}$ (second column) and their logarithms.

Cost	Multiplicative residual	log (base e)
1278	1.084	0.081
724	0.879	−0.129
809	0.912	−0.093
920	0.976	−0.024
772	0.938	−0.064
877	1.101	0.096
1064	1.000	0.000
865	1.053	0.052
961	0.953	−0.048
856	1.105	0.100
1293	1.021	0.021
717	0.981	−0.019
648	1.037	0.036
Median		1.000454
Average		0.000796
Standard deviation		0.072318

Figure 14.2 Multiplicative residuals $e_{\times i}$ and their log (neperian).

It clearly appears that the multiplicative residuals are centered around 1 (here the median is the important characteristic, as explained below), which is logic. It is possible to make on the log $e_{\times i}$ the same analysis that were done on the additive residuals.

It is also interesting to discover what the Gauss' hypotheses mean for the true residuals $e_{\times i}$. These hypotheses are relative to log $e_{\times i}$: all the computations about the multiplicative residuals are made on these values.

We refer here to Gauss' hypotheses which will be mentioned in the next part. These hypotheses were proposed in order to allow to use for the multiplicative formula the results which were demonstrated for the additive (bilinear) formula; one of the most important one is that, for the whole population, the deviations[1] should be distributed according to the normal law. If we want – but this is not compulsory – to use these results here, it means that the $e_{\times i}^*$ should be distributed according to the same law.

[1] We use the word "deviation" instead of "residual" for the population.

The Distribution of the $e_{\times i}$

We know that $\ln e_{\times i}$ follows a normal distribution, centered around 0 (if the Euclidian metric was used for establishing the formula, otherwise the center is called m) and a standard deviation σ. In the example given $\sigma = 0.072318$.

Generally speaking, as we saw it, the average m is – whatever the metric – so close to 0 that it can be neglected for practical computations.

Referring to Chapter 3, one can say that the $e_{\times i}$ themselves follow a log-normal distribution given by the expression:

$$\frac{1}{e_{\times i}\sigma\sqrt{2\pi}}\exp\left(-\frac{(\ln e_{\times i})^2}{2\sigma^2}\right)$$

This could be embarrassing because, as it can be seen in Chapter 3, this distribution is rather complex and should be difficult to handle for practical cost estimates. However – and fortunately – σ is generally very small (do not forget that it is the standard deviation of the log of the multiplicative residuals: even a residual of 1.5 – which means that the cost value is at 50% of the dynamic center! – gives a Neperian log of 0.4) and the distribution appears simple, as illustrated with the values of the example.

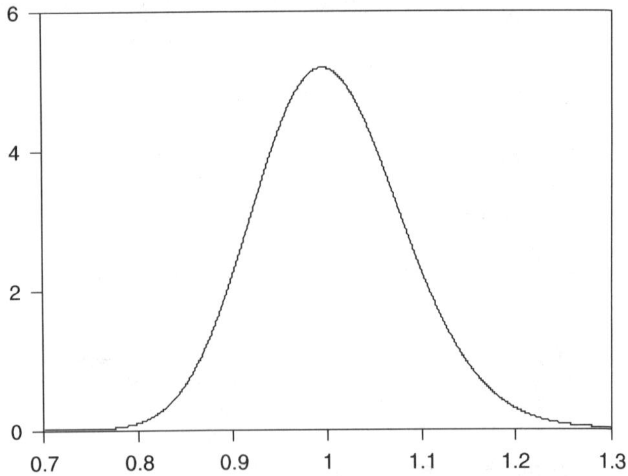

Figure 14.3 The distribution of the $e_{\times i}$ with $\sigma = 0.0723$.

This distribution has the following characteristics:

- It is, as expected, nearly centered around 1. More exactly its mode (the $e_{\times i}$ value corresponding to the maximum of the distribution) is equal to $e^{-\sigma^2}$ (0.995 for our example), its mean to $e^{\sigma^2/2}$ (1.003 for our example) and its median to 1.
- The variance is given by $s^2 = e^{\sigma^2} \times (e^{\sigma^2} - 1)$: 0.00526 for our example, to which correspond a standard deviation of 0.0725.
- The skewness by $(e^{\sigma^2} + 2) \times (e^{\sigma^2} - 1)$: 0.0157 for our example, very close to 0.
- The kurtosis by $e^{4\sigma^2} + 2e^{3\sigma^2} + 3e^{2\sigma^2} - 3$: 3.0854 for our example.

This distribution, for practical purposes, can therefore be very well **approximated by a normal distribution** with mean 1 and standard deviation $s = \sqrt{e^{\sigma^2} \times (e^{\sigma^2} - 1)}$, where σ is the standard deviation of the log of the multiplicative residuals. This will make the computations easier!

As σ is always a small number, an approximate value of s can be found with the power series: as $e^{\sigma^2} \approx 1 + \sigma^2 +$ etc., one can write $s \approx \sigma$. For our example this gives a value of 0.723 instead of 0.725, very close indeed. Therefore, for practical computations, we can very well use this approximation.

14.3.2 An Interesting and Important Comment

In the example, one can see that the $e_{\times i}$ are small and centered around 1. It is then possible to develop them in a Taylor polynomial:

$$e_{\times} = 1 + \varepsilon$$

where ε is a random variable centered around 0. Therefore one can write:

$$y = \hat{y} \times e_{\times} = \hat{y} \times (1 + \varepsilon) = \hat{y} + \hat{y} \times \varepsilon$$

which means that *the multiplicative formula automatically produces an additive deviation which is proportional to the cost.* This is a satisfactory result for cost studies: the deviations in these studies are known in percentages, which means that their value is proportional to the cost; the subject is mentioned in Chapter 9.

So the multiplicative formula automatically solves this problem.

14.3.3 Looking at the Additive Residuals

Although the formula is defined as multiplicative, it quite possible – and it is recommended at least to be able to compare both solutions – to recompute, once the formula is found, the residuals as additive. The purpose of this computation is to compare them with the residuals computed from an additive formula.

For the example, the results are given in Figure 14.4.

The second column is a copy of the table given in Section 1 of Chapter 13, reproduced here for convenience; for the additive formula the deviations average is 0, the average absolute relative residuals being 4.81%. When the formula becomes multiplicative, the deviations average – computed here in an additive way – is 2.56, the average absolute relative residuals being 5.89%. From this point of view, the multiplicative formula is, for this example, slightly inferior to the additive formula.

These computations were made just to illustrate that the only way one can compare formulae is to compute the residuals the same way.

Additive formula		Multiplicative formula	
Cost	Additive residual	Cost	Additive residual
1278	63.64	1278	98.12
724	−72.52	724	−99.81
809	−67.42	809	−79.32
920	−17.51	920	−23.27
772	−24.52	772	−51.81
877	88.49	877	80.08
1064	−35.76	1064	−0.37
865	19.10	865	42.95
961	−24.02	961	−48.03
856	100.94	856	81.10
1293	−3.11	1293	25.42
717	−20.82	717	−14.38
648	−6.48	648	22.68
Average	0		2.56
Average of absolute deviation	4.81		5.89

Figure 14.4 Comparing additive residuals on both the formulae.

14.4 Are Multiplicative Residuals Interesting?

Donald MacKenzie in a recent paper[2] made a lot of regressions using a multiplicative formula $\hat{y} = b_0 x^{b_1}$ – where x quantifies the mass – and found that the residuals "appear to be log-normally distributed and proportional to cost magnitude for a wide range of space hardware box types". This is a good reason to use this type of multiplicative formula when the mass is used as the cost driver.

[2] Donald MacKenzie. Cost-estimating relationship regression variance study.

Part V

Building a Specific Model

The first four parts of this book was developed in order to prepare this one.

In Parts I and II, data were collected (this is the sample) and then analyzed.

Then a convenient way was investigated in order to replace – in the sample – all the data by something much more convenient to be used. The information available in the sample was split into two parts of equal importance:

1. The (dynamic) center of the data on one hand (Part III).
2. The spread of the information around this center on the other hand (Part IV).

The sample is now well understood and the information conveniently presented to the cost analyst.

The important part is now the capacity to answer the question: how can these information be used for estimating the cost of a new product not belonging to the sample? This new product will be considered as being part of the whole population from which the sample is drawn.

This Part V deals with this question.

Part Contents

This chapter is an important one, for both theoretical and practical points of view. It explains how the results found in the sample can be extrapolated to the whole population the cost analyst is interested in.

This will allow him/her afterwards to estimate the cost of any new product belonging to the population, which implies computing a confidence level for his/her estimate.

This chapter summarizes all the results found up to now: it can be considered of a summary of what must be known in order to build a specific model.

It insists on the decisions the cost analyst has to make, in supplement to the computations.

15 From Sample to Population

Summary

Up to now we studied only the sample and the results we obtained are only valid for this sample.

It is time to see now what can be said about the population as a whole, which means for any object of the population. After all, our purpose is to be able to estimate any object of the population ...

The chapter starts by developing the principles on which are based the extrapolations of what was found on the sample to the population. It first reminds the cost analyst that the shape of the relationship between the dependent variable and the parameters or cost drivers is a choice he/she has to make. From the sample we want then to compute estimators of the coefficients which appear in the formula giving the dynamic center of the cost distribution for the whole population. The qualities expected for these estimators are mentioned.

The important question is: how far can be the estimators derived from the sample to the true value of the coefficients? Two solutions are generally proposed, based on hypothesis testing on one hand, confidence interval on the other hand, both being the different faces of the same coin.

Then the way the perceived relationships – in the sample – can be extrapolated to the population is investigated: two solutions are possible:

1. The classical solution is illustrated with the correlation coefficient: if a correlation has been found in the sample between two variables, what can be said about these variables for the population?
2. The modern approach, based on the Bootstrap, or its "little brother" the Jackknife.

In order to introduce the solutions on a simple case, the principles are first applied to the simple problem of a cost distribution with no cost driver: this is a basic idea, for instance, for the opinion surveys. One shows how, classically, the center of the cost distribution – always for the population – and its spread can be estimated. The important "t" variable is explained. Then the modern approach is illustrated and the results of both approaches compared.

The case of a distribution using one cost driver is then presented, followed by the case involving several quantitative parameters and qualitative parameters.

15.1 The Principles

This chapter is dedicated to the population. Let us remind our definition of the population: the population is the set – potentially infinite – of all objects that can be part of the product family we previously defined. This product family is – in principle – supposed to be homogeneous, the size being the only characteristic which distinguishes the products inside the population. However we can accept a few inhomogeneities between the products, these few inhomogeneities being taken care of by a small set of variables, quantitative or qualitative. The names of these variables, including the size, are called $V_1, V_2, \ldots, V_j, \ldots, V_J$.

A product to be estimated, drawn out from the population, is defined by its values of the variables. These values are called[1] $X_1, X_2, \ldots, X_j, \ldots, X_J$. From these values we want to be able to estimate its cost.

The question is: How can we do that?

15.1.1 About the Population

Everything starts from a belief!

We believe that there is a set of variables, of which number is $J + K$ from which a fully determined – which means that the cost is totally determined, with no "deviation" – relationship between the values of the variables and the cost, called Y:

$$Y = F(X_1, X_2, \ldots, X_j, \ldots, X_J, X_{J+1}, \ldots, X_{J+K};\ B_0, B_1, \ldots)$$

where $B_0, B_1, \ldots$ are constants. These coefficients are also called the "characteristics" of the population, because, *with the function F*, they characterized this population.

Note that the set of the variables may be larger than the set of the variables known in the sample.

Such a belief is based on the fact we think that costs do *not* come by chance; if the costs are deterministic there should be such a relationship, which may include the type of machines, the names of the operators, the management of the company, etc.

What Is the Shape of the Relationship F()?

It is very simple to answer this question: we do not know! As well as we do not know the list of the variables which should be included in it.

Theoretically it should be possible, if we knew the whole population, to find out a polynomial that would fit exactly with all the data (this is generally called "curve fitting") and there are powerful algorithms (see Ref. [10]) to do that. But we are not interested in this solution which could be far too complex for practical purposes.

The first decision is therefore to limit, in the study, the number of variables to a reasonable small set. It is not yet the set of the variables which will be really included in the formula: this set will be a subset of the theoretical set of variables

[1] We use capital letters for any object extracted from the population, small letters being reserved to objects which are part of the sample.

(see Chapter 16). To make the process very clear, there are three sets of parameters:

1. The theoretical set, from which everything could be described with no deviation.
2. The set we decide to use in the studies.
3. The set which will be eventually included in the formula (this will be the object of another decision).

For the simplicity of notation we will keep the same variables names X_1, X_2, ..., X_j, ..., X_J for all sets, although it is clear that the sets of variables we consider can be a very small set of the "theoretical" variables. Let us consider here the second set only.

Now we know that, due to this decision, some "fluctuations", due to the unknown variables which should have been included, will occur and confuse a little bit (we hope!) the issue.

The relationship $F()$ can therefore be split in two terms:

1. A function $\hat{Y}(X_1, X_2, \ldots, X_j, \ldots, X_J)$ which will be called the "*dynamic center*" of the cost distribution for the whole population.
2. Deviations, called E, from this function. These deviations can be expressed as additive or multiplicative – or as any other form – to $\hat{Y}$; they will be represented for convenience as additive, but the reader can convert it to other forms. The *distribution of these deviations* – which are NOT "residuals", because they result from a deliberate choice – will be called Ψ.

The conclusion of this introduction is that the model we are looking for is a set of two things: the function $\hat{Y}$ and the distribution Ψ.

What can be said at this stage about the function $\hat{Y}$? It is the subject of **a second decision.**

The relationship $\hat{Y}(X_1, X_2, \ldots, X_j, \ldots, X_J)$ is selected on an *a priori* basis.
The benefit is getting something easy to handle.
The price to pay is some loss of information.

Choosing *a priori*[2] the type of relationship does not mean that several relationships will not be tried until we are happy with the results. At that time, we want to insist on the following points:

- Choosing the wrong relationship – and the linear relationship so often used is quite often a wrong relationship – is the first cause of poor cost estimates.
- It is always highly recommended to check if the price we pay – for loosing information, due to the selection of a small set of parameters – is not too large. This is the second cause of poor relationships.

These are not the only causes: we will discover other ones in due course.

The relationship $\hat{Y}(X_1, X_2, \ldots, X_j, \ldots, X_J)$ includes some constants, or coefficients, called $B_0, B_1, B_2, \ldots$ which are of course unknown and will remain forever unknown. What can be done about them?

15.1.2 What Are We Going to Do with Our Sample? Looking for "Estimators"

The data contained in our sample (our observations) will be used to get **approximate values** of the coefficients $B_0, B_1, B_2, \ldots$

[2] Obviously this choice will be based on the study of the sample. Nevertheless it is eventually a choice.

"Approximate" is a very important adjective: the exact values of these coefficients will never be known. In order to know them, we should know the information for any possible object of the population,[3] which is, by definition, impossible, as the population is infinite (and if it finite but large it will be too costly).

In order to explicitly recall the reader that the values we obtain from the sample are just approximations of the real values, we will use the following symbol[4] $\hat{B}_0, \hat{B}_1, \hat{B}_2, \ldots$, the little "hat" reminding what they really are (estimates of true values).

Of course, as you may expect, we need an answer to the question: **how close are these $\hat{B}_0, \hat{B}_1, \hat{B}_2, \ldots$ from the true values** $B_0, B_1, B_2, \ldots$? This is one of the most important question when using our data for estimating the cost of a new product: if these values are far away from the true values, we can expect that our estimates will be completely wrong. This question will therefore come out quite often in the following section.

The beautiful thing with statistics is that this distance can often be estimated! How is that possible? As you will discover, the whole logic to estimate this distance is based on the following principle (the adverb "randomly" being the key word):

> Our sample was randomly selected among the available data. As this sample was randomly selected, another one could have been selected as well.

There are nowadays two solutions to estimate this distance:

1. The "classical" approach, which was discovered by Carl Friedrich Gauss and is purely analytical, but needs **very stringent hypotheses.**
2. The "modern" approach, which is more recent (about 1980) and palliate the drawbacks of the first one.

Randomly selecting a sample is, for the "classical" approach, not sufficient: the *population* from which it is selected must also have some properties (these are the hypotheses just mentioned). These properties will have to be carefully examined in order to check if they apply to the population we are interested in. Consequently classical statistics offer sometimes a limited help in our domain (cost estimating). What can be said at this stage is the following:

- Elementary statistics can compute this distance if some hypotheses – and sometimes very severe ones – are met. And these hypotheses are rarely verified in our domain.
- More advanced statistical concepts (such as "non parametric" statistics[5]) are often much more suited to our domain.
- In some cases we will have to find out another way to compute these distances which does not require these hypotheses. Recent developments in statistics allow to do that; they will be explained later on in Section 15.3.2 of this chapter under the name of "modern approach".

This section is primarily devoted to these "elementary statistics": understanding them is a prerequisite for understanding most of the manuals on the subject and more advanced methods.

The numerical values of the constants, $b_0, b_1, b_2, \ldots$ computed from the information we get in the sample, are logically called "**estimators**" of the true values $B_0, B_1, B_2, \ldots$

[3] Such a knowledge is called a "census".

[4] Which is very common in books of statistics.

[5] The word "non parametric" clearly refers here to statistics, not to "parametric cost estimating" as it was defined in Chapter 1 of Volume 1.

15.1.3 What Are the Qualities Expected for an Estimator?

In order to be really useful, an estimator should present some qualities that are briefly commented here; all these qualities are not compulsory to practically used estimators, but they must be known. Let B be a coefficient – or a characteristic – of the population and $\hat{B}$ an estimator of B, estimator "extracted" from the sample.

These qualities are given below.

Lack of Bias

In order to define what is meant by this term, let us suppose we can select K *different samples* of the same size I (the number of products included in the sample). From these samples we compute – using here *the same algorithm* – different values of our estimator; let us call them $\hat{B}^{(1)}, \hat{B}^{(2)}, \ldots, \hat{B}^{(k)}, \ldots, \hat{B}^{(K)}$. We will say our estimator is unbiased – it should be said to be correct: "the algorithm we use for computing the estimator is unbiased" – if the average value of this set $\hat{B}^{(1)}, \hat{B}^{(2)}, \ldots, \hat{B}^{(k)}, \ldots, \hat{B}^{(K)}$ is mathematically equal to the true value of the coefficient:

$$\bar{\hat{B}} = B$$

(this notation is standard: a flat hat represents the average value). Of course, as we only have one sample, we cannot check that and it has to be demonstrated mathematically. For instance it can be shown that, in order to estimate the average value of a population Z (its arithmetic mean $\bar{Z}$), the sample mean $\bar{z}$ is an unbiased estimator; but the sample variance s, which can be used for estimating the variance S of the population, is a biased value. We will return to that in the next section.

Efficiency

The idea here is: let us assume we have several ways (several different algorithms) to estimate B. These N *different algorithms* will give us several values for $\hat{B}$ from the same sample. Let us call them ${}_{(1)}\hat{B}, {}_{(2)}\hat{B}, \ldots, {}_{(n)}\hat{B}, \ldots, {}_{(N)}\hat{B}$. Suppose we can do all these computations on K *different samples*; the first algorithm will give a set of K values, labeled for instance ${}_{(1)}\hat{B}^{(1)}$ for sample 1, ${}_{(1)}\hat{B}^{(2)}$ for sample 2, etc. ...; the second algorithm will also generate another set of K values, labeled ${}_{(2)}\hat{B}^{(1)}$ for sample 1, ${}_{(2)}\hat{B}^{(2)}$ for sample 2, etc. ... It is possible to compute the variance[6] of the values provided by each algorithm on the different samples; for algorithm n we have

$$\text{var}({}_{(n)}\hat{B}) = \frac{1}{K}\sum_{k=1}^{K}({}_{(n)}\hat{B}^{(k)} - {}_{(n)}\bar{\hat{B}})^2$$

The algorithm which gives the estimator with the least variance will be said the most efficient. It can be shown that, for instance, the sample mean $\bar{z}$ is the most efficient estimator of the population mean $\bar{Z}$ (this does not mean that this mean is the best value estimator of the "center" of the population).

Consistency

Consistency is an "asymptotic" property. An estimator is said to be consistent if, when the sample size I grows indefinitely, the value of the estimator trends towards

[6] The term is defined in the next chapter.

the value of the true characteristic:

$$\lim_{I \to \infty} \hat{B} = B$$

Sufficiency

An estimator is said to be sufficient if no other algorithm may provide more information about the characteristic we are interested in. In other words the estimator contains all the information the sample may provide – about the characteristic it is an estimator of. This quality is rather hard to demonstrate; but it has been proven that – if the population follows a "normal" distribution with a known variance S – that the sample mean $\bar{z}$ is a sufficient estimator of the population mean $\bar{Z}$.

Robustness

Robustness has two definitions which depend on what we are interested in. This deserves some explanation. An estimator is based on an algorithm using the observed values of the sample; it is possible to define robustness when looking at the algorithm (its logic), or when looking at the values (of course the quality always refers to the algorithm, but this distinction helps explain what we are looking for):

- If we look at the logic of the algorithm: an algorithm, in order to give a reliable value for the estimator, is based on some assumptions (for instance, quite often, it assumes that the population from which the sample is drawn is "normal"). It is said to be robust if its qualities are rather insensitive from any departure from these assumptions; as it can be expected, the stronger the assumptions it is based on, the less robust it will be.
- if we look at the data, an estimator is said to be robust if it is insensitive to a small change of one observed value. Of course we want the estimator to be sensitive to "normal" changes of the values, but not too much. It was the purpose of the search of "outliers" to find out the samples values which may change to a large extent the value of the estimator. The median was shown to be a very robust characteristic, whereas the arithmetic mean was discovered as not robust at all.

In conclusion an estimator is said to be robust if it is insensitive either to small departures on the assumptions it is based on, or to the data it uses.

All these qualities are interesting[7] from a theoretical point of view. From a practical point of view, two qualities are especially important. The lack of bias is important but not fundamental. Robustness, from the cost estimator point of view, is certainly the most important quality. This is the reason why it has been so much investigated in Part III.

What Is the Logic for Establishing the Qualities of an Estimator?

Estimators of the coefficients we are looking for are computed by some mathematical algorithm which delivers a value. How is it possible to discuss the quality of such a value?

---

[7] Rao ([46], p. 314) devotes 20 pages to the "minimum variance unbiased estimation". The interested reader will find there many interesting theorems.

The logic refers to the principle which has just been mentioned: the data from which the values are computed constitute our sample; this sample is assumed to be drawn randomly from the population. **Theoretically** we could therefore have "drawn" another sample: in other words, if it were possible, we could have observed the cost of different products – from the same population or product family – or even the cost of the same products made under different circumstances. These costs may have given values different from the ones we have: they would constitute another sample. We – always theoretically – could have repeated the process several times and so get different samples.

The qualities of the descriptors should normally be based on the different results observed from these different samples: if all these samples would produce about the same values for the descriptors, we would estimate their quality very high; otherwise we will consider them carefully before using them!

As it is impossible, we have to find out another way to quantify the quality of our estimators. We already mentioned there are several of them. Whatever the method, it is always be based on the following hypothesis:

Sample values are randomly extracted from the population we study.

15.2 How to Get Values for Our Estimators from the Sample?

Two methods may be proposed: a theoretical one and a practical one.

15.2.1 The Method of Maximum Likelihood

This is the theoretical method, proposed by Fisher. It is based on the following hypothesis: the population from which the sample data were drawn is parametric; this means that the shape of its distribution is known (for instance it can be normal or χ^2 or anything else) but its characteristics (the constants which characterize it) are not. Let us call ξ this distribution.

Let us suppose for the simplicity of the text that just one characteristic – let us call it π is unknown. In order to have an estimator $\hat{\pi}$ of it, we draw a sample of size I: $x_1, x_2, \ldots, x_i, \ldots, x_I$.

The idea is the following one: the probability of drawing the particular sample from a population of which distribution is $\xi(x; \pi)$ is equal to the product of the probability of drawing each x_i. As this probability is equal to $\xi(x; \pi)$, the probability of drawing this sample is given by:

$$L(\pi) = \prod_i \xi(x_i; \pi)$$

which is a function of π (as all the x_i are known). As we actually drew this sample, we look for a value of π which maximizes $L(\pi)$; this value will be the estimator $\hat{\pi}$ we are looking for.

As an example, suppose we study a population of a variable X and we know that the distribution of these X follows a normal law. It can be for instance the time required to go by car from point A to point B: we know, from experience, that this time follows a normal law. Suppose we also know the standard deviation S of this distribution: also from experience, we know that, in this area, the driving time has a standard deviation $S = 10\,\text{min}$. What we are looking for is the average time $\bar{X}$ for this travel. In order to get this value, we do 10 times the travel and observe 10 different values $x_1, x_2, \ldots, x_{10}$.

We are therefore looking for the value of $\bar{X}$ which maximizes the function:

$$L(\bar{X}) = \prod_i \frac{e^{-\frac{(x_i - \bar{X})}{2S^2}}}{S\sqrt{2\pi}}$$

The computation is not difficult: this function is maximum for $\Sigma(x_i - \bar{X}) = 0$. This means that the estimator of $\bar{X}$ is given by:

$$\hat{\bar{X}} = \frac{\sum x_i}{I}$$

which is the arithmetic mean of the sample! This computation explains why the sample mean is so often used for estimating the center of a distribution. But remember the hypotheses: the distribution follows a normal law and the standard deviation is known.

15.2.2 The Practical Method: The Plug-in Principle

Practically we are looking for two things:

1. The formula giving the value of the dynamic center of the distribution of the costs in the population we are studying.
2. The distribution Ψ of the deviations around this dynamic center.

The study of the sample revealed that the dynamic center was a computed function of the variables $\hat{y}(x_1, x_2, \ldots, x_j, \ldots, x_J)$ with computed coefficients $b_0, b_1, \ldots$, and that the distribution ψ of the residuals around this dynamic center was a computed distribution.

The plug-in principle is the following one: we **decide** that:

1. The function giving the dynamic center of the population has *the same form* as it had in the sample:

$$\hat{Y}(X_1, X_2, \ldots, X_j, \ldots, X_J) = \hat{y}(x_1, x_2, \ldots, x_j, \ldots, x_J)$$

   This means that we decide that:

$$\hat{B}_0 = b_0$$
$$\hat{B}_1 = b_1$$
etc....

2. The distribution Ψ of the deviations inside the population has the *same shape* as distribution ψ of the residuals in the sample, with, however, different characteristics.

This is the basic of the results we will use for estimating the distribution of the costs in the population, and therefore to estimate the cost of any product belonging to this population.

As the distribution $\hat{y}$ is already computed from the sample, the only problem we still have to solve is to establish the characteristics of the distribution Ψ.

There are two ways to solve the problem:

1. Estimating the characteristics of the distribution Ψ around the static or the dynamic center of the distribution of the costs in the population.
2. Estimating the reliability of the estimators $\hat{B}_0, \hat{B}_1, \ldots$ from which the position of the center can be computed.

These two ways are the two faces of the same coin: they express the same thing. Generally the second one is preferred because it is easier to use, but the first one probably reveals more about the distribution of the cost.

15.3 Extrapolating One Characteristic from the Sample to the Population

This section deals with an important question: we found one characteristic value for the distribution of the cost values inside the sample (for instance its arithmetic mean, or the correlation between the size and the cost, etc.). What does this mean for the population as a whole? Can we extrapolate this value to the population? How far could be the value observed in the sample from the value of the same characteristic for the population?

How Do We Compute the Value of the Characteristic in the Sample?

The characteristic we are interested in for the population is theoretically computed by an algorithm: for instance the arithmetic mean for the whole population is defined, for an infinite population, as:

$$\bar{Y} = \int Y g(Y) \mathrm{d}Y$$

where $g(Y)$ is the continuous distribution of the values Y.

The way it is computed in the sample uses the same algorithm; this is called the "plug-out principle" and is the counterpart of the "plug-in principle".

15.3.1 What Are We Looking for?

We found a value for characteristic A in the sample of size I, which may be – and is often in the cost domain – small. This sample comes from a random drawing (this is our basic hypothesis) inside the population. Is this value representative for the whole population?

Maybe we have been lucky in our random drawing, or we have been unlucky, or the sample is a fair representation of the population: we do not know. In the current life, your experience is a great help to qualify the representativity of a sample. For instance if you observe the speed of 10 cars on a freeway on which the speed is limited to 130 km/h and found a set of values such as 50, 65, 58, ... with an average of

62 km/h, can you conclude that all the people drive at this average speed on this freeway? Your experience contradicts this conclusion.

It is exactly the same in the cost domain, except we have generally no experience about the family we are studying. The only thing we can do then is to turn to computation.

In the language of statistics, we want to know if the value we found in the sample is "statistically significant", which means it can reasonably be applied to the population; the word "reasonably" has been mentioned on purpose: the statistician can never say "the result can be applied to the population". More exactly the property can be applied to the population with a given "level of confidence", but we cannot be completely sure about the value.

Let us take an example – on which we will return: we found, in the sample, a correlation of 0.7 between one causal variable and the dependent variable. The statistician can maybe say: there is – with a probability to be wrong of 0.05 (this is the level of confidence) – really a correlation between these variables in the population – the correlation is statistically significant – but we cannot be sure it takes the value 0.7.

Once we have found the value of an estimator, the next question we ask therefore is: How reliable is this value $\hat{A}$? This will of course become a crucial question when we will use it for estimating the cost of a new product.

Two techniques, closely related, have been developed for answering this question; they are called "hypothesis testing" on one hand, and "confidence interval" on the other hand.

1. *Hypothesis testing* starts from a assumption: we think (this is our hypothesis) that the true value of the characteristic A has a preconceived value – let us call it A_0. We just want to check if our idea is right or wrong. In order to do that we draw a sample from which we compute a value a; according to the plug-in principle we write $\hat{A} = a$. We have now to answer the question: does this value confirm or infirm our hypothesis?
2. *Confidence interval* does not require a preliminary assumption. It aims to answer the question: "How far can be the true value A from our estimator $\hat{A}$?"

Both tests are based on the hypothesis: the sample was randomly drawn from the population.

Both tests are the two faces of the same coin: what you see depends on the face you look at.

15.3.2 Hypothesis Testing

What we are interested in is the value of A, supposed to exist, for the whole population. In order to achieve this goal, the first strategy is the following one:

1. We make an hypothesis about the characteristic of the population we are interested in: let us say we assume that A has the value A_0. This hypothesis is generally called $\boldsymbol{H_0}$, the alternative hypothesis being called $\boldsymbol{H_1}$.
2. As we are not sure about this value, we draw a sample of size I and, from this sample, using for instance the plug-in principle, we obtain an estimator of which value is $\hat{A}_{(I)}$ (the index I reminding us about the size of the sample it is computed from).

The question is now: **does this value $\hat{A}$ validates our hypothesis or not?**

Suppose this value $\hat{A}_{(I)}$ is close to A_0; we may reasonably think that it is a good hint that the hypothesis is validated. But we know that our sample was randomly selected: so this fortunate result might very well be a happy result and the true value of A might be very different from A_0, this result being just an artefact!

Suppose now it is quite different from A_0. Can we make the conclusion that hypothesis $\boldsymbol{H}_0$ is not true? Maybe, but we can also make the same statement: the true value of A may be close to A_0 and this unfortunate result is only due to bad luck.

This example shows that, from the values of the sample, we can make two different mistakes:

1. Rejecting $\boldsymbol{H}_0$ when it is true, just because the result was bad (due to bad luck). This is generally called Type I error.
2. Accepting $\boldsymbol{H}_0$ when it is wrong, just because the result was good (due to luck). This is generally called Type II error.

Can we be more specific and quantify the risk of making an error?

The Classical Approach

The classical approach requires to make **hypotheses** about the population.

The logic of the quantification will be based on the following reasoning: as the sample is randomly selected and due to the fact we know something about the population, we must be able to compute the probability of, if hypothesis $\boldsymbol{H}_0$ is true, obtaining the result $\hat{A}_{(I)}$ from a population in which $A = A_0$.

This quantification will therefore be based on two assumptions:

1. the sample is really randomly selected,
2. we know something about the population.

So what we need to know about the population is enough information to be able to *compute* this probability. Going from this information to this probability may be extremely difficult – mathematically speaking – and this is, in most cases, impossible. It is however possible in a few cases, the most frequent one being the assumption that the population follows a normal distribution. This is a very strong assumption rarely met in practice in the domain of cost . . . Nevertheless it is still often being assumed – generally implicitly – in order to be able to compute something.

As testing hypothesis allows to introduce important concepts, and as it is mentioned in several manuals, we will go on with the subject, even if its application for cost purposes is limited. The theory is explained in this section; examples will illustrate it in the following ones.

It is important at this stage to understand that the computation will never be able to give a final and definite answer to the question: "Is $\boldsymbol{H}_0$ true?" (or false). The computation can only give a probabilistic answer: the probability to make an error – in accepting or rejecting $\boldsymbol{H}_0$ – is so much. This probability is called the "level of confidence" and is symbolized by the Greek letter α.

α is generally given *a priori*: when we say "I want, when I base my decision on the result of this sample, to make a mistake less than 10% of the time", it means that we want $\alpha > 0.9$. It is quite possible to compute, when we found a value $\hat{A}_{(I)}$, the exact probability of making such an error; authors call this probability value the P-value. Due the limited use of hypothesis testing in the domain of cost, it is probably not worth the effort.

Reasoning on H_0

So, we have drawn a sample of size I from our population of which distribution is known and

- made the hypothesis called H_0,
- chosen a confidence level, let us say 0.9.

From the assumptions already mentioned, we are able to draw a curve giving the probability of obtaining – if H_0 is true – the value $\hat{A} = a$ from our sample. This curve is reproduced on Figure 15.1. Obviously the mode of this distribution (corresponding the greatest probability) is equal to $\hat{A}$. This curve is *the theoretical distribution* of a, distribution we should observe if we were able to draw a lot of samples of size I from the population (assuming of course that $A = A_0$).

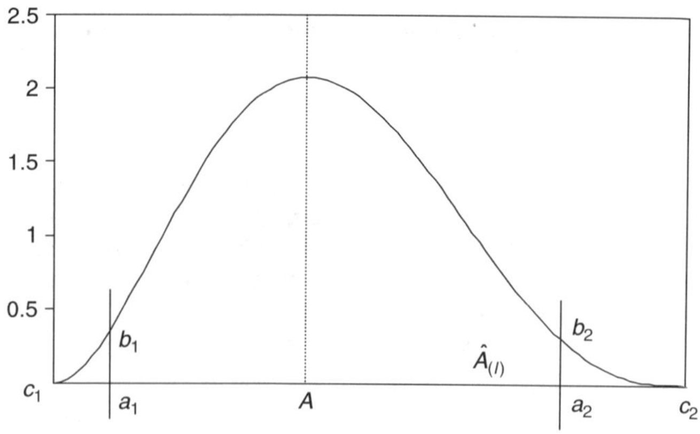

Figure 15.1 Probability to compute from the sample a value $\hat{A}_{(I)}$ for the estimator, given hypothesis $\mathbf{H_0}$ $(A = A_0)$: theoretical distribution of a.

Two small lines are drawn on the figure, corresponding to a_1 and a_2. These values are computed in such a way as the area under the probability curve on the left of a_1 is equal to 0.05 and the probability on the right of a_2 is also equal to 0.05. As the total area under the probability curve equals 1, it means that the area between a_1 and a_2 equals 0.9.

What are our conclusion if we observe a value a in the sample?

1. If $a_1 < a < a_2$: the sample does not invalidate the hypothesis H_0 and we keep the value A_0. The probability to make a wrong decision (a Type II error) is 0.1.
2. If $a < a_1$ or $a > a_2$: now we consider that the sample invalidate hypothesis H_0 which therefore we reject. The probability to make a wrong decision (a Type I error) is also 0.1.

The Modern Approach

The modern approach uses the Jackknife[8] (if the sample size is large enough) or, more frequently, the Bootstrap.

[8] The Jackknife – which does not allow repetitions and is therefore limited to the "creation" of N different samples – is the "small brother" of the Bootstrap; this can be mathematically demonstrated.

The Bootstrap is able to directly draw the distribution of $\hat{A}_{(I)}$ without making any hypothesis on the distribution of the cost in the population. The idea is to "extract" all the information available in the sample. In the classical approach the sample is only used to compute the value of the characteristic: a. This is a very poor use of the information! In order to extract all the information the Bootstrap "says": this sample was randomly drawn from the population; this means that another sample could have been drawn as well. It then simulates other samples[9] of the same size I by randomly selecting values – with repetition – from the sample.

A large number of samples – all of them could have be drawn – can so be generated (400 is a usual number); each one is called a **replicate** of the sample. For each sample characteristic a is computed. From these 400 values of a, its distribution curve can be computed – the BETA distribution is frequently used, due to its high flexibility, for this purpose – and the interval $\lfloor a_1, a_2 \rfloor$ also computed. After that, the reasoning is the same.

15.3.3 Confidence Interval *g*

Looking for the confidence interval is another way of solving the problem.

The idea is to say: we observe in the sample a value a for the characteristic we are interested in and ask the question: How far could be the true value A for the same characteristic in the population? This distance is defined by a "confidence interval".

As usual in statistics, the confidence interval cannot be determined in a definite manner: the only thing we can look for the interval in which A has a given probability – for instance 0.9 (called the level of confidence) – to be. Obviously the better the desired level of confidence, the larger the interval: at one extreme, the probability of finding A in the interval $a \pm \infty$ is equal to 1.

In this section, we do not make any hypothesis about the true value of the characteristic. We know an estimate $\hat{A}_{(I)} = a$, computed from our sample of size I. The question is now: keeping $\hat{A}_{(I)}$ constant, where can be A?

It is impossible to answer directly this question: it has to be rephrased: is it possible that A be far away from $\hat{A}_{(I)}$? After all $\hat{A}_{(I)}$ was computed on a sample from the population and we may guess that A should not be too different from $\hat{A}_{(I)}$.

What is the meaning of "not too different?"

To quantify this assertion, we must compute, for different values of A, the probability of finding $\hat{A}_{(I)}$, the logic being that if A is very different from $\hat{A}_{(I)}$, then the probability of finding $\hat{A}_{(I)}$ is very small. The probability distribution of $\hat{A}_{(I)}$ can be drawn and this curve will tell us, how far can be A from $\hat{A}_{(I)}$.

This curve is the counterpart of the one we studied in the previous paragraph:

- In the previous section, we kept A constant, equal to A_0, and looked for the probability in finding different values for a.
- In this section, we keep a constant and compute the probability distribution of the values of A which could produce such a value a in the sample.

---

[9] The mathematical theory of the Bootstrap has been made.

15.3.4 Introducing the Standard Error of an Estimate

Suppose we were able to draw several samples from the population. They will generate a set of K estimators called $\hat{A}_{(I)}^{(1)}, \hat{A}_{(I)}^{(2)}, \ldots, \hat{A}_{(I)}^{(k)}, \ldots, \hat{A}_{(I)}^{(K)}$. It is possible to compute the mean and the standard deviation of this set:

$$\hat{A}_{(I)} \quad \text{and}^{10} \quad \sum_k \left(\frac{\sum (\hat{A}_{(I)}^{(k)} - \bar{\hat{A}}_{(I)})^2}{K} \right)^{\frac{1}{2}}$$

This standard deviation of the set of estimators is precisely what is called the standard error of the estimator. Do not confuse:

- The standard deviation s (in small letters) of the values of the sample.
- The standard error of the estimate. This standard error refers to the estimator and consequently to the population. For this reason it will be labeled $S\hat{E}$; the little hat comes from the fact this value is in fact an estimate of the true SE, the one we could find if the number K of samples could grow indefinitely.

How can $S\hat{E}$ be Estimated?

It may seem strange that we can, from just one sample, estimate the value of the characteristic we are looking for and the standard deviation of the estimate! Nevertheless there are two major ways to estimate $S\hat{E}$.

The Classical Way

The classical way, which analytically solves the question, is mathematically impossible if some hypotheses are not made, as we saw it in the previous section.

The logic is the same.

The Modern Way

It so happens that the Bootstrap – and to a lesser extent the Jackknife – does create the samples we need: they therefore solve the problem in a natural way.

15.4 Extrapolation of the Perceived Relationships from the Sample to the Population

This section illustrates the concepts introduced in the previous section.

When studying the relationship between two variables, it appeared that an important characteristic was the Bravais–Pearson correlation coefficient. We will use two logics: the test of hypothesis for the classical approach, the confidence interval for the modern approach.

[10] The divider is not really K but the number of degrees of freedom of the expression.

15.4.1 The Classical Approach

A correlation was found in the sample between the dependent and the causal variables; its value was quantified in the "Bravais–Pearson correlation coefficient".

The data for this example are shown in Figure 15.2.

Name	Cost	Mass(kg)
A	1278	6.83
B	724	2.18
C	809	3.80
D	920	4.55
E	772	2.18
F	877	2.11
G	1064	4.67
H	865	2.81
I	961	2.55
J	856	1.68
K	1293	6.30
L	717	1.98
M	648	1.25

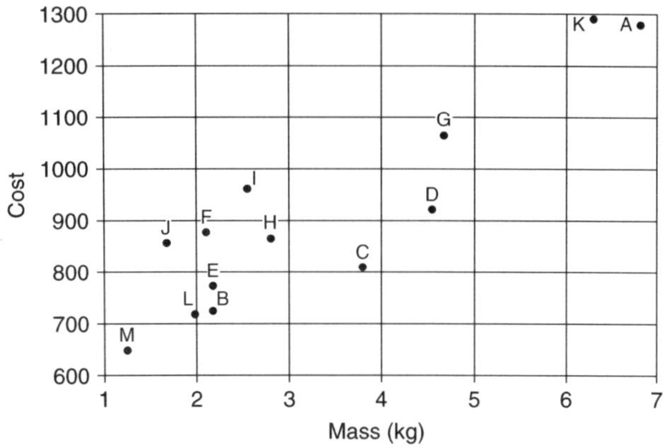

Figure 15.2 The graph of the sample data for the example.

In the example, the value $r_B = 0.901$ was found in accordance with the graph.

Can we infer from this value, observed in the sample, that there is really a correlation between these variables for the whole population?

Let us make the hypothesis – called $\boldsymbol{H_0}$ – that the sample was drawn (randomly) from a population in which there is really no correlation at all between these variables. The relationship we thought have found is then just due to chance; in fact both variables are independent. Does the correlation we observed in the sample validate this hypothesis?

If this hypothesis is true, one can expect to find, in the sample, a correlation coefficient r_B close to 0. However the random sampling of the observations may produce, just by chance, a coefficient different from 0. In such a case, we could conclude that the correlation inside the whole population is different from 0 (this is called a *"second type" error*) whereas it is not.

How can we make a conclusion? The best way is to study, in the case of this hypothesis $\boldsymbol{H_0}$, the distribution we could observe for r_B if we were able to draw several samples of the same size I (13 in this example).

Let us assume it is possible. If the hypothesis $\boldsymbol{H_0}$ is true, how the value of r_B would be distributed?

Such a distribution is extremely difficult to compute without a hypothesis on the distribution of the data in the whole population. We will therefore make a very strong hypothesis about it: this distribution is "binormal" which means that both V_0 and V_1 are normally distributed. In our sample, the distribution is not normal for the observed couples in the sample, but this is irrelevant.

In the case of this hypothesis, the distribution $f(r_B)$ of r_B can be computed. It is given by:

$$f(r_B) = \frac{1}{B\left(\frac{1}{2}; \frac{I-2}{2}\right)} (1 - r_B^2)^{\frac{(I-4)}{2}}$$

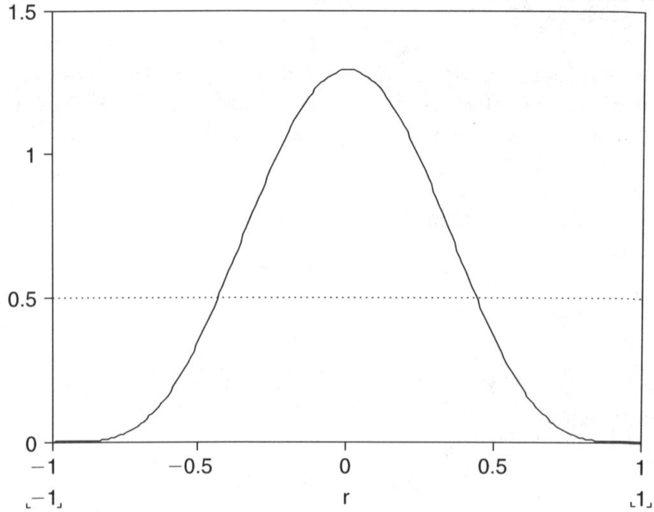

Figure 15.3 The distribution of r_B for $I = 4$ (dotted line) and $I = 13$ (full line).

where $B(p, q)$ is given by $\Gamma(p)\Gamma(q)/(\Gamma(p + q))$, with $\Gamma(t) = \int_0^\infty e^{-x}x^{t-1}\,dx$ (see Chapter 3). It looks a bit complex, but you will not have to compute it.

Note that this distribution is not defined for $I < 4$. For $I = 4$, r_B follows a uniform distribution: all the values are equally probable. For $I > 4$ the distribution has a bell shape. It is represented in Figure 15.3 for $I = 4$ and $I = 13$ (our example).

For large I the distribution becomes close to the normal one (with a mean equal to 0 and a standard deviation equal to $1/\sqrt{I-1}$).

Let us give now a level of confidence of 95% (both sides). It is not too difficult to compute the interval values of r_B for which the area under the distribution curve will be equal to 0.95, which means that the area outside this interval will be 0.025 on both sides: the interval is $(-0.554, 0.554)$; this is illustrated in Figure 15.4. If the sample provides us with a value outside this interval, we can say – with a level of confidence of 95% – that there is really a correlation between both variables inside the whole population; this is the case in our example. Consequently for the population studied, we can reject hypothesis H_0.

As a matter of fact, as Figure 15.4 shows it, the level of confidence is much higher than 95%: computation shows – this is the P-value already mentioned – that it reaches the level of 99.998%. Practically, we are sure there is such a correlation!

Pay attention to the fact that the method "demonstrates" – or not – that there is a correlation between the variables inside the whole population, but does not guarantee at all the level of correlation equals the one we found in the sample values.

The comment we can make is only that the correlation coefficient found in the sample is "statistically significant".

The previous computation was made in the H_0 hypothesis: we assume that there was no correlation inside the population and use the sample value in order to check this hypothesis, with the added hypothesis that the population was binormal.

Is this hypothesis about binormality reasonable? We do not know the population, so we cannot have a direct judgment. However we have a sample and we may

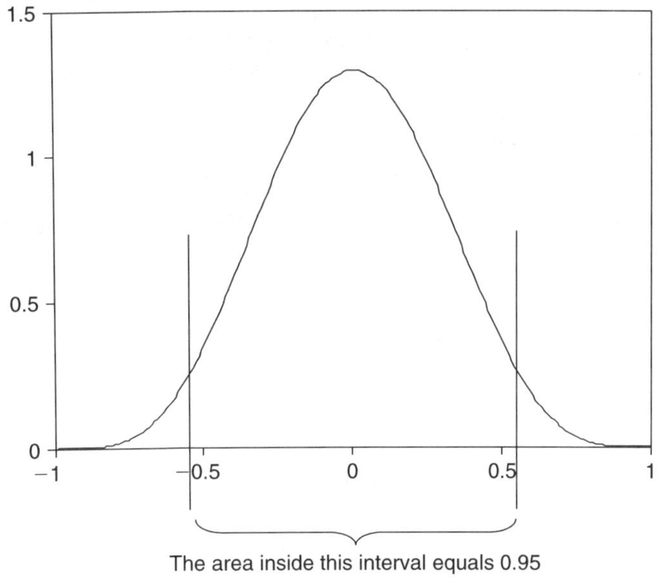

Figure 15.4 Using the distribution of r_B for our example.

ask the question: can we consider that this sample was drawn from a binormal population? We present below two tests:

1. The first one is based on the characteristics of a normal distribution: it skewness equals 0 and its kurtosis equals 3. The characteristics for the sample are:
   - for the dependent variables: skewness = 0.850, kurtosis = 2.734,
   - for the causal variable: skewness = 0.840, kurtosis = 2.424.

   As the sample was randomly drawn, we cannot, even if it is drawn from a binormal population, expect that its distributions will be both exactly normal: small deviations from normality can be expected. How small? Using the logic developed upwards (about testing hypothesis), computations show that, for a sample of size 13, we cannot reject the hypothesis that the population is binormal if (given an average level of confidence):
   - both skewness are lower than 0.858,
   - both kurtosis are in the interval [1.869, 4.114].

   Both conditions are nearly fulfilled here, even if the skewnesses are a bit high.
2. The second one is due to Kolmogoroff and Smirnoff: a value is computed which should be – always for a sample of size 13 and an average level of confidence – lower than 0.229. For both variables the computed value is equal to 0.115.

Both tests are positive here; note that we cannot be sure that the population from which the sample is drawn is binormal: the only thing we can say is: "we cannot reject the hypothesis it is binormal" and this is different from saying: "the population is binormal". Due to the result of the tests, we therefore conclude that the fact there is a correlation in the population is statistically significant, even if we are not really sure about the value the Bravais–Pearson correlation coefficient may have in the population.

Is it possible to make the same computation starting from another hypothesis, for instance $r_B = \rho$ in the population? The computation is possible, but difficult (Ref. [50], p. 136). An approximation of the distribution of r_B gives:

- A mean equal to $\rho - (\rho(1 - \rho^2))/2I$.
- A standard deviation equal to $(1-\rho^2)/\sqrt{I-1}$.

Let us apply these formulae to our example, starting from the hypothesis $\rho = 0.9$ in the population. In such a case, the computation shows that the distribution of r_B in a sample of size 13 would have a mean value equal to 0.893 and a standard deviation equal to 0.055. The value found in the sample – 0.901 – does not invalidate this hypothesis.

Can we hope to get a value equal to 0.95 for the population? Then the distribution of r_B for the sample would give a mean equal to 0.946 and a standard deviation equal to 0.028. The value found in the sample is at 1.607 standard deviation of the mean value; this gives a level of confidence (P-value) equal to 5.4% (one side only), leaving little hope that the correlation in the population is really equal to 0.95.

This discussion illustrates the classical approach of using a sample statistic for determining the value of a characteristic of a population. Do not forget the hypothesis:

- The population is binormal.

15.4.2 The Modern Approach

The modern approach uses the Bootstrap and does not make any hypothesis about the population. As previously said, we start by making 400 replicates, randomly selected with repetitions, which are randomly prepared as indicated in Figure 15.5.

Replicate no.						
1		2		3		....
1278	6.83	1278	6.83	724	2.18	
724	2.18	724	2.18	724	2.18	
809	3.8	1278	6.83	809	3.8	
920	4.55	920	4.55	648	1.25	
772	2.18	772	2.18	772	2.18	
877	2.11	877	2.11	877	2.11	*etc ...*
1064	4.67	1064	4.67	1064	4.67	
865	2.81	717	1.98	865	2.81	
961	2.55	961	2.55	961	2.55	
856	1.68	856	1.68	856	1.68	
1293	6.3	1293	6.3	1293	6.3	
717	1.98	717	1.98	648	1.98	
648	1.25	648	1.25	648	1.25	
Correlation						
0.901		0.943		0.88		

Figure 15.5 Illustration of the Bootstrap procedure.

For each replicate, the Bravais–Pearson correlation coefficient is computed; then its average and its standard error are computed. The results are:

- coefficient of correlation: 0.871,
- standard error: 0.010.

Obviously, due to the fact that the establishment of the replicates is random, if you make a second computation, you will not get exactly the same values. For instance, making 4 Bootstrap computations gives the following set for the coefficient: 0.883, 0.877, 0.876, 0.879. There is nothing bad about that, as long as the values stay in a reasonable range, which is the case if the number of replicates is high enough.

The nice thing about this modern approach is that it gives immediately the standard error of the coefficient of correlation.

The Jackknife is the "little brother" of the Bootstrap: it also makes replicates of the sample, but without repetition. This means that, with a sample of size I, it can make only I such replicates of size $I - 1$. This is not a problem when the sample size is large, but it is so rare in the domain of cost that the Bootstrap is generally preferred.

15.4.3 Conclusion

It is of course impossible to decide, between the classical approach and the modern approach, which one is the right one! However one can notice that the values are rather close and that the standard error is small. The analytical value is at 3 standard error of the result computed by the Bootstrap; due to the rather small size of the sample and the hypothesis made to compute analytically the coefficient, such a difference is not abnormal.

15.5 The Case of One Variable (No Causal Variable)

This section deals with populations described by one variable only, which will be called Y. This variable takes values we call y in the sample and Y in the population.

This variable can be anything; it can be, for our purpose:

- The dimension of a machined part, measured on several hundred identical parts.
- The cost or the duration of any activity which is repeated several times, as it occurs frequently in production.
- The cost of the same product made by different manufacturers.
- Etc.
- Including the residuals mentioned in Part IV.
- Or the specific cost of several different products belonging to the same product family. The specific cost is the product cost divided by its size; the most frequent example is given by the cost per kilogram (or the cost per cubic meter, etc.) so frequently – without any justification – used by many people.

This comment about the specific cost will deserve some explanation.

The Center Value Is an Important Information

If you are looking for a reliable forecast, the center value of the population is something important in cost analysis because it used as a reference value.

For the statistician, the center value of a distribution is also a very important concept, for two reasons:

1. It is the simplest "model" of a population and the easiest computation which can be done from the sample values.
2. It is far more efficient to study the distribution of the values around the center value than to study it globally.

In the following sections, we will introduce variables in order to get a better model. We will see when preparing the quality tests of the models that **the reference** is always the simplest "model" given by the center value. In other words the quality tests always try to answer the question: does the introduction of a variable really improve our "reference" model?

Consequently it is necessary to be able to estimate, from the sample values, this center value for the whole population.

15.5.1 Introduction

We study a population, supposed to be infinite in size because the values we are interested in are continuous (such as $ or €), described by one variable Y. Let us remind the reader that capital letters are used for anything related to the population, small letters for anything related to the sample drawn from this population.

The distribution – called $\Phi()$ – of the values Y taken by this variable may be unknown or known; both cases will be discussed. If it is unknown we will see in the following pages if we must nevertheless make some hypothesis about it in order to be able to estimate any characteristic.

What we are interested in is this distribution $\Phi()$, for estimating purposes. Therefore we would like to determine some characteristics of this **population**. As indicated in Part I:

> The distribution of the values of one variable is described by a few set of characteristics:
> - its center $\bar{Y}$ or $\tilde{Y}$, or anything else, which is probably the most important,
> - its standard deviation S, or anything else, which is the second in order of importance,
> - and its shape (skewness Γ_1 and kurtosis Γ_2).

In an attempt to get some information about these characteristics, we drew a sample of size I from this population; the values are labelled $y_1, y_2, \ldots, y_i, \ldots, y_I$ according to our usual notation scheme.

The characteristics of the **sample** distribution, named $\varphi()$, are called:

- Its arithmetic mean $\bar{y} = \Sigma_i y_i / I$.
- Its median $\tilde{y}$.
- Its standard deviation $S_y = [\Sigma_i (y_i - \bar{y})^2 / I]^{0.5}$.
- Its skewness and kurtosis given by (the name of the variable is not repeated in the symbol of the skewness and kurtosis in order to simplify the notation and

because these characteristics are rarely used):

$$\gamma_1 = \frac{\sum_i (y_i - \bar{y})^3}{Is^3}$$

$$\gamma_2 = \frac{\sum_i (y_i - \bar{y})^4}{Is^4}$$

The question we try to answer to in this chapter is:

What can be said about the population from the sample values?

Two approaches are possible:

1. The classical approach which generally requires strong hypotheses for answering the question.
2. The modern approach which is much more efficient.

15.5.2 The Classical Approach

The classical approach is purely mathematically oriented: it tries, by making hypotheses and computing from them, to establish some properties of the population.

The way these properties are demonstrated is the inverse one: starting from hypotheses about the population, the mathematician tries to forecast properties for any sample (not specifically the one we know) drawn from this population. From these properties, some characteristics of the population are inferred.

The Center of the Distribution Φ

What can be said about the center of the distribution Φ when we know the center of the sample distribution φ?

As it was seen in Chapter 4, there are two main measures of the center of a distribution: the (arithmetic) mean $\hat{Y}$ and the median $\tilde{Y}$, other ones being rarely used.

The Arithmetic Mean

Characteristics of the Arithmetic Mean The arithmetic mean has very interesting properties (the first three ones were demonstrated by Carl Friedrich Gauss):

- This value has the maximum likelihood if the population distribution is normal. This should not be a surprise for you because Gauss precisely built this "normal" law in order to get this property.
- The value is not biased, which means that $\mathcal{E}(\bar{y} = \bar{Y})$.
- From all the linear (in the y_i) estimators it is the one which has the minimum variance.
- This property has been afterwards extended to all unbiased estimators, linear or not.

- Jerzy Neyman, Egon S. Pearson and Abraham Wald demonstrated that it was the more globally precise of all estimators, biased or not (globally means whatever the value $\bar{y}$, as other estimators could be more precise locally, which means for a particular value of $\bar{y}$).

Note that all these properties are demonstrated for the normal curve. The distributions we have to work with are not always normal.

The Stein's Paradox If you have to estimate the mean of one population distribution, the best choice is of course to use the sample arithmetic mean.

If you have to estimate the means of two independent population distributions, the same result stands.

The Stein's paradox (Pour la Science no. 11) starts when the number K of means to estimate exceeds 3. Suppose you have to estimate the specific costs – assumed to be constant – for several different and independent product families, named A (making a trench), B (building a construction), C (manufacturing a bicycle), etc. You draw three samples from these families and compute their arithmetic means: $\bar{y}_A, \bar{y}_B, \bar{y}_C, \ldots$

According to Stein, your best estimate for the product family $\hat{Y}_A$ for instance (the same is true for the other estimates) is not $\bar{y}_A$ but something such as (this is the James–Stein theorem):

$$\hat{\bar{Y}}_A = \bar{\bar{y}} + c_A(\bar{y}_A - \bar{\bar{y}})$$

where $\bar{\bar{y}}$ is the general arithmetic mean: $[\bar{\bar{y}} = (\bar{y}_A + \bar{y}_B + \bar{y}_C)/3]$ and c_A a factor smaller than 1. This equation means that:

- if $c_A = 0$, then all estimators have the same value: the general mean $\bar{\bar{y}}$,
- if $c_A = 1$, then $\hat{\bar{Y}}_A = \bar{y}_A$, the usual estimator,
- if $0 < c_A < 1$, then $\hat{\bar{Y}}_A$ "regresses" toward the general mean.

James and Stein propose the following value for c_A (for $K > 3$):

$$c_A = 1 - \frac{(K-3) \times s_A^2}{\sum_{k=1}^{K} (\bar{y}_k - \bar{\bar{y}})^2}$$

where K is the number of product families and s_A^2 the variance of the values in the sample A. One can observe from this equation that:

- The larger the number of product families, the more $\hat{\bar{Y}}_a$ regresses toward the general mean.
- The smaller the standard deviation inside a sample, the less $\hat{\bar{Y}}_a$ regresses towards the general mean for this product family (this sounds logical).
- In order to avoid a strong regression towards the general mean, $\Sigma(\bar{y}_k - \bar{\bar{y}})^2$ should be as large as possible, which means that all the individual means $\bar{y}_k$ should be as different as possible (the equation has a problem if all the $\bar{y}_k$ are identical).

The strange thing about this theorem is that the product families can be completely different. Also note that it seems to favour the regression towards the mean we criticized in Chapter 9: the truth is decidedly difficult to discover, but the question here is of a different nature.

As the mean is an important characteristic of the distribution Φ, our attention must be, at this stage, devoted to establish its values from the sample values. We start here by establishing some properties which are true whatever the shape of the population distribution. Afterwards we will see how these properties can be improved if the shape of the population is known.

What Can be Said About the Mean $\bar{Y}$ When the Distribution $F()$ Is Unknown? If nothing is known about the distribution of the population, the information which can be given is rather general, as the reader may expect.

In order to understand how it is developed, suppose we are able to draw a large number of different samples of size I from the population. Sample number k delivers a set of values that we call, in order to clearly distinguish them $_{(k)}y_1, _{(k)}y_2, \ldots, _{(k)}y_i, \ldots, _{(k)}y_I$, the index placed in front of a value referring to the sample number k. It is possible to compute the mean value $_{(k)}\bar{y}$ of this sample.

The set of samples then produces a set of arithmetic means that we call $_{(1)}\bar{y}$, $_{(2)}\bar{y}, \ldots, _{(k)}\bar{y}, \ldots, _{(K)}\bar{y}$.

Let us now compute the mean of all these values $_{(k)}\bar{y}$; this is called the mathematical expectation of these $\bar{y}$ and written $\mathscr{E}(_{(k)}\bar{y})$. It could also be written, because it is a mean of means, $\bar{\bar{y}}$. The important result – which can be demonstrated – is that the mathematical expectation of the mean of all these means is equal to the mean of the population.

$$\mathscr{E}(_{(k)}\bar{y}) = \bar{Y}$$

If we return now to our unique sample, we can say we get a first result, not very strong one maybe, but nevertheless usable: the population mean $\bar{Y}$ should be "in the vicinity" of the sample mean $\bar{y}$.

How Far Could be $\bar{Y}$ from $\bar{y}$? If we know nothing about the population (this distribution could be completely "strange"), the only piece of information we can add is probabilistic in nature (Ref. [50], p. 267):

$$\lim_{k\to\infty} {}_{(k)}\bar{y} \to \bar{Y}$$

We are not going to discuss here the way the series $_{(k)}\bar{y}$ converges towards $\bar{Y}$: it is a pure mathematical subject which has a limited interest for the cost analyst. Only the consequence of this convergence is of interest to him/her. The consequence is based on the central limit theorem:

> The theorem "central limit"
>
> If $y_1, y_2, \ldots, y_k, \ldots, y_K$ is a set of random variables with the same expectation $\bar{Y}$ and standard deviation S, then
>
> $$\frac{1}{\sqrt{K}}\left(\frac{y_1 + y_2 + \cdots + y_k + \cdots + y_K - K \times \bar{Y}}{S}\right) \underset{K\to\infty}{\longrightarrow} N(0,1)$$

This is called **the law of large numbers.** Let us express it in our case (we know – this is our basic hypothesis when studying a sample – that the values are drawn from the same population): if I values $y_1, y_2, \ldots, y_I$ are drawn from a population of which mean is $\bar{Y}$ and the standard deviation is S, when $I \to \infty$ (which means when the sample size grows indefinitely), then:

$$\frac{\bar{y}-\bar{Y}}{S/\sqrt{I}} \longrightarrow N(0,1)$$

In other words, the mean $\bar{y}$ follows, when I is large, a normal distribution centered on $\bar{Y}$ (the population arithmetic mean) with a standard deviation $S/\sqrt{T}$ (the population variance divided by the square root of the sample size).

This law is often expressed the following way: **if a variable (here $\bar{y}$) is the sum of a large number of small causes, then its distribution is normal.** This is quite frequent in human activity: such an activity is the sum of hundred additive micro-activities; consequently it is reasonable to consider that the cost of an activity is distributed according to a normal law. It explains what was said earlier that the cost of one activity is a random variable and what is measured when we measure it is just a cost among a lot of possible costs distributed according to a normal law.

Note that this is the theoretical distribution of the $\bar{y}$, theoretical in the sense that it cannot be observed as it would require to draw several samples of the same size from the same population, which is something which is excluded. *It has nothing to do with the distribution of the sample.*

We learnt something about the distribution of the arithmetical mean $\bar{y}$ of a sample of size I: the distance of this mean to the population mean will increase with the population standard deviation S and will decrease with the sample size. All sounds to be logical.

But this does not give us any information about how far could a particular $\bar{y}$ value be from $\bar{Y}$, until we know something about S. We then have to wait until an estimate of S is found out.

Nevertheless $\bar{y}$ can be used as an estimator of $\bar{Y}$ but *we do not know how reliable it is* if the population standard deviation S is unknown.

What Can be Said About the Mean $\bar{y}$ When the Distribution $\Phi()$ Is Known but Not Normal? Knowing the distribution $\Phi()$ means that we know its standard deviation S, plus maybe its skewness Γ_1 and its kurtosis Γ_2. Of course the arithmetic mean $\bar{Y}$ is not known, otherwise no sample would be necessary! This situation is not as theoretical as one may think: we may know, from experience, that the standard deviation of the tolerances of a machine tool or of the specific cost could have a given value and could be slightly skewed. We draw a sample in order to get a good idea of its mean.

We use here the same technique: we compute, from what we know about $\Phi()$, some characteristics of the distribution of the ${}_{(k)}\bar{y}$. Afterwards we will see what we can infer from the sample for the population.

The knowledge of S, Γ_1 and Γ_2 allows to get a better understanding of the distribution of the ${}_{(k)}\bar{y}$. It can be demonstrated (Ref. [50], p. 267) that:

- The variance of this distribution is given by $\text{var}({}_{(k)}\bar{y}) = S^2/I$. This result is logic: if we draw just one sample from a population of which variance S^2 is small, the values $y_1, y_2, \ldots, y_i, \ldots, y_I$ we draw should very rarely be far away from $\bar{Y}$ and so will be their mean.
- Its skewness is given by $\gamma_1({}_{(k)}\bar{y}) = (\Gamma_1(Y))/\sqrt{I}$.
- And its the kurtosis by $\gamma_2({}_{(k)}\bar{y}) = 3 + ((\Gamma_2(Y) - 3)/I)$.

Let us return to what we can infer from the sample characteristics.

An estimate of the population mean is still given by $\hat{\bar{Y}} = \bar{y}$ and we know the "standard error" of this estimate $s\hat{e} = S/\sqrt{I}$, plus the fact that it follows, if the sample size is large enough, a normal distribution. Compared to the preceding section "What can be said about the mean $\bar{Y}$ when the distribution $F()$ is unknown?", we can now be more specific and give an interval estimation, at least if the sample size is large enough: we know that in this case the distribution of the $\bar{y}$ is nearly normal, which means that 90% of the time (this is the confidence level) the true value $\bar{Y}$ will be in the interval $[\bar{y} - 1.165 \times s\hat{e}, \bar{y} + 1.165 \times s\hat{e}]$, or 95% of the time in the interval $[\bar{y} - 1.96 \times s\hat{e}, \bar{y} + 1.96 \times s\hat{e}]$.

Note that the variance of the distribution of the ${}_{(k)}\bar{y}$ decreases with the sample size; it follows that the standard error will decrease with the square root of the sample size. Therefore we can expect that the accuracy with which $\bar{Y}$ will be estimated will decrease only slowly with the sample size: in order to double the accuracy (which will mean to divide by 2 its confidence interval) the sample size will have to be multiplied by 4: the accuracy is costly!

Also note that, when I increases indefinitely, then $s\hat{e} \underset{I\to\infty}{\rightarrow} 0$. This confirms that $\bar{y}$ then converges towards $\bar{Y}$ and quantifies the speed of this convergence.

What can be inferred from the knowledge of the skewness and the kurtosis of the distribution of the ${}_{(k)}\bar{y}$? It is obvious that, if the sample size I grows indefinitely, then:

$$\gamma_1({}_{(k)}\bar{y}) \to 0 \quad \text{and} \quad \gamma_2({}_{(k)}\bar{y}) \to 3$$

which shows that the distribution of all the means we can compute from various samples of large sizes is normal (as its skewness becomes 0 and its kurtosis becomes 3). This confirms and quantifies what was said in the previous section.

About the Standard Error

Suppose we are interested in the means $\bar{y}$ of several samples drawn from the population of which we known the standard deviation S.

These means will have an expected value $E({}_{(k)}\bar{y}) = \bar{Y}$ and a certain spread. This spread can be quantified by the $\bar{y}$ standard deviation. This spread is called the "standard error" of the mean we can observe from one sample drawn from this population.

It is noted $s\hat{e}(\bar{y})$ and has the same dimension as $\bar{y}$: if $\bar{y}$ is given in €, $s\hat{e}(\bar{y})$ will also be in €.

Notice that:

- $s\hat{e}(\bar{y})$ has nothing to do with the standard deviation s of the values y_i of the sample.
- It is also NOT an estimate of the population standard deviation.

From what is written in the previous page, if the population standard deviation S is known, then

$$s\hat{e}(\bar{y}) = \frac{S}{\sqrt{I}}$$

What Can be Said About the Mean $\bar{Y}$ When the Distribution $\Phi()$ Is Normal? Here again we start by computing what is the distribution of the $\bar{y}$ under this hypothesis.

Now all the values y_i sampled from the population are random variables with a normal distribution. So is their sum; consequently, **the mean $\bar{y}$ follows a normal distribution** whatever the sample size, of which it can be demonstrated (Ref. [50], p. 273) that the standard deviation is still $S/\sqrt{I}$.

Are not we in the same situation, as the standard deviation of the population S has to be known? Fortunately not because it can be shown that:

$$t = \frac{\bar{y} - \bar{Y}}{\frac{s}{\sqrt{I-1}}}$$

where s is the sample standard deviation, follows a Student distribution with $I - 1$ degrees of freedom.

The demonstration of this very important formula – very important because it does not depend on the population standard deviation – rests on the fact that the Student distribution[11] is defined the following way: if A represents a random variable following a normal law $N(0,1)$ and B a random variable following a χ_q^2 distribution[12] (with q degrees of freedom: see Chapter 2), then the ratio $A/\sqrt{(B/q)}$ follows a Student distribution with q degrees of freedom.

It so happens (from what has been said in the second paragraph of this section) here that $[(\bar{y} - \bar{Y})/(S/\sqrt{I})]$ follows a $N(0,1)$ distribution and that $(I \times s^2)/(S^2)$ follows a χ_{I-1}^2 distribution (see below); in the division of these two quantities, S and I disappear!

Returning now to what can be inferred from the values of our sample, it is now possible to give a confidence interval for $\bar{Y}$ which is **based only on the sample values:**

$$\bar{y} - t_{I-1,\alpha/2}\frac{s}{\sqrt{I-1}} < \bar{Y} < \bar{y} + t_{I-1,\alpha/2}\frac{s}{\sqrt{I-1}}$$

where $t_{I-1,\alpha/2}$ is the value of the Student distribution with $I - 1$ degrees of freedom which lets on the right and of the left $\alpha/2$ of the total area, $1 - \alpha$ being the level of confidence.

The area on the left of $-t_{\alpha/2}$ and on the right of $t_{\alpha/2}$ represents 2.5% of the total area under the curve (the distribution is symmetrical): the area external to the interval therefore represents 5% of the curve. This is this result which allows to say that $\bar{Y}$ will be inside this interval 95% of the time (Figure 15.6).

Let us take an example: we observed a sample of size $I = 10$, for which $\bar{y}$ takes the value 15 and s the value 2. Let us choose a confidence level $1 - \alpha = 95\%$. For the Student curve with 9 degrees of freedom $t_{0.025} = 2.262$ which gives $t_{\alpha/2}(s/\sqrt{I-1})$ 1.508; one may say that $\bar{Y}$ will be inside the interval [13.492, 16.508] 95% of the time.

Note that, compared to the previous section when the population standard deviation was supposed to be known, the formula uses the value $I - 1$ and not I anymore. This comes from the fact we lost one degree of freedom when computing s.

About the Degree of Freedom A characteristic of a distribution is a random variable, as its value will change from one sample to another sample. This variable is computed from the sample values; if it happens that this computation involves several coefficients

[11] The Student distribution is presented in Chapter 3.
[12] The χ^2 distribution is presented in Chapter 3.

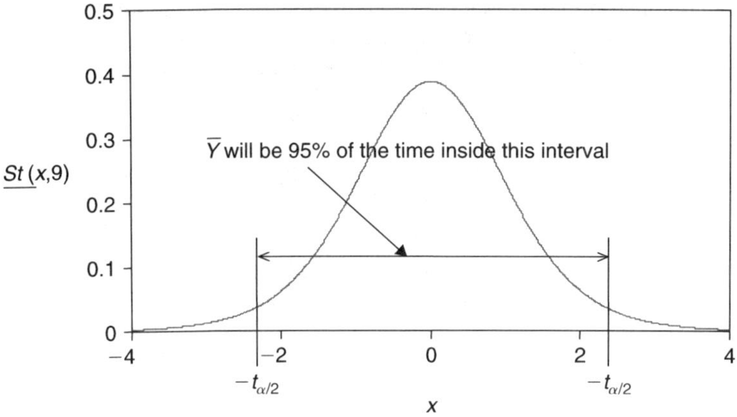

Figure 15.6 The Student distribution for 9 degrees of freedom and $\alpha = 0.05$.

also computed from the sample values, these coefficients "absorb" some information of the sample. Then the characteristic gets less information from the sample.

The degree of freedom of a random variable is equal to the number of data available to compute it, less the number of the coefficients included in its formula. For instance the sample mean computed on a sample of size I has I degrees of freedom, whereas the variance – which uses the sample mean – only has $I - 1$.

About the Variable *t*

The variable t is used quite often in the manuals about formulae building.

It is often said, for getting a reliable estimate, that t must be at least equal to 2. Where does this rule come from?

The following table gives the values of $t_{\alpha/2}$ for several degrees of freedom ν – equal, as previously mentioned, to $I - 1$. It is clear from this table that:

- Even with a small sample, a reasonable level of confidence of 90% (both sides) can be reached as soon as t is in the vicinity of 2; of course this level of confidence improves (this means that a larger sample standard deviation s becomes acceptable) if the degrees of freedom increase. For instance a value of t equal to 2 for 10 degrees of freedom puts the level of confidence to 92%.
- But if you need a better level of confidence of 95% (both sides), the sample must be rather large.

ν	$\alpha/2 = 0.05$	$\alpha/2 = 0.025$
5	2.02	2.57
10	1.81	2.23
20	1.73	2.10
30	1.70	2.04
∞	1.65	1.96

Student distribution: Values of t for obtaining the level of confidence $1 - \alpha$ (both sides) as a function of the degree of freedom ($\nu = \infty$ corresponds to the normal distribution).

The conclusion is that, if $t \geqslant 2$, then even with a small sample the level of confidence of 10% (both sides) is easily obtained. This is quite sufficient in practice in the domain of cost: the rule comes from this computation.

The Median

Unfortunately the classical approach does not help for finding the confidence interval for $\tilde{Y}$ if the population distribution is not normal.

We limit therefore ourselves to simple rules.

What Can be Said About the Median $\tilde{Y}$ When the Distribution $\Phi()$ Is Normal? If the distribution $\Phi()$ is normal, then the median is equal to the mean: $\tilde{Y} = \bar{Y} \ldots$ Then the median can be estimated equal to $\bar{y}$.

The variance of this estimate can be approximated[13] by $\pi/2\ S^2/I$.

About the Spread of the Distribution $\Phi()$

We try to answer here the same question that we asked about the population distribution center: what is, from the information we have in the sample, the best guess for S?

What Can be Said About S When the Distribution F() Is Unknown?

Here again we suppose that it is possible to draw a lot of samples from the population. The variance of sample k is called ${}_kS^2 = (1/I)[\Sigma({}_ky_i - {}_k\bar{y})^2]$.

It can be demonstrated,[14] as we did for the mean, that ${}_ks^2$ converges towards S^2 and that the mathematical expectation of ${}_ks^2$ is given by $E(s^2) = [(I - 1)/I]S^2$. Therefore an unbiased estimate of S^2 is given by

$$\hat{S}^2 = \frac{I}{I-1} s^2 = \frac{\sum_i (y_i - \bar{y})^2}{I-1}$$

which is very frequently used. The division by $I - 1$ instead of I comes from the fact that we lost one degree of freedom for computing s.

Using s^2 as an estimate of S^2 would introduce a bias equal to s^2/I.

The fact that the estimate of S is not biased is made at the detriment of its accuracy. For instance Rao[15] shows that: $[\Sigma_i(y_i - \bar{y})^2]/(I + 1)$ could also be used as an estimate of S^2; this estimate has a smaller variance, but it is biased. This shows that the compromise between the lack of bias and the accuracy is rarely obvious and depends on what is looking for.

What Can be Said About S When the Distribution $\Phi()$ Is Known but Not Normal?

By "known", we mean here that we know the value of S and the fourth central moment U_4 of $\Phi()$.

---

[13] Ref. [50], p. 275.
[14] Ref. [50], p. 269.
[15] Ref. [56], p. 316.

Can we give more information about this estimate? One can establish[16] that the variance of ${}_k s^2$ is given (this equation requires the knowledge of the moment U_4 of the population distribution) by

$$\operatorname{var}({}_k s^2) = \frac{I-1}{I^3}\left[(I-1)U_4 - (I-3)S^4\right]$$

which does not really helps the cost analyst!

What Can be Said About S When the Distribution Φ() Is Normal?

It can be shown that the classical estimate $\hat{S} - s\sqrt{I/(I-1)}$ is very slightly biased. Theoretically one demonstrates that:

- If the population mean $\bar{Y}$ is known, then

$$\hat{S} = \sqrt{\frac{I}{2}}\frac{\Gamma\left(\frac{I}{2}\right)}{\Gamma\left(\frac{I+1}{2}\right)}\sqrt{\frac{\sum_i (y_i - \bar{Y})^2}{I}}$$

  is an unbiased estimator of S, with a minimum variance. Notice that the last term of the expression if not the standard deviation of the sample, because $\bar{Y}$ is used instead of $\bar{y}$.
- If the population mean is not known, then

$$\hat{S} = \sqrt{\frac{I}{2}}\frac{\Gamma\left(\frac{I}{2}\right)}{\Gamma\left(\frac{I+1}{2}\right)}s$$

  is an unbiased estimator of S, with also a minimum variance.

The following question is now: how far could be this estimate $\hat{S}$ of the true value S? The answer can be done by finding a confidence interval of S (the population variance) around this value $\hat{S}$ or around s (the sample variance); this second solution is easier: it has been demonstrated that

- If the population mean $\bar{Y}$ is known, then

$$\frac{I}{S^2}\sqrt{\frac{\sum_i (y_i - \bar{Y})^2}{I}}$$

  follows a distribution discovered by I. J. Bienaymé and now known under the term "χ^2 distribution with I degrees of freedom" and consequently[17] frequently referred to as $\chi^2_{(I)}$.

[16]Ref. [50], p. 270.

[17]Ref. [49], p. 139.

- If the population mean is not known, then $I(s^2/S^2)$ follows a $\chi^2_{(I-1)}$ distribution (we loose one degree of freedom due to the use of s).

These expressions are nice but a bit too complex for the cost analyst. Let's then concentrate on the second point, which is the only interesting point to us (as we never know the population mean $\bar{Y}$), and let us try to get an approximate confidence interval, interval which will be sufficient for our studies.

As $I(s^2/S^2)$ follows a $\chi^2_{(I-1)}$ distribution – distribution which has a mean equal to $I-1$ and a variance equal to $2(I-1)$ – a sufficient approximation for practical purposes is given by:

$$I\frac{s^2}{S^2} = (I-1) \pm k\sqrt{2(I-1)}$$

where k depends on the level of confidence you want (generally speaking, as previously mentioned, k in the vicinity of 2 is selected)

or
$$\frac{s^2}{S^2} \approx 1 \pm k\sqrt{\frac{2}{I}}$$

as soon as I is large enough (we do not need a precise value of this confidence interval!). Consequently

$$S^2 \approx \frac{s^2}{1 + k\sqrt{\frac{2}{I}}} \approx s^2\left(1 \pm k\sqrt{\frac{2}{I}}\right)$$

$$S \approx s\left(1 \pm k\sqrt{\frac{1}{2I}}\right)$$

which gives an easy way to get an approximate value of the confidence interval of S. $\sqrt{s^2/2I}$ is the "standard error" of S around s.

About the Skewness and the Kurtosis of the Distribution $\Phi()$

The skewness and the kurtosis are computed in the sample.

The classical approach is unable to compute what these values could be for the population when the values for the sample are computed. The only information which is available is given for the:

- skewness $\mathscr{E}(\gamma_1) = 0$ with a variance of $6/I$,
- kurtosis $\mathscr{E}(\gamma_2) = 3$ with a variance of $24/I$.

This information is not really helpful for the cost analyst.

15.5.3 The Modern Approach

The classical approach is a powerful one, as it is able to demonstrate very important properties. However it needs several hypotheses which may sometimes be difficult to accept.

This modern approach was briefly introduced in Section 15.4.2. Its purpose is NOT to improve the value of the required characteristic, but to easily compute its standard error. We will limit the discussion here to the population mean and variance, but the method can very well be used for the skewness and the kurtosis, as the example in Section 15.4.2 illustrates.

Two modern approaches are possible: the Jackknife and the Bootstrap, the former being the "little brother" of the second one. Both are based on the generation of several replicates of the sample, which are defined as the samples which could have been drawn, assuming that the sample was randomly drawn. Starting from a sample of size I:

- The Jackknife generates I replicates of size $I - 1$, by removing, in each one, just one product at a time.
- The Bootstrap generates a large number of replicates of size I, by randomly selecting, with replacement, I products among the I available. The number K of replicates is let to the cost analyst, 400 being a common one.

The Jackknife requires a rather large value of I in order to get a sufficient number of replicates. As it is rarely the case in cost estimating, we will concentrate here on the Bootstrap.

Once the replicates have been generated, the required characteristic – here the mean and the standard deviation of the sample – is computed for each. At the end of this process, we get 400 – if this figure was selected – different values of this characteristic. From these values its distribution can be computed; this means for us the center (here the arithmetic mean is always used), the standard deviation, the skewness and the kurtosis.

The validity of this method has mathematically been proved (Bootstrap): see, for instance Ref. [31].

About the Bootstrap

The general procedure starts with the randomly created replicates. For each replicate k the characteristic z is computed on the replicate values; let us call it ${}_{(k)}z$.

Then the estimator of the characteristic Z for the whole population can be taken as the z value from the original sample, or as the mean $\bar{z} = (1/K)\sum_k {}_{(k)}z$ of all ${}_{(k)}z$. But, more important its standard error is classically computed as:

$$\hat{se}_Z = \left\{ \frac{\sum_k ({}_{(k)}z - \bar{z})^2}{K - 1} \right\}^{\frac{1}{2}}$$

About the Jackknife

The logic is the same, but the standard error of the estimator is computed in a different way:

$$\hat{se}_Z = \left\{ \frac{I-1}{I} \sum_k ({}_{(k)}z - \bar{z})^2 \right\}^{\frac{1}{2}}$$

where I is the sample size.

These formulae can be applied to any characteristic.

15.5.4 Comparing the Approaches

Let us do that on an example.

Suppose you get this set of values in your sample:

1278
724
809
920
772
877
1064
865
961
856
1293
717
648

What can we say, from this sample about the mean and the standard deviation of the whole population from which this sample was (randomly) drawn?

The results are given in Figure 15.7.

	Mean of the population		Standard deviation around the mean	
	Estimated	Standard error	Estimated	Standard error
Classical computation	906.462	55.708	192.979	37.846
Jackknife	906.462	53.523	191.927	41.908
Bootstrap	908.742	52.929	181.52	37.065

Figure 15.7 Comparing the solutions.

How do the standard error compare? This is a satisfactory result. It shows that each time the classical computation is valid, the Jackknife and the Bootstrap compute about the same values. But if the hypotheses on which the classical computation is based are not fulfilled, then the result of this computation may be wrong.

15.6 The Case of Two Variables (One Parameter)

The two variables are the dependent variable on one hand, a causal variable – generally related to the object size – on another hand.

The presentation is here limited to the classical approach: the modern approach does not differ from what has previously been said.

Example

In this section, we will use the example defined in Figure 15.2.

15.6.1 Extension to the Population of the Perceived Relationships in the Sample

The analysis of the sample revealed some correlations (mainly Bravais–Pearson, Spearman) between both variables (the causal and the dependent) and we must ask the question: are these correlations valid for the whole population?

This question was dealt with in Section 15.4 of this chapter, for both the classical and the modern approaches.

Now, in the sample, we found that the distribution of the cost – called φ – could conveniently be described as the sum of a "dynamic center" and the distribution – called ψ – of the residuals around this dynamic center.

The question is now: How can be use this information for the whole population?

15.6.2 Using Additive Deviations for Studying the Distribution of the Cost

We are now dealing with the population as a whole and from the understanding we have about the population, and/or the results we found in the previous section, we **assume** that the distribution Φ of the dependent variable Y can be described by the sum of two terms (capital letters are always used for values of the population):

A *dynamic center* $\hat{Y}$ which is related to the causal variable X by a linear relationship:

$$\hat{Y} = B_0 + B_1 X$$

A *random deviation*, that will be noted E_+, around the dynamic center. Its distribution is called Ψ.

So that a particular value of the dependent variable can be written:

$$Y_i = B_0 + B_1 X_i + E_{+i}$$

All terms B_0, B_1 and the distribution of E_+ are unknown. The purpose of studying the sample was to get some information about these terms.

The reader probably already knows the results of the classical approach. Mathematically speaking these results are quite nice. We will nevertheless reproduce them, not so much for them than for recalling the hypotheses on which they are based: the purpose is to let the reader decide on his/her right to use these results in the practical situation he/she works for.

No demonstration is given, as these results appear in most books on statistics (see for instance Refs [38], [50], . . .).

Assumptions About the Population

The classical approach is unable to make any statement without making some assumptions about the population from which the sample is drawn and for which we expect to estimate some of its characteristics.

Not all assumptions are required to demonstrate everything: the assumptions which are strictly required will be stated in due course for each property.

Assumptions[13] that can be made on the population

1. The average of Ψ is 0 (the mathematical expectation of E_+ is 0).
2. The distribution Ψ does not depend on X (assumption called "homoscedasticity") and its variance, called S^2_{E+} or simply[14] S^2, does exist.
3. Deviations E_{+i}, E_{+j} relative to two values X_i, X_j are mutually independent.
4. The distribution Ψ is normal.
5. The distribution of X has a finite average $\bar{X}$ and a finite variance S^2_X.
6. No *a priori* information is available on B_0 and B_1.

The Dynamic Center of the Population

Estimating B_0 and B_1

The best thing we can do is to consider that the values we computed from the sample can be used as estimators of B_0, B_1 and of the distribution of E_+. We then write (this is the "plug-in principle"):

$$\hat{B}_0 = b_0$$
$$\hat{B}_1 = b_1$$

This is an important statement: it shows we are confident in the values we get in our sample and in the computations we made for finding b_0, b_1. The values we get from our particular sample are reminded here:

- When we decide to have an intercept different from 0:

$$\hat{B}_1 = \frac{\sum_i (x_i - \bar{x})(y_i - \bar{y})}{\sum_i (x_i - \bar{x})^2} = b_1$$

$$\hat{B}_0 = \bar{y} - b_1 \bar{x} = b_0$$

- When we decide to have an intercept equal to 0 (there is no $\hat{B}_0$ anymore):

$$\hat{B}_1 = \frac{\sum_i x_i y_i}{\sum_i x_i^2}$$

[13] These assumptions are usual in all the statistical books.

[14] When no confusion is possible.

From now on, we will study the first case only, the reader being able to convert, if necessary, the result for the second case.

Now, even if we are confident in the work we did up to now, we know that these values were computed from *one* sample of *size I* and that forces some constraints in the confidence we may have in these values.

The purpose of this section is to quantify this level of confidence. *Quantifying this level of confidence is a statistical property*: it has no meaning for the particular sample we have; we return to this question later on in this chapter. What we can do at this stage is to derive the statistical properties of the $\hat{B}_0$ and $\hat{B}_1$. If we were able to draw a lot of samples of the same size from the population: what would be the distribution of these values?

The Distributions of the $\hat{B}_0$ and $\hat{B}_1$

$\hat{B}_0$ and $\hat{B}_1$ are values computed from a particular sample; in order to avoid any confusion between these particular values and the values you could observe from different samples, we will call them $\bar{\hat{B}}_0$ and $\bar{\hat{B}}_1$. As this sample was supposed to be randomly drawn, we may have drawn a different sample: as a matter of fact, if we were able to draw several samples of size I, we would certainly compute different values for them.

The question is then: what confidence can we have about the values we have just computed? Or, if you prefer another way of saying the same thing: what is the distribution of $\hat{B}_0$ and $\hat{B}_1$?

What we can do is then to estimate the distribution of these values, distribution we could observe if we were able to draw many samples of the same size. As a matter of fact, we are not interested in the distribution of these random variables, but in the distribution of the differences – which are also random variables – of $B_0 - \hat{B}_0$ on one hand, of $B_1 - \hat{B}_1$ on the other hand, the purpose being to answer the question: are these differences small enough for considering we get a good knowledge of the population?

As usual the distribution of these variables is studied by computing their center, their standard deviation, their skewness and their kurtosis.

The Center of the Distributions of $B_0 - \hat{B}_0$ and $B_1 - \hat{B}_1$ If Assumption 1 is accepted, then both centres equal 0.

This property, which can also be written $\mathscr{E}(B_0 - \hat{B}_0) = \mathscr{E}(B_1 - \hat{B}_1) = 0$ where $\mathscr{E}$ means "expected value", is often called "lack of bias" because it means that the average values of $\hat{B}_0$ and $\hat{B}_1$ equal B_0 and B_1, which are the true values of the coefficients:

$\hat{B}_0$ and $\hat{B}_1$ – are unbiased linear estimators of B_0 and B_1

"linear" meaning that they are both linear functions of the y_i, "unbiased" meaning that their expected values are equal to B_0 and B_1.

Never forget it is a statistical property: *it does not say anything* about the particular values of the $\hat{B}_0$ and $\hat{B}_1$ you extrapolated from the sample.

The Variances of the Distributions of $B_0 - \hat{B}_0$ and $B_1 - \hat{B}_1$ As B_0 is a fix quantity, the variance of $B_0 - \hat{B}_0$ is equal to the variance of $\hat{B}_0$, the same being true for $\hat{B}_1$; it is therefore easier to simply speak about the variances of $\hat{B}_0$ and $\hat{B}_1$.

If Assumptions 1, 2 and 3 are accepted, then the variances of the distributions of $\hat{B}_0$ and $\hat{B}_1$ which could be observed on a lot of samples of size I, are given by:

$$\text{var}(\hat{B}_1) = \frac{S_{E+}^2}{\sum_i (x_i - \overline{x})^2}$$

$$\text{var}(\hat{B}_0) = \frac{\sum_i x_i^2}{I\sum_i (x_i - \overline{x})^2} S_{E+}^2$$

their covariance being given by:

$$\text{cov}(\hat{B}_0, \hat{B}_1) = \frac{-\overline{x}}{\sum_i (x_i - \overline{x})^2} S_{E+}^+ = -\overline{x} \times \text{var}(\hat{B}_1)$$

In order to prepare the case with several quantitative variables, it is interesting to mention here the matrix "stenography". Let us first consider the product:

$$\|{}^+x\|^t \otimes \|{}^+x\| = \begin{Vmatrix} 1 & 1 & \dots & 1 & \dots & 1 \\ x_1 & x_2 & \dots & x_i & \dots & x_I \end{Vmatrix} \otimes \begin{Vmatrix} 1 & x_1 \\ 1 & x_2 \\ \dots & \dots \\ 1 & x_i \\ \dots & \dots \\ 1 & x_I \end{Vmatrix}$$

This product of a matrix of type $\Re^{2\times I}$ by a matrix of type $\Re^{I\times 2}$ gives a matrix of type $\Re^{2\times 2}$: this is a square symmetric matrix with two rows and two columns. It can be developed as:

$$\|{}^+x\|^t \otimes \|{}^+x\| = \begin{Vmatrix} I & I\overline{x} \\ I\overline{x} & \sum_i x_i^2 \end{Vmatrix}$$

Let us now compute the inverse of this matrix. In order to do so, we need first to compute its determinant. It is given by:

$$\det = I\sum_i x_i^2 - I^2\overline{x}^2 = I \times \left(\sum_i (x_i - \overline{x})^2 \right)$$

Therefore the inverse of $\|{}^+x\|^t \otimes \|{}^+x\|$ will be given by:

$$(\|{}^+x\|^t \otimes \|{}^+x\|)^{-1} = \begin{Vmatrix} \dfrac{\sum_n x_i^2}{I\sum_i (x_i - \overline{x})^2} & -\dfrac{\overline{x}}{\sum_i (x_i - \overline{x})^2} \\ -\dfrac{\overline{x}}{\sum_i (x_i - \overline{x})^2} & \dfrac{1}{\sum_i (x_i - \overline{x})^2} \end{Vmatrix}$$

If you compare the elements of this matrix with the previous relationships, you understand why the matrix:

$$\text{cov} = (\|^{+}x\|^{t} \otimes \|^{+}x\|)^{-1} \times S_{E+}^{2}$$

is called the "covariance matrix" of the data observed in the sample: the diagonal elements give the variances of $\hat{B}_0$ and $\hat{B}_1$, and the other element (this matrix is symmetrical) gives their covariance: it is a useful stenography!

Can We Say More About the Distribution of $\bar{B}_0$ and $\bar{B}_1$? Yes, if new hypotheses are introduced.

These hypotheses are first hypothesis numbered 4, second[15] that the x values are fix and that only the y values were random. This means that we are supposed to be able to prepare several products described by the same x values and that we observed their costs; these costs are random values. This is certainly unrealistic in the cost domain. However Johnston (Ref. [34], p. 29) shows that this constraint can be relaxed.

Given these hypotheses, and replacing in the formula giving $\hat{B}_0$ and $\hat{B}_1$ all the y_i by their value:

$$y_i = B_0 + B_1 \times x_i + E_{+i}$$

It is clear that $\hat{B}_0$ and $\hat{B}_1$ are only functions of these E_{+i} and are therefore random numbers with a normal distribution.

Consequently it can be said that $\hat{B}_0$ and $\hat{B}_1$ both follow normal distributions with the variances computed upwards. In other words $(B_0 - \hat{B}_0)/(\text{var}^{0.5}(B_0))$ and $(B_1 - \hat{B}_1)/(\text{var}^{0.5}(B_1))$ follow a normal $N(0,1)$ distribution.

The conclusion is therefore that the skewness of the random variables $B_0 - \hat{B}_0$ and $B_1 - \hat{B}_1$ is 0 and that their kurtosis is equal to 3.

Can we be more specific?

Unfortunately both variables include the term S_{E+}^2 which is unknown. But we will see in Section "The distributions of the $\hat{B}_0$ and $\hat{B}_1$" that an estimate of it can be computed, let us call it $\hat{S}_{E+}^2$, and that $(I-2)/(S_{E+}^2) \times \hat{S}_{E+}^2$ follows a χ_{I-2}^2 distribution. Therefore, as explained in Section 2.2.1 of this chapter, it appears that:

$$\frac{B_0 - \hat{B}_0}{S_{E+}} \times \frac{\left(\sum_i x_i^2\right)^{0.5}}{\left(I \times \sum_i (x_i - \bar{x})^2\right)^{0.5}} \times \sqrt{\frac{I-2}{\frac{I-2}{S_E^2} \times \hat{S}_{E+}^2}} = \frac{B_0 - \hat{B}_0}{\hat{S}_{E+}} \times \frac{\left(\sum_i x_i^2\right)^{0.5}}{\left(I \times \sum_i (x_i - \bar{x})^2\right)^{0.5}}$$

and

$$\frac{B_1 - \hat{B}_1}{\hat{S}_{E+}} \times \frac{1}{\left(\sum_i (x_i - \bar{x})^2\right)^{0.5}}$$

follow a Student distribution with $I - 2$ degrees of freedom. The conclusion in Section 5.2.1 can therefore be applied.

[15] This was an hypothesis made by Gauss.

This explains why the "t" values, defined as: $\hat{B}_0/\hat{\text{var}}^{0.5}(\hat{B}_0)$ and $\hat{B}_1/\hat{\text{var}}^{0.5}(\hat{B}_1)$, are often computed. As previously explained, they should be about greater that 2 to get some confidence (level of confidence 90%) about the estimated values.

Properties of the Estimators $\bar{B}_0$ and $\bar{B}_1$

$\hat{B}_0$ and $\hat{B}_1$ are the Best Linear Unbiased Estimators An interesting result, due to Gauss and Markoff is that, if hypotheses 1, 2 and 6 (above cited assumptions) are satisfied, then $\hat{B}_0$ and $\hat{B}_1$ are the **best linear unbiased estimators** of B_0 and B_1; all the words are important. Gauss and Markov never demonstrated that these values were the best unbiased estimators, but that they were the best *linear* (linear meaning they linear in x_i and y_i) unbiased estimators. It is quite possible that other estimators – non linear and/or biased – could be better estimators.

Malinvaud (Ref. [38], p. 95) insists on the fact that hypothesis 4 is not used in the demonstration: it does assume nothing about the distribution of the deviations in the population. He adds that "contrary to the belief sometimes held, the assumption of normality of the deviations E_+ inside the population is not of basic importance for the theory".

This demonstration is a bit long, but available in many books on statistics.

If Assumptions 1, 2, 3, and 5 Are Valid, Then $\hat{B}_0$ and $\hat{B}_1$ Are Consistent This means (see Ref. Malinvaud [38], p. 88) that when $I \to \infty$ (the sample size increases indefinitely), then the variances of both $\hat{B}_0$ and $\hat{B}_1$ tend to 0, which means that:

$$\hat{B}_0 \to B_0$$
$$\hat{B}_1 \to B_1$$

This is an interesting result, but which, as usual, does not prove anything about the particular values you got from the particular sample you are interested in.

If Assumptions 1, 2, 3 4 and 6 Are Valid, Then $\hat{B}_0$ and $\hat{B}_1$ Are Sufficient This nice property is demonstrated by Malinvaud (Ref. [38], p. 99). Note that now the hypothesis of normality is required.

Studying the Deviations Around the Dynamic Center for the Population

In Section "The dynamic center of the population" we studied the first part of the decomposition of the cost distribution in the population: its dynamic center. We found, without forgetting we had to make some hypotheses, interesting results.

It is time now to study the second part of it: the distribution of the deviations E_+ around this dynamic center. This is also important because, in order to practically apply the results found in the previous section, we need an estimate of their variances S^2_{E+}.

About the Center of the Distribution of the E_+

This center cannot be estimated from the sample: it is an hypothesis we have to make: it is hypothesis 1:

$$\mathscr{E}(E_+) = 0$$

Of course the fact that the sum of the residuals in the sample is, in the classical approach using the standard linear regression, equal to 0 is a hint in this direction, but is does not prove anything about the individuals E_+.

Estimating the Variance of the E_+ Around this Center

When studying a sample, the word "residual" was used: it designated the information that could not be included in the dynamic center, due to observation errors, or due to lack of luck in the sampling process, or due to missing variables, or due to the wrong choice of the moving center, etc.

For the population, assuming – this is our hypothesis – a linear relationship between the variables, we will use the terms "deviation": in the population as a whole, there is no "error" in the value of the variable Y and there is no sampling bad luck. Therefore the differences between the "true" dynamic center (up to now we have only been able to find an estimate of it) may only be due to:

- scattering due to human activities (if you are working on cost) or to inconsistent cost accounting system,
- lack of variables.

The differences are then true deviations, the word "residual" being inappropriate.

It can be noted that the variance of the residuals (in the sample then) depends only – whatever the number of variables – on the variance of the cost and the correlation between the cost and the causal variables. This was established in Part IV.

Studying the variance of this distribution is a statistical property: it assumes we were able to draw a lot of samples of the same size from the population. After that, some computations are required.

These computations require making an hypothesis: this hypothesis, called 2 (assumption cited above), is that the distribution of E_+ around its center is normal with a constant variance called S^2_{E+}.

At this stage we can investigate two points: is this hypothesis 2 realistic or not? Can we estimate this variance?

Checking Hypothesis 2 We would like this to be true, as the relationships we wrote assume it is. Can we check it? As the only information which is available is the information contained in the sample, we have to use it to confirm it or not. As usual when dealing with a sample, the answer can never be a definite yes or no, but either "it is very likely so" or "it is unlikely", the likeliness depending on the required level of confidence.

Technically the problem can be described as the following one: from the distribution of the residuals e_{+i} in the sample, can we infer they were drawn from a normal distribution of the E_+?

Different authors built what is called tests of normality, some of them are simple (such as just plotting the data on a normal probability paper) or rather complex (Ref. Sachs [49], p. 322). Two tests are presented here.

Hypothesis H_0 is: the distribution of E_+ is normal.

Test on Skewness and Kurtosis It is well known that the normal distribution presents a skewness equal to 0 and a kurtosis equal to 3. Departure from these values may witness a lack of normality.

How much departure is allowed? As usual it depends on the level of confidence you require. Tables were prepared and are available in statistical text books (Ref. Sachs [49], p. 326).

Test of Kolmogoroff–Smirnoff The test of Kolmogoroff–Smirnoff is logic and interesting: the idea is to compare the distribution of the e_{+i} with what could be expected if they were drawn from a normal distribution. This test works well for small samples, which is quite often the case in cost estimating. It involves the following steps:

- The arithmetic mean $\bar{e}_+$ and the standard deviation s_{e+} of the residuals are computed: this is rather easy if the standard least squares regression was used, as then $\bar{e}_+ = 0$.
- Then the data (residuals) are centered and scaled: let us call them as usual ${}_{cs}e_{+i}$ (this will allows to use the standard normal curve with mean 0 and standard deviation 1):

$$ {}_{cs}e_{+i} = \frac{e_{+i} - \bar{e}_+}{s_{e+}} $$

- Data (residuals) are grouped into classes of equal size; let w be this size. Refer to Chapter 2 for deciding on the width w of the classes:

$$ w \cong \frac{R}{1 + 3.32 \times \log_{10} I} $$

  where $R = e_{\max} - e_{\min}$ represents the range and I the total number of data points.
- The center of class k is referred to as c_k and the number of data point for this class as I_k.
- Then, for each class, this number is compared to what it should have been if the distribution were normal.

From this comparison, a characteristic is computed and compared to tabulated values (Ref. Sachs [49], p. 330).

Estimating the Variance S_{E+}^2 S_{E+}^2 is the variance of the true cost values in the population (what we call the "deviations") around their dynamic center.

The variance of the e_i is in the sample, as usual, computed as:

$$ \frac{\sum_i e_{+i}^2}{I} $$

It can be used for estimating the variance of the population by computing its expectation. The result, of assumptions 1, 2 and 3 are considered as valid, gives:

$$ E\left(\sum_i e_{+i}^2\right) = (I-2) \times S_{E+}^2 $$

The factor $(I - 2)$ comes from the number of degrees of freedom of this distribution when the intercept is different from 0. It should be changed to $(I - 1)$ if the intercept is forced to 0.

An estimate of the population variance is therefore given by:

$$ \hat{S}^2 = \frac{\sum_i e_{+i}^2}{I-2} $$

It can be demonstrated (see Ref. Theil [56], p. 114) that this value is an unbiased estimator of S^2 and corresponds to the maximum likelihood estimator (see Ref. [56], p. 126).

How Accurate is this Variance Estimate $\hat{S}^2_{E+}$? With the same hypotheses, one can establish that the ratio:

$$\frac{I-2}{S^2_{E+}} \times \hat{S}^2_{E+}$$

has a χ^2_{I-2} distribution. From this result a confidence interval for S^2_{E+} can be computed, for a given confidence level.

Example

Using the example mentioned at the beginning of this section, we find the following vector of the residuals:

$$\vec{e}_+ = \begin{Vmatrix} 14.644 \\ -69.328 \\ -148.08 \\ -112.891 \\ -21.328 \\ 90.747 \\ 18.98 \\ 7.99 \\ 130.272 \\ 113.212 \\ 83.217 \\ -56.112 \\ -51.323 \end{Vmatrix}$$

from which we compute $s_{e+} = 83.805, \hat{S}_{E+} = 91.105$ and the variances–covariances matrix:

$$\begin{Vmatrix} 2988 & -712.259 \\ -712.259 & 215.887 \end{Vmatrix}$$

and therefore

$$\hat{B}_0 = 572.972 \text{ with a variance equal to } 2988$$
$$\hat{B}_1 = 101.081 \text{ with a variance equal to } 215.887$$

The "t" values of these estimates are given by:

$$t_{\hat{B}_0} = 10.481$$
$$t_{\hat{B}_1} = 6.88$$

They give a satisfactory level of confidence about these estimated values.

15.6.3 Using Multiplicative Deviations

Using multiplicative deviations is a simple application of the additive deviations.

Starting from the formula $y = b_0 x^{b_1}$ for the sample, we get for the population:

$$\hat{Y} = \hat{B}_0 \times X^{\hat{B}_1}$$

with $\hat{B}_0 = b_0$ and $\hat{B}_1 = b_1$. Their variances are given by:

$$\operatorname{var}(\ln \hat{B}_0) = \frac{\sum (\ln x_i)^2}{I \times \sum \left(\ln x_i - \frac{1}{I} \sum \ln x_i \right)^2} \hat{S}^2_{\ln}$$

where

$$\hat{S}^2_{\ln} = \frac{\sum_i (\ln \hat{y}_i - \ln y_i)^2}{I - 2}$$

is the estimated variance of the deviations of the cost logarithms around their dynamic center in the population and:

$$\operatorname{var}(\hat{B}_1) = \frac{\hat{S}^2_{\ln}}{\left(\ln x_i - \frac{1}{I} \sum \ln x_i \right)^2}$$

If $\ln \hat{B}_0$ follows a normal distribution with the computed variance, then $\hat{B}_0$ follows a log-normal distribution of which the standard deviation is given by:

$$\left\{ e^{\operatorname{var}(\ln \hat{B}_0)} \times \left(e^{\operatorname{var}(\ln \hat{B}_0)} - 1\right) \right\}^{\frac{1}{2}}$$

15.7 Using *J* Quantitative Variables

Using several quantitative variables is very similar – for the classical as well the modern approach – to the case of two variables. The only difference is the general use of matrices. This section just introduces the subject.

The set of data in the sample is therefore represented by two matrices: one is the vector $\vec{y}$ of the values taken for different products by the dependent variable Y, the second one being the set of the values taken by the causal variables $V_1, V_2, \ldots, V_j, \ldots,$

V_P for the same products; if the intercept is not forced to 0 (which is the general case), a column of 1, corresponding to a constant V_0 must be added:

$$\vec{y} = \begin{Vmatrix} y_1 \\ y_2 \\ \cdots \\ y_i \\ \cdots \\ y_I \end{Vmatrix} \quad \|^+x\| = \begin{Vmatrix} 1 & x_{1,1} & x_{1,2} & \cdots & x_{1,j} & \cdots & x_{1,J} \\ 1 & x_{2,1} & x_{2,2} & \cdots & x_{2,j} & \cdots & x_{2,J} \\ \cdots & \cdots & \cdots & \cdots & \vdots & \cdots & \vdots \\ 1 & x_{i,1} & x_{i,2} & \cdots & x_{i,j} & \cdots & x_{i,J} \\ \cdots & \cdots & \cdots & \cdots & \vdots & \cdots & \vdots \\ 1 & x_{I,1} & x_{I,2} & \cdots & x_{I,j} & \cdots & x_{I,J} \end{Vmatrix}$$

Let us remind for the sake of clarity, the conventions:

- There are J causal variables (or "parameters"), numbered from 1 to J, the current number being j. In the matrix $\|^+x\|$ a column – except the first one which bears the number 0 – is attributed to a causal variable. The variables are here either quantitative or qualitative.
- There are I products, numbered form 1 to I, the current number being i. In the matrix $\|^+x\|$ a line is dedicated to a product.
- The indexes always follow the rule $x_{\text{product,variable}}$ or $x_{\text{line\_number,column\_number}}$.
- Small letters always refer to the sample values, capital letters being reserved for the population.

In the analysis of the sample, it was showed (Part III) that the distribution of the dependent variable could be split into two terms:

- A **dynamic center** of which – in the present linear case – formula was established as:

$$\hat{y}_i = b_0 + b_1 x_{i,1} + b_2 x_{i,2} + \cdots + b_j x_{i,j} + \cdots + b_J x_{i,J}$$

  where the b_0 could take different values as a function of the qualitative variables.
- A set of "**residuals**" which took a value e_{+i} for product I, if the additive formula is used.

The advantage of this formulation is to replace a complex distribution, depending on several variables, by just a formula, plus the distribution of one new variable.

We limit the presentation here to the classical approach; the modern approach does not differ from what is previously explained.

The data are now represented by matrices.

15.7.1 Extension to the Population of Various Concepts

The hypothesis we start from is that there is, *in the population*, an approximately true linear relationship between the dependent variable and the J causal variables. The dynamic center of the dependent variable in the population, which expresses this relationship, is then assumed to be:

$$\hat{Y} = B_0 + B_1 X_1 + B_2 X_2 + \cdots + B_j X_j + \cdots + B_J X_J$$

As the relationship between the cost Y and the causal variables is not perfect, we write that for a particular object of the population:

$$Y_i = B_0 + B_1X_{i,1} + B_2X_{i,2} + \cdots + B_jX_{i,j} + \cdots + B_JX_{i,J} + E_{+i} = \hat{Y}_i + E_{+i}$$

$$Y_i = \overleftarrow{X}_i \otimes \vec{B} + E_{+i}$$

where the $X_{i,1}, X_{i,2}, \ldots, X_{i,j}, \ldots, X_{i,J}$ are the values of the parameters for this particular object , $\vec{B}$ being the set of the B_j (with a first line B_0 if the intercept is not forced to 0) and E_{+i} the deviation of the cost Y_i for this particular object from the value of the dynamic center $\hat{Y}_i$ always for this object. It must be clear that for a particular object of this population E_{+i} is not random: it has a definite value, which of course we do not know.

When we consider now all the objects in the population, the only thing we can say about the E_+ is that this variable has a random behaviour because we are completely unable to forecast its value. From our perspective, it seems to randomly fluctuate from one object to another one.

For the distribution of this variable we can make hypotheses, the first one being that its expected value $\mathscr{E}(E_+) = 0$, meaning that the values of E_+ observed on a large number of objects drawn from the population is equal to 0.

In the sample, we also admit that there is a similar relationship and we write for each object of the sample:

$$y_i = b_0 + b_1x_{i,1} + b_2x_{i,2} + \cdots + b_jx_{i,j} + \cdots + b_Jx_{i,J} + e_i = \hat{y}_i + e_{+i}$$

$$y_i = \overleftarrow{x}_i \otimes \vec{b} + e_{+i}$$

and we were able, when trying to minimize the sum $\Sigma_i e_{+i}^2$, to establish that:

$$\vec{b} = (\|^+x\|^t \otimes \|^+x\|)^{-1} \otimes \|^+x\|^t \otimes \vec{y}$$

no hypothesis being required. From that an estimate of $\vec{B}$ is considered (this was called the "plug-out" principle):

$$\hat{\vec{B}} = \vec{b}$$

15.7.2 What Can be Said About This Estimate $\vec{B}$?

The variances of the coefficients are given by the diagonal elements of the variances-covariances matrix:

$$\text{var}(\vec{b}) = S_{E+}^2 \times (\|^+x\|^t \otimes \|^+x\|)^{-1}$$

where S_{E+} is the standard deviation of the deviations from the formula for the whole population; it can be estimated by:

$$\hat{S}_{E+}^2 = \frac{\vec{e}_+^{\,t} \otimes \vec{e}_+}{I - J - 1}$$

where $\vec{e}$ represents the vector of the I residuals relative to the I products (and therefore $\vec{e}_+^t \otimes \vec{e}_+ = \Sigma_i e_{+i}^2$) and J the number of quantitative parameters; the "minus 1" at the denominator comes from the presence of the intercept: it would disappear if the intercept is forced to be 0 (in such a case the data matrix has no column of 1). **These equations need to make several hypotheses**; these hypotheses were previously discussed.

How Far Is This Estimate from the True Value?

We are interested in the difference or error $\hat{\vec{B}} - \vec{B}$ between our estimate and its true value. In order to compute it, we replace in the formula giving $\hat{\vec{B}}$ the vector $\vec{y}$ (the set of all the costs of our sample, which are the "true" costs of our products) by its definition:

$$\hat{\vec{B}} = (||^+x||^t \otimes ||^+x||)^{-1} \otimes ||^+x||^t \otimes (||^+x|| \otimes \vec{B} + \vec{E}_+)$$

which can be written since

$$(||^+x||^t \otimes ||^+x||)^{-1} \otimes (||^+x||^t \otimes ||^+x||) = ||1||$$

$$\hat{\vec{B}} = \vec{B} + (||^+x||^t \otimes ||^+x||)^{-1} \otimes ||^+x||^t \otimes \vec{E}_+$$

or

$$\hat{\vec{B}} - \vec{B} = (||^+x||^t \otimes ||^+x||)^{-1} \otimes ||^+x||^t \otimes \vec{E}_+$$

Of course we cannot compute this expression because we do not know $\vec{E}_+$. However this expression shows that, whatever $\vec{E}_+$, the error $\hat{\vec{B}} - \vec{B}$ is proportional to the matrix $(||^+x||^t \otimes ||^+x||)^{-1}$ we already met when studying the multi-collinearities problems. We noticed that this matrix, if it is ill conditioned, could produce large values which will damage the variances and covariances of $\hat{\vec{B}}$. The interesting point here is that – always in the presence of multi-collinearities – it can also seriously damage the error $\hat{\vec{B}} - \vec{B}$. Of course we do not know – always because $\vec{E}_+$ is unknown – for which component of $\hat{\vec{B}}$ this happens, but it certainly happens! This explains, as we noticed it when studying the multi-collinearities problems, that some coefficients may be "illogical", even it is not always so obvious as in the example we used.

What Are the Properties of the Estimate $\hat{\vec{B}}$

Under the same ordinary least squares (OLS) hypotheses already mentioned when using one parameter only, the same properties can be demonstrated. Computations are not difficult using the matrix notations.

The variances–covariances of the estimators are given by:

$$\text{var}(\hat{\vec{B}}) = (||^+x||^t \otimes ||^+x||)^{-1} \otimes S_{E+}^2$$

an estimate of S_{E+} being given by:

$$\hat{S}_{E+}^2 = \frac{(\vec{y} - \hat{\vec{y}})^t \otimes (\vec{y} - \hat{\vec{y}})}{I - J - 1}$$

15.8 Using Qualitative Parameters

As already mentioned, qualitative parameters are most often required for finding the dynamic centres. Before using them, we must nevertheless check if they are useful or not.

When dealing with the sample, we found that, in order to do that, we had to check if the important constraint (the relationship between the dependent variable and the quantitative causal variable(s) is the same whatever the modalities of the qualitative parameters) is fulfilled or not. Suppose the test was positive and that, consequently, the equation giving the dynamic center of the sample includes qualitative parameters.

In order to use this equation for the population as a whole, we have now to check is the positive result we got for the sample was not an artifact; if the test is positive this equation will deliver estimates for the coefficients usable for the whole population.

In order to make the check for the sample, we made three analysis:

1. In the first one, called α, we do not care about the qualitative variables. We will look for the dynamic center. The euclidian norm of the residuals is named, for a reason which is explained in Chapter 11 χ^2_α.
2. In the second one, called β, we consider that the constraints are fulfilled and we dealt the normal way with the qualitative variables. The euclidian norm of the residuals is named χ^2_β.
3. In the third one, called γ, we consider that the constraints are not fulfilled (the relationship between the cost and the quantitative variable(s) may depend on the qualitative variables) and we consider we have different sub-families depending on the qualitative variables; each sub-family is dealt with independently. The euclidian norm of the residuals is named χ^2_γ.

Let us call, as usual, J the number of quantitative parameters, I the number of products and C the number of used modalities for all the qualitative parameters; to each modality corresponds a different intercept (if we are using linear equations).

We now have to make hypotheses in order to go on. These hypotheses are quite familiar:

- All the residuals computed in the three analysis are independent.
- Their distribution is normal with mean equal to 0 and the same variance (this is the classical homoscedasticity hypothesis).

The fact that these hypotheses are familiar does not mean they are always satisfied. They are reasonable and, if they were not, the BOOTSTRAP could help solve the problem but we are only interested in the concepts here without entering into too much complexity.

Distribution of the Euclidian Norms of the Residuals Vectors

Assuming these hypotheses are valid, the three $\chi^2_\alpha, \chi^2_\beta$ and χ^2_γ then follow, according to its definition, χ^2 distributions of which the number of degrees of freedom df_α, df_β, and df_γ must be computed:

$$\begin{aligned} df_\alpha &= I - J - 1 \\ df_\beta &= I - J - C \\ df_\gamma &= I - C \times (J + 1) \end{aligned}$$

What we win in the Euclidian norms of the residuals when we go from α to β (from computing the dynamic center with no qualitative parameter to computing it with all the qualitative parameters) can be written as:

$$\chi^2_{\alpha-\beta} = \chi^2_\alpha - \chi^2_\beta$$

because – due to an important property of the χ^2 distribution – it also follows a χ^2 distribution, of which the number of degrees of freedom is equal to $df_{\alpha-\beta} = C - 1$.

On the same way, what we win in the euclidian norms of the residuals when we go from β to γ (from computing the dynamic center with all the qualitative parameters to computing C independent dynamic centres) can be written as:

$$\chi^2_{\beta-\gamma} = \chi^2_\beta - \chi^2_\gamma$$

which also follows a χ^2 distribution, of which the number of degrees of freedom is equal to $df_{\beta-\gamma} = C \times (J + 1) - C - J$.

Testing a First Hypothesis

We may have noticed in the sample that the gain $\chi^2_{\alpha-\beta}$ is important.

This means that the different intercepts we compute when using the qualitative parameters are different for the sample. We want to be sure it is not an artifact: is it reasonable to expect the same situation for the population?

What we can do in order to check it is to consider and test this $\boldsymbol{H_0}$ hypothesis: let us suppose that all the intercepts are the same in the population. Is it realistic, then, to compute the gain $\chi^2_{\alpha-\beta}$ in the sample?

Using one of the definitions of the *F*-distribution (see Chapter 3), we know that the ratio

$$F_{\alpha-\beta} = \frac{\dfrac{\chi^2_{\alpha-\beta}}{df_{\alpha-\beta}}}{\dfrac{\chi^2_\alpha}{df_\alpha}}$$

follows an *F*-distribution with $[df_{\alpha-\beta}, df_\alpha]$ degrees of freedom.

If all the intercepts are equal (this is the $\boldsymbol{H_0}$ hypothesis), we expect that this random variable takes a value equal to 0 for the population (we gain nothing in considering the intercepts are different). If it is different from 0 for the sample, we can compute, for a given level of confidence, if we can or not reject the null hypothesis.

Working on an Example Let us take the same example as the one which was presented in Chapter 11. The following values were computed: $\chi^2_\alpha = 119904.39$ (with $df_\alpha = 28$ degrees of freedom) and $\chi^2_\beta = 19268.89$ which gives a $\chi^2_{\alpha-\beta}$ with $df_{\alpha-\beta} = 2$ degrees of freedom.

$F_{\alpha-\beta} = 11.75$ consequently follows an F-distribution with [2,28] degrees of freedom. Looking at the *F*-distribution tables, we find that, given the levels of freedom and the $\boldsymbol{H_0}$ hypothesis (all intercepts are identical):

- The probability that a value equal or higher than 19.5 be found is 0.05.
- The probability that a value equal or higher than 9.45 be found is 0.1.

The computed value for the sample (11.75) has therefore a low – but not negligible – probability to be found given the H_0 hypothesis. The conclusion is that this hypothesis can be rejected with a low risk: the intercepts can be considered as different.

Testing a Second Hypothesis

Given the conclusions about the first test, it is also interesting to check if different formulae for the dynamic center have to be considered or not. This is equivalent to check if it is reasonable or not to consider that the slopes (in the case of linear relationships) are equal?

The null hypothesis H_0 is now: all the slopes are identical (this is the usual way to deal with qualitative variables). In order to make this test, we made the analysis γ and found $\chi^2_\gamma = 2719.64$. Here again the gain is important and amounts to $\chi^2_{\beta-\gamma}$ with $df_{\beta-\gamma} = 2$ degrees of freedom. Does this value favour accepting or rejecting hypothesis H_0?

We compute now:

$$F_{\beta-\gamma} = \frac{\dfrac{\chi^2_{\beta-\gamma}}{df_{\beta-\gamma}}}{\dfrac{\chi^2_\beta}{df_\beta}}$$

As $df_\beta = 24, F_{\beta-\gamma} = 10.31$ follows an F-distribution with [2,24] degrees of freedom. Looking at the F-distribution tables, we find that, given the levels of freedom and the H_0 hypothesis (all the slopes are identical):

- The probability that a value equal or higher than 19.5 be found is 0.05.
- The probability that a value equal or higher than 9.45 be found is 0.1.

We are here in a controversial situation: the probability to find a value of 10.31 under the hypothesis of all the slopes being identical is not far from 0.1. It is up to the cost analyst to decide on what to do. Due to the fact that considering that the slopes are identical is simpler, we would prefer not to reject the hypothesis.

Note: Generally speaking, when using qualitative variables a different result is expected: we expect to find out that hypothesis H_0 cannot be clearly rejected which means that the constraint "all the slopes are equal" is acceptable. The example was chosen in order to show that the contrary may happen and that therefore the test is important and that the decision is always in the hand of the cost analyst.

16 Building the Model

Summary

This chapter closes this part dedicated to the construction of a specific model.

It first tries to motivate the reader of the interest of building such a model.

Of course model building assumes that data are available, that they have been analyzed in order to discover any potential problem, that correlations have been studied, etc.

It also assumes that the difference between "sample" and "population" has been clearly understood.

Once everything has been studied, model building is a rather easy task. The role of the model builder is then to make **several decisions**.

All decisions are based on what is expected for the formula: When it is going to be used? What is the precision which is looked for? Is the information available in terms functional or physical?, etc.

The first decision refers to the variables that should be included in the model. Some comments are made in this chapter, in addition to everything which has been already said.

The second decision is the kind of formula which has to be selected. Among the infinite number of types of formulae, the best compromise has to be found between a complex formula which completely describes the data and the simple one which is easy to handle.

The third decision is the choice of the metric to be used. This choice depends on several things from an *a priori* knowledge of the scattering of the data.

When the formula has been built, its quality should be checked and quantified, one of the reasons for this quantification being to be able to compare different solutions. The second one, maybe more important, is to communicate the information to potential users.

We insisted several times on the fact that several characteristics of a model have to be *decided* by the cost analyst: data analysis helps a lot in this decision-making process, but nevertheless the real decision is in the hand of the model builder. For instance one may quote:

- What are the variables which have to be selected?
- How many variables should be kept?

- What form should take the formula?
- What metric should be used?

The procedure is sometimes called "specification analysis".
This chapter starts with some comments about the so-called specific cost.
It goes on with establishing some criteria for specification decision.

16.1 Why Should We Build a Specific Model?

This is obviously the first question to answer to. After all many people seem to be quite happy with the use of the specific cost. Why should we then bother making something more complex?

About the Specific Cost

The "specific cost" is the cost per unit of size: the cost per kilogram for mechanical items, the cost per meter or square meter in the building industry, the cost per cubic meter for excavation, the productivity (the inverse of the man*day per instruction) for software, etc.

Many people believe in the stability of the specific cost – inside a product family of course, but sometimes also across product families, for product manufactured by some similar technology.

In the vicinity of one data point, the use of the specific cost may seem reasonable, unless of course a change of technology is required. Mathematically this could be demonstrated with Taylor's polynomial: in the vicinity of this data point (defined by its size), the cost for an increment in size is given by the relationship:

$$\text{cost}(\text{size} + \delta\,\text{size}) = \text{cost}(\text{size}) + \frac{\partial \text{cost}}{\partial \text{size}} \times \delta\,\text{size}$$

However, there is no reason to believe that the derivative (∂cost/∂size) (which can be called the "incremental cost") should be equal to the ratio (cost/size), as Figure 16.1 geometrically illustrates.

Of course the incremental cost is more difficult to compute than the specific cost: this may be the reason why the second one is more used than the first one! However, if you want to use the specific cost for estimating purposes, checking it first is a good practice.

The specific cost is largely used by many people, particularly by managers as a quick test for making a judgment about a quotation. It is also sometimes used by project managers for budgetary purposes. The reason may be that it is easy to understand (the human mind thinks linearly!) and very easy to use.

The fact it is largely used, plus the fact it is potentially wrong, deserve some comment.

Where does the use of this specific cost come from? The specific price is deeply engraved in our culture: since our first childhood purchases, we learnt that apples, meat, butter, etc. are sold at a specific price. Why is it so? Certainly not because their specific cost is constant (we will return to it), but because it is convenient: imagine how shopping would be difficult if the specific cost would depend on the quantity you buy! It sometimes happens that retailers propose goods at a specific cost which

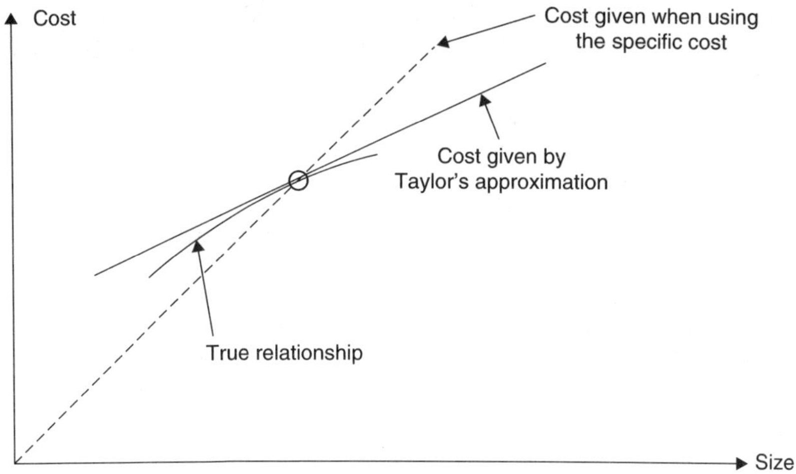

Figure 16.1 The specific cost is generally not the incremental cost.

decreases with the bought quantity, but it seems to be more an appeal than a true application of the variation of this cost.

It is not difficult to indicate, not to demonstrate, why the specific cost cannot be constant: suppose you produce a piece of material weighting 1 kg and that the cost is broken down into two components:

1. The raw material costing, let us say, €10.
2. The machining of this part, costing €100.

The specific cost is obviously 110 €/kg. Suppose now you double all the dimensions of this piece of material. The new mass will amount to $2^3 = 8$ kg and the cost can still be broken into two parts:

1. The raw material, which will cost €80, assuming that the cost of the raw material increases linearly with the mass (it is not exactly the case in industrial procurement: the larger the quantity you buy, the better you are able to negotiate the price, but let us forget this effect here).
2. The machining of the part. What is machined is not the mass but the surface of the part; a first order of magnitude of the cost, assuming that the thickness of metal to be removed is the same, could be proportional to the surface: €400.

The cost of this new piece now amounts to €480 and its specific cost shifts from 110 to 60 €/kg. Note that the cost changes with the mass at the exponent of 0.7, not far from the laws sometimes used by the people working in the engineering business.

This is a first "explanation" of the decrease of the specific cost with the size. But there are others.

A second "explanation", already mentioned when the "correction-by-constant formula" was discussed, deals with the production of something. For machining a part, or making a trench, or making anything such as boring a hole in your wall, two costs have to be considered:

1. A cost for preparing the work, which can be considered, once again it is just a first order of magnitude, as independent of the size of the work or slightly dependent on it.

2. A cost for making this work, which can be considered as proportional to its size, if the size is properly defined (for instance the surface for machining a part, or the volume for excavation).

This result in a non-linear behavior of the total cost.

A third, more basic, explanation was given in Chapter 13 of Volume 1: we show that when the size is increased, for example, the proportion of "low" technologies in the product increases faster than "high" technologies. As low technologies are cheaper than high technologies, the product cost CANNOT grow linearly with the size.

A fourth "explanation" is linked to the industrial practice: if you have to dig a trench of 10 m^3 and another one of 10 000 m^3, you will certainly not use the same machine to do it. For the second one you will use a much more powerful and efficient machine: it will produce the trench at a lower specific cost than the first one.

It is not too difficult to convince people that the specific cost should at least be limited to a product family: after all the cost per kg of apples is different from the cost per kg of pears.

It is more difficult to convince them that the relationship between cost and size is not linear.

When discussing the subject with managers,[1] they eventually agree on the second "explanation". For instance they take the example of the taxi driver who charges first a fix price, and adds to it a price proportional to the distance – and quite often the time – of the journey.

Most of the people know this effect and, in order to keep the specific cost they believe in, decide to use different specific costs for different sizes: this is a first step towards true relationships.

What is the risk of quoting at a fix specific cost? The risk is not difficult to assess: **you loose all the time!**

Let us use Figure 16.2 as an example: the small circles represent the data points, the full line the average specific cost which may be computed from them, the dotted line the true change with the size of this specific cost. Both lines intersect for a size *M*. If you decide to use the average value as the specific cost for your quotations:

- If you quote for a project size smaller than *M*, you will win the contract (your competitors will use a higher specific cost) but you will lose money on it, because your cost is too low.
- If you quote for a project size larger than *M*, you will loose the competition (your competitors will use a smaller specific cost).
- Only if you quote for a project of size in the vicinity of *M* you can expect to win the contract.

Two conclusions at this stage:

1. Check, with the methods developed in this book, how does the cost change with the size.
2. If you want to use a constant specific cost, do it. But at least use a constant specific cost inside a limited range and be careful.

[1] It is true, in many companies, that the cost of materials appear to the project managers to have a constant specific cost. The procurement department negotiates with the manufacturer a price which depends on the quantity the company is going to buy for the following year. Once this is done, this department "resells" the materials to the project managers at a fix specific cost, whatever the quantity they need: for them it appears that the specific cost is constant – at least during a year and probably much longer if the company always buys about the same quantity per year.

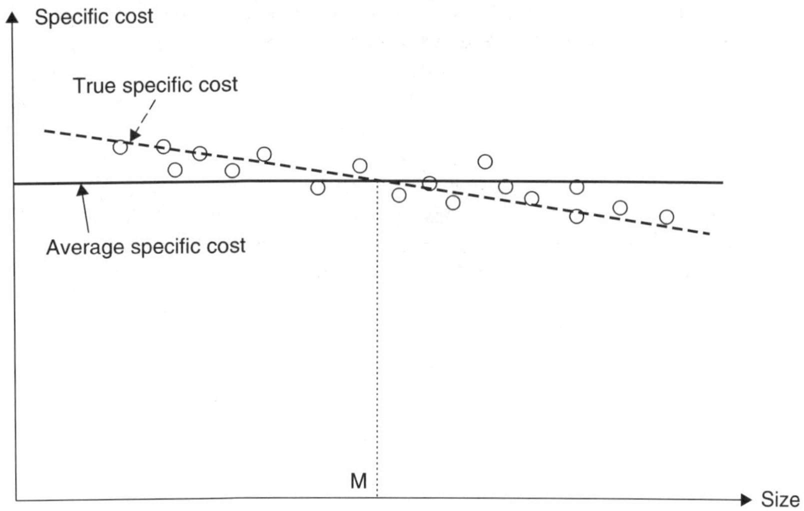

Figure 16.2 The danger of using a fix specific cost.

16.2 How Many Variables?

If the specific cost is not considered reliable enough to make a cost estimate, the idea is then to create a "specific" model which means a model that can be used for a dedicated product family.

The first question to answer to is then: How many variables should be kept for building such a model? And what variables, if there is a choice?

The answer to this question is sometimes linked with the type of formula which will be selected. It may happen that the "best" variables for building an additive formula are not the best one for building a multiplicative one. It means that the choice of the variables may have to be made several times, with different formula types.

We assume here a linear relationship is investigated.

Qualitative Comments

Before any choice might be done, some qualitative comments can be made:

1. The formula must make sense. What we mean is that it has to be acceptable from an engineer's point of view. We already saw this problem when dealing with collinear variables (Chapter 6). Another point of view to be considered is when engineers strongly believe that such or such variable should be included: the formula has to be used by people; it is sometimes better to build a formula slightly less satisfactory than the one the cost analyst would prefer, than to build one that the potential users would reject.
2. This question has a more fundamental background: must we use any preliminary information that does not appear itself when working with the sample? We believe so. As it was several times reminded, a sample is just (supposed to be) the result of a random drawing from a population and it is quite possible that an

interesting phenomenon, due to bad luck, escaped from the drawing process. After all engineers may have good reason to consider that one variable must be part of the formula. It is always better of course to check their statement, if we can.

3. Although it is obvious, it must be recalled that the variables which are included in the formula must be known at the time the formula will be used. It serves no purpose to build a "perfect" formula which cannot be used.
4. Functional or physical variables? This question was already mentioned but is worth repeating. Functional variables are variables which describe the product from the user's point of view: What can he/she expect from the product (examples: power, flow, load, etc.)? Physical variables describe the product from the engineer's, the manufacturer's, point of view: mass, number of parts, materials, etc. The cost analyst should remember that:
   - the formula must fit the needs of the people who uses it,
   - the user of the formula may be more interested, depending on his/her needs at a particular time, in the capacity of making trade-off analysis than on just getting a cost. Trade-offs analysis have to be based mainly on functional variables.

   Functional characteristics are especially useful in the early phases of a project, physical ones becoming more interesting when some definition of the product is already available. Therefore the choice of the variables depends on the time the cost estimate with the formula will have to be done. Generally speaking it is often a good idea to build two different formulae, one using functional characteristics, the other one physical characteristics.
5. How many variables should be included in the formula? Keeping in mind what has already been said, there is always a trend to add as many variables as possible, with the hope they will improve the quality of the cost forecast. This is sometimes true, if variables are not collinear. But one may always ask the question: Is it worth it to keep variables which bring only a marginal improvement of the quality, as measured for instance by the R^2?

   This point is strongly correlated with the number of data points you have. It must not be forgotten that a specific cost model is a set of two things: a formula giving the dynamic center plus the distribution of the deviations around this dynamic center. The variance of this distribution is estimated from the sum of the squares of the residuals in the sample, divided by $I - J$, where I represents the number of data points and J the number of causal variables, quantitative or qualitative (including the intercept, if any). When J becomes closer and closer to I, then this variance increases very quickly. This means that you will get imprecise coefficients or even that you will not be able to quantify the model quality.

   We all know that, in the cost domain, the number of data points is always limited. A practical rule of thumb is therefore to get, if it possible, at least 5 data points per variable. This means that if you want to use 2 variables, try to get at least 10 data points. This rule can be alleviated when the number of variables increases, but not too much. That being said, you may have to work with less data points but be careful about the formula you get, except if you are completely confident in the cost values of the database you use.
6. For deciding about the variables to be included as "cost drivers", one might think it is a good idea to get some information about the manufacturing process. Of course it is difficult to oppose to this formulation. Nevertheless as the title of the book reminds it, the purpose of a specific model is to be able to cost estimate before the operations sheets ate prepared; this means from the product description (functional or physical). However, it is sometimes a good idea to

check the consistency of the detailed cost estimating process by building models from the information available in these sheets; some people even think that one the primary purpose of specific – or even general – models is to do it and also to "debug" database. Why not, of course? And models are very powerful to do it, but it is a slightly different problem.

Let us comment a bit more about this subject, because it sets a limit to the specific – or even the general – modelization process. Suppose you have a set of costs for products you consider as similar enough to be considered as belonging to a product family. You try to build a model from these costs but it does not work properly. Investigating the question you discover that some products were machined on a new, very efficient, machine because it was available, whereas other ones were made on more conventional machines. You also discover that small products were machined several at a time because their size allows it; etc. The obvious conclusion, if we are to build one model from data, is that the manufacturing process should be about the same or to add other parameters.

7. The number of parameters to be included in a formula depends also on the homogeneity of the product family. In a very homogeneous one, one variable is enough (the product size), but it is rather rare in the cost domain.

   Using qualitative variables is an interesting subject which should be considered. If you do it, check about the right use of these variables.
8. Try to avoid, as much as possible, subjective variables. Refer to the comments we made in Chapter 1 about this subject.

16.2.1 A Simple Selection: The "Stair Case" Analysis

A simple procedure for deciding about the variables to be included in the formula consists in computing all possible formulae, using from one to all variables. If the number of variables does not exceed three, this solution can still be done manually although it represents already the computation of seven formulae. If more variables are existing, this becomes time consuming: it is then possible to automatize the process.

The procedure involves 6 steps:

1. The cost analyst selects the formula he/she thinks is suitable.
2. He/she selects the criterion from which the procedure will compare the results: the procedure has to make some choices and it must "know" on which criterion the choice has to be based.
3. The procedure then, *using just one variable*, makes all the computations and, according to the selected criterion, decides what is the parameter which gives the best results.
4. Now using two variables by adding, one at a time, a new variable to the already selected variable, it determines what is the best couple, always according to the same criterion.
5. It then searches for the third important variable, then the fourth, etc.
6. Until it cannot improve the results.

The procedure is fast and efficient. It can be improved by starting, at step 3, by using two variables simultaneously: it may happen that, in a couple, no variable is particularly interesting but that, as a couple, both variables are a good choice. But this is rather rare, but we already saw it happen.

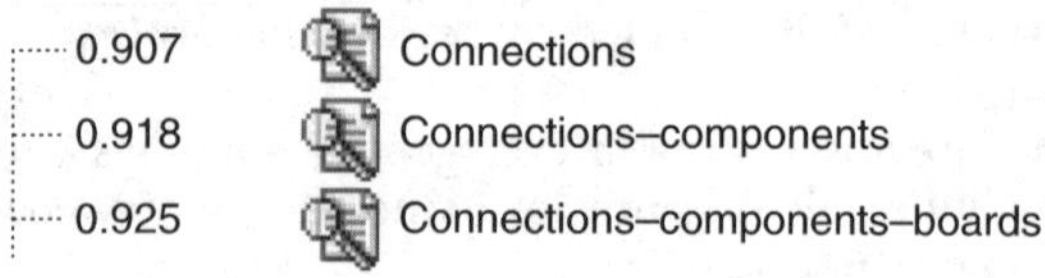

Figure 16.3 A stair case analysis.

This can be easily implemented. The result, for the example we previously used, detecting the R^2 as the criterion, appears in Figure 16.3:

- The most interesting parameter is the number of connections: used alone, it provides an R^2 equal to 0.907.
- The second more interesting is the number of components: added to the number of connections, the R^2 goes to 0.918.
- The third one is the number of boards (R^2 goes to 0.925).
- Whereas the use of the mass gives a very marginal increase, as R^2 goes only to 0.927 only (and do not forget that the mass is strongly correlated to several other variables).

From this result it is possible to conclude that, assuming of course that all the parameters are known at the time the estimate is needed, the mass can be discarded.

16.2.2 A Logic Approach Based on "Partial" Regressions

This section illustrates what can be done for parameters selection. It is also based on the increase of the coefficient of determination R^2, but on a more "in depth" analysis. It uses the step-by-step analysis already developed in Chapter 6, about visualization.

We have to select quantitative parameters. This requires computing how much each parameter does contribute to the R^2. The procedure consists to work in 3 steps:

1. A linear regression (ordinary least square, OLS) is made for y on all the variables – the intercept being considered as a variable – except one, let us say x_j (one column of the $\|{}^{+}x\|$ matrix), which is therefore removed from this matrix. The new matrix is called $\|{}^{+}x_{[\cdot,j]}\|$.
2. Then a regression is made for x_j on the same "reduced" matrix.
3. And eventually the residuals of these two regressions are simultaneously compared. These residuals are obviously "parents" as they are based on variables on the same matrix. Suppose the residuals are about the same: this means that x_j contains information which could be very useful for explaining the residuals of y.

Introducing Some Notations

For making the computations, in order to make the presentation as clear as possible, we need to introduce some notation.

We investigated in Chapter 3 two linear regressions for each variable V_j:

- $\vec{y}$ on $\|{}^{+}x_{[\cdot,j]}\|$ on one hand. We call $\vec{e}(y, x_{[\cdot,j]})$ the residuals of this regression: the symbols between parenthesis reminds us where these residuals originate (the vector and matrix notations are removed from this parenthesis in order to alleviate the notation). Note that the coefficients $b_{k,x[\cdot,j]}$, related to variables V_k ($k \neq j$), of this regression are not generally equal to the coefficients of the full regression, up to now called b_j; it is the case only if all variables are orthogonal. The **multiple correlation coefficient** of this regression is noted R_j.

- $\vec{x}_k$ on $||^{+}x_{[\cdot,j]}||$ on the other hand, and we call its residuals $\vec{e}(x_k, x_{[\cdot,j]})$. This is the regression of parameter j on all the other parameters. We found that:
  - the slope of this regression is equal to b_j which is the coefficient of the full regression of $\vec{y}$ on $||^{+}x||$,
  - and its residuals are the residuals of the full regression of $\vec{y}$ on $||^{+}x||$.

Now we have to investigate how much these residuals are related.

Theil investigated ([56], p. 171) the correlation of these residuals $\vec{e}(y, x_{[\cdot,j]})$ and $\vec{e}(x_k, x_{[\cdot,j]})$. He called r_k this **partial correlation coefficient** and demonstrated the interesting relationship:

$$1 - R^2 = \left(1 - R_j^2\right) \times \left(1 - r_j^2\right)$$

From this relation, it is clear that the contribution of variable V_j to the coefficient of determination is given by:

$$R^2 - R_j^2 = r_j^2 \times (1 - R_j^2)$$

One Example

Let us work on the example presented in Chapter 6. We have five variables: the intercept, the mass, the number of connections, the number of components and the number of boards (Figure 16.4).

A regression of y on all the variables leads to $R^2 = 0.927$. The following figure presents the results of the computations on partial regressions.

Omitted variable (j)	R_j^2	r_j^2	Contribution $R^2 - R_j^2$
Intercept	0.166	0.901	0.751
Mass	0.925	0.025	0.002
Number of components	0.908	0.205	0.019
Number of connections	0.883	0.374	0.044
Number of boards	0.920	0.089	0.007

Figure 16.4 The multiple and partial correlations for the example.

What conclusions can be drawn from these computations?

1. The contributions follow exactly the hierarchy found in the simple solution. The advantage of this logic solution is that the intercept is added.

2. This intercept explains a lot of the residuals in the full regression. This probably comes from the fact that the slopes are rather small.
3. A high value of r_j^2 which quantifies the correlation between the residuals $\vec{e}(y, x_{[\bullet,j]})$ and $\vec{e}(x_j, x_{[\bullet,j]})$ is an interesting feature: it shows that the residuals on y when parameter j is not selected can largely by explained by this parameter, which is therefore probably a very important parameter: large r_j^2 must be looked at.
4. When R_j^2 is large (which is another way to look at the correlations between parameters), one cannot expect an important contribution of parameter j to the R^2: the formula given on the previous page quantifies it and the example illustrates.
5. When r_j^2, the correlation between the residuals are small; this means that parameter j does not really "explains" the variations of y. This is the case, for instance, of the mass.
6. You may have noticed that the sum of the contributions, here 0.823, is not equal to R^2. This comes from the fact that, in the computation of R^2, the number of products and the number of parameters should be taken into account in order to produce the "adjusted R^2":

$$R^2_{\text{adjusted}} = R^2 - \frac{J-1}{I-J}(1-R^2)$$

16.2.3 The "Press" Procedure

This procedure was proposed[2] by D. M. Allen. It is based on the deletion of products.

When product i is deleted from the observations, the new value of the residual for this product i is given by:

$$e^*_{i\,[i,\bullet]} = \frac{e_i}{1-h_{i,i}}$$

where $h_{i,i}$ is the diagonal element of the HAT matrix, corresponding to product i. The "press" value, which stands for "prediction sum of squares", is given by:

$$\text{press} = \sum_i e_i^{*2}$$

should be as small as possible, the logic being that the residuals should be as small as possible, even when one data point is deleted.

Allen recommends to make all the regressions, using one, then two, etc. variables and then to keep the solution which gives the less *press*. Of course the amount of computations is rather high!

[2] Allen. Technical Report No. 23. Department of Statistics, University of Kentucky, 1971.

16.2.4 The Residual Variance Criterion

Suppose we have two sets of causal variables and that we make regressions on both sets; one set is the correct one, the other one being the incorrect one. Theil showed ([56], p. 543) that, given some hypotheses, on the average, the residual variance estimator of the incorrect set exceeds that of the correct one.

He concludes that this justify the selection of the set with the smallest residual variance estimate. This is an interesting result, but which is true only on the average (which means the mathematical expectation) and therefore says nothing about your particular case study.

16.2.5 What to Do with a Limited Set of Data?

We already insisted several times about adding variables to get as most as possible a complete description of the products gathered for preparing a specific model.

The first thing to do is obviously to get as many products as possible in the product family for which you want to build a model. But there are circumstances, which are not rare, where the number of products is limited. You should nevertheless get a good description of these products; consequently you may have several variables you would like to use in order to "explain" the cost.

A good rule of thumb already mentioned is to get about at least five products per parameter. You could, mathematically speaking, build a model including as many products as you have parameters. The inconvenience is that you do not have the faintest idea of the validity of this model.

Another solution is to use a step-by-step analysis:

1. First build the formula for the dynamic center of the cost distribution with just one variable and no more than two coefficients, whatever the formula type. This variable should be related to the products size:

$$y_i = f(b_0, b_1, x_{i,1}) + e_{i+}^{(1)}$$

2. Then compute the residuals around this dynamic center. These residuals should be, whatever the formula, the additive e_{i+}.
3. Try afterwards to correlate these residuals with another variable; as you have as many residuals you have products, this remains possible. For doing so you compute the correlation between these residuals and all the remaining variables and select the one which presents the highest correlation (do not forget that you can establish, if it is needed, a correlation between a quantitative variable – the residuals – and a qualitative parameter).
4. Establish the formula giving the dynamic center of these residuals according to this second variable:
   - if this variable is quantitative, you get a second relationship such as:

$$e_{i+}^{(1)} = g(b_3, b_4, x_{i,2}) + e_{i+}^{(2)}$$

- if it is qualitative, you may group your products into sub-families, as far as the qualitative variable is concerned and decide about a constant for each modality of the qualitative variable:

$$e_{i+}^{(1)} = m_1, \text{ or } m_2, \text{ or } \ldots + e_{i+}^{(2)}$$

5. As you were using additive residuals, you may now write if variables V_1 and V_2 are not correlated:

$$y_n = f(b_0, b_1, x_{i,1}) + g(b_2, b_3, x_{i,2}) + e_{i+}^{(2)}$$

You can of course go on, "explaining", one after the other the residuals with new variables. This procedure is of course a little time consuming, although modern software can prepare that very easily. The main advantage of it is that you keep a complete visibility of the process and you see immediately how adding a variable allows to improve the quality of the model.

If V_1 and V_2 are correlated, you should, before going to step 4, "decorrelate" them. For doing that, follow the procedure described in Chapter 2 under the name "step-by-step" analysis:

- You establish the formula giving the dynamic center of y as a function of V_1 as, for instance, if the correlation is linear:

$$x_{i,2} = c_0 + c_1 x_{i,1} + z_i$$

  where z_i is the residual of the operation.
- Then you establish the link between V_1 and V_2:

$$e_{i+}^{(1)} = g(b_3, b_4, z_i) + e_{i+}^{(2)}$$

Using such a formula is not difficult: you just have to follow the procedure backwards.

16.3 What Kind of Formula?

This is an important step when developing the formula to find out the "dynamic center" of the cost distribution in the sample.

Introduction

Throughout the chapters of this volume, we mentioned analyzing the data with the purpose of building a model consisting in the sum of one formula (the "dynamic center") and the distribution of the residuals.

But a cost-estimating model might be more complex than that: it can include several formulae, organized in a chain, each formula computing the value of an intermediate variable that will be entered in the following formula. If the variables are independent and can be computed from other independent variables, this solution presents no difficulty at all.

But the situation might be more complex, as the following example illustrates: suppose you are interested in the development cost of products belonging to a product family. You have data and would like to use them; however these data are "polluted" by another variable, let us call it "experience", which is a continuous variable. You can of course attribute an experience value to each product (from 0, meaning no experience at all, to 1, meaning the designers will just have to copy an existing design), but doing so you force the model to apply to the variable "experience" the same law as the other variables (for instance a law in power), whereas the influence of the experience may not follow at all such a law.

Another solution is possible; it consists in several steps. Suppose you decide that you want to build a model based on what the cost should be for a development when no experience is available; then the level of experience will be used to reduce this "0 experience development cost" to the expected cost taking into account this level. The steps are then the following ones:

1. You build a scale of experience, for instance from 0 to 1. This scale – which, to start with, is very subjective – gives examples of what is meant by several levels of experience. It is then applied to each product in the database.
2. You "guess" what the relationship for taking into account the level of experience should be, for instance, a linear or sinusoidal relationship going from 1 when no experience is available, to 0.1 (or any other figure you can estimate when interviewing people who met this situation before) when a full experience is available.
3. You use this formula to *normalize* the cost to a 0 experience cost.
4. You then create a dynamic center, where the experience does not appear any more, linking the normalized cost to other variables. This dynamic center generates a set of residuals.
5. You compute the correlation between the level of experience and these residuals. Ideally no correlation should exist, but it is generally not the case immediately. Analyzing this correlation should suggest how to improve the relationship mentioned in step 2; as you are dealing with one variable only, the use of the graph is the right tool to be used. This analysis may suggest:
   - checking the level of experience attributed to each product: after all, you may have been too optimistic or pessimistic for such or such product;
   - rebuilding the scale or, preferably, "reguessing" the relationship mentioned in step 2.
6. You restart the process several times until a satisfactory output appears.

This procedure allows to combine subjective ideas and computations in a way that is clear: computations allow to remove a lot of subjectivity from the ideas.

Building a Formula

It has been repeated that the selection of the formula type is a decision, as no algorithm can be used to automatize it. The choice may be based on previous experience, preconceived ideas or testing different kinds of formulae. Nevertheless it must not be forgotten that a formula can very well fit with the data available in the sample but this does not prove it is really valid for the whole population.

For this reason we always recommend to check the formula type with experienced people: if, for example, a linear formula is selected, does linearity make sense with such or such variable? It can very well happen that an interesting potential formula can be:

$$\hat{y} = b_0 + b_1 x_1 + b_2 x_2^{b_3}$$

which is not easy to establish (looking at the residuals on a preliminary formula might help) but quite possible.

The cost analyst should never forget that the mathematical treatments are there to establish the values of the formula coefficients, not to decide about the shape of the relationship.

Now, let us remind the cost analyst that there are four "basic" formulae:

1. *Linear, or additive*: This formula is very basic but rarely the best one when dealing with cost; however, it must be considered for small product sizes:

$$\hat{y} = b_0 + b_1 x \quad \hat{y} = b_0 + b_1 x_1 + b_2 x_2 + \cdots$$

2. *Multiplicative*: This formula can be chosen as soon as the product size is not small:

$$\hat{y} = b_0 \times x^{b_1} \quad \hat{y} = b_0 \times x_1^{b_1} \times x_2^{b_2} \times \cdots$$

3. *With a correction by a constant*: This formula synthesizes both linear and multiplicative formulae; it should therefore be considered when the product sizes goes from small to large values:

$$\hat{y} = a + b_0 \times x^{b_1} \quad \hat{y} = a + b_0 \times x_1^{b_1} \times x_2^{b_2} \times \cdots$$

4. *Exponential*: This formula may give better results in some circumstances, but should be used with extreme care outside the domain of definition of the product size:

$$\hat{y} = b_0 \times b_1^{x} \quad \hat{y} = b_0 \times b_1^{x_1} \times b_2^{x_2} \times \cdots$$

The cost analyst must also not forget that variables are there in order to palliate inhomogeneities in the product family. In this respect, the use of qualitative variables is, when dealing with cost, very often compulsory.

16.4 Selecting the Metric

If your data are not scattered (you are lucky!) you may use any kind of metric. In such a case use the simplest one: the linear regression (OLS).

But it is rarely the case in the cost domain: selection of the metric becomes important when the data start to be scattered, let us say when R^2 becomes less than 0.9, which means the correlation coefficient less than 0.95. It becomes critical when this R^2 is less than 0.8 (or the correlation coefficient less than 0.9).

The question is dealt with in Part III. Generally speaking:

1. consider using the metric defined by the median;
2. if you want to use the standard linear regression, do not forget to correct for its bias;
3. if the range of your product sizes is large (let us say larger than 3 and, at fortiori, 10) the metric defined by a ratio should be preferred, the most common one being:

$$\left| \frac{y_i - \hat{y}_i}{\hat{y}_i} \right|^{\alpha}$$

with α generally equal to 2.

4. if a few data points are far from the main set, you may prefer, if you want to keep them in the formula, to use the metric defined by the biweight.

And never forget that building a formula must be preceded by a careful data analysis, a search for potential outliers and the resolution of problems which may be caused by multicollinearities.

16.5 Quantifying the Quality of the Formula

16.5.1 Introduction

No algorithm can tell the cost analyst about the real, or predictive, quality of a cost model. The cost analyst is the only person who may say that "the model does make sense", which is eventually the most important thing about a cost model.

Once it has been decided that the model does make sense, algorithms can help give some information about the precision with which future estimates can be made. These algorithms are all based on the analysis of the residuals; this is the reason why this analysis is an important step in establishing a formula: the smaller the residuals, the better the estimates.

Therefore the standard error of an estimate is an important piece of information about the quality of a specific model. It can be directly computed, as it is explained in Part IV of Volume 1.

However, it is time consuming and one would prefer to get a synthetic information about this quality. A very good information is provided by the variance of the formula coefficients; these variances are of course strongly correlated to the standard error of the estimate. They have been computed in details.

What about having just a number which could quantify the quality of a specific model and which could easily be used for comparing different formulae? This section is devoted to the search of such a number.

16.5.2 Numbers Directly Based on the Residuals

There are a lot of numbers which can be built. This section proposes several ones and you can certainly build your own: the best numbers are the ones you are used to.

Numbers Directly Based on the Sample Values

One could start by computing the sum of the residuals $\Sigma_i e_{+i}$ or their average absolute value $\Sigma_i |e_{+i}|/I$. However, in the cost domain (as previously mentioned) relative residuals are more indicative than their values, especially if the cost range is large. We therefore turn towards relative values.

The easiest number to compute is the total percentage – if it is expressed in percentage – of relative residuals (TRR) (sometimes called "total bias" but this appellation has not our preference) defined as:

$$\text{TRR} = 100 \times \sum_i \frac{y_i - \hat{y}_i}{\hat{y}_i} = 100 \times \sum_i \frac{e_{+i}}{\hat{y}_i}$$

where e_{+i} are the additive residuals. A very low value tells that the built dynamic center goes very well "in the middle" of the data points, at least for our purpose.

But this number is difficult to understand, because it depends on the number of data points on one hand and because negative values can very well be compensated by positive values, giving an illusion a high quality model: a dynamic center can very well go exactly in the middle (the value of the TRR is then 0) of the data points but the model may have a very low predictive quality.

The simplest value we prefer is the ARR or "Average of Relative Residuals" defined as (in percentage):

$$\text{ARR} = \frac{100}{I}\sum_i \frac{|e_{+i}|}{\hat{y}_i}$$

This value is very easy to understand: if you find a value of 5%, it means that, on the average, your data points are at 5% of the dynamic center. There is no need to be trained on statistics to understand it. It should nevertheless be checked with a view on the graph of the residuals: a small value may hide the fact that you have a lot of very small values and a few high ones (are not these data outliers?): the sign test may also help discover that.

The drawback is that it has no statistical property: it is, generally speaking, very difficult to compute with absolute values. For this reason statisticians prefer to work with the e_{+i}^2: we will find them in the next sections.

Numbers Based on Estimated Values for the Population

You want to be sure that the model you built is satisfactory for the whole population: if it is satisfactory for the sample, it is a cause of satisfaction, but what about the population?

You have no way to check if the dynamic center you computed goes exactly through the distribution of the costs for the whole population. We established in Chapter 15 that, given some hypotheses, the coefficients of the formula have no bias. This is fine, but it is a statistical property which does not say anything about the *particular formula* you just built.

So we are not completely sure about the center (static or dynamic) of the cost distribution for the population.

However, we saw that the variance of the distribution of the deviations, always given some hypotheses, around this center can be estimated by:

$$\hat{S}_{E+}^2 = \frac{\sum_i e_{+i}^2}{I - J - 1}$$

where I is the number of data points, J is the number of parameters (not including the intercept). Some model builders, in order to work with relative deviations, prefer to use:

$$\frac{\sum_i \left(\frac{e_{+i}}{\hat{y}_i}\right)^2}{I - J - 1}$$

from which the "standard error" of the deviations, also called the standard error of the estimate which means the same thing, can easily be computed as:

$$100 \times \sqrt{\frac{1}{I-J-1}\sum_i \left(\frac{e_{i+}}{\hat{y}_i}\right)^2}$$

in percentage.

16.5.3 The Coefficient of Determination R^2

The R^2 is one of the most frequently used global test about the quality of a specific model. Many cost analysts look first at this "coefficient of determination" for quantifying the quality of the formula.

This R^2 is easy to understand using the additive residuals; this is one of the reason we advocated, whatever the way the residuals are computed, to always, at the end of the process, compute the additive form of the residuals.

Let us remind the notations: the dynamic center of the cost distribution is given by $\hat{y}$. For a particular product of the database we write:

$$y_i = \hat{y}_i + e_{+i}$$

when the residuals e_{+i} are defined as additive. An important value is given by $\Sigma_i e_{+i}^2$.

For a Linear Formula Using the Ordinary Regression Analysis (OLS)

Using the standard linear regression, $\Sigma_i e_{+i}^2$ is precisely the quantity we tried to minimize and from the minimization process[3] the values of b_0 and b_1 were computed (in the case of one parameter only):

$$b_1 = \frac{\sum_i (y_i - \bar{y})(x_i - \bar{x})}{\sum_i (x_i - \bar{x})^2}$$

$$b_0 = \bar{y} - b_1 \bar{x}$$

Once the regression is made, what remains is given by:

$$\sum_i e_{+i}^2 = \sum_i \left(y_i - b_0 - b_1 x_i\right)^2$$

[3] For simplicity purposes, the discussion here is based on one parameter only. It can obviously be extended to any number of parameters.

As this value is not dimension free and as we wanted – as we are used to when computing correlation coefficients – to get a value between 0 and 1, with 1 being synonymous to something good – we may define R^2 as:

$$R^2 = 1 - \frac{\sum_i e^2_{+i}}{\sum_i (y_i - \bar{y})^2}$$

This expression does not require any hypothesis and therefore is always valid. It can be used for any formula, built from any metric, as soon as the additive residuals were computed (which does not mean that the sum of their squares was minimized by the metric: this is something different).

Let us get another expression of this formula. One can write:

$$\sum_i e^2_{+i} = \sum_i (y_i - \hat{y}_i)^2 = \sum_i [(y_i - \bar{y}) - (\hat{y}_i - \bar{y})]^2$$

In the second expression the first parenthesis is the difference between the cost values and their average, the second being, because (this is valid only for the linear regression):

$$\bar{\hat{y}} = b_0 + b_1\bar{x} = (\bar{y} - b_1\bar{x}) + b_1\bar{x} = \bar{y}$$

and therefore $\bar{\hat{y}} = \bar{y}$, the difference between the nominal cost and their average. Let us develop this second expression:

$$\sum_i [(y_i - \bar{y}) - (\hat{y}_i - \bar{y})]^2 = \sum_i (y_i - \bar{y})^2 + \sum_i (\hat{y}_i - \bar{y})^2 - 2\sum_i (y_i - \bar{y})(\hat{y}_i - \bar{y})$$

Fortunately enough (this is only true also for the linear regression), the third term of the second expression is equal to 0, as the reader may prove it by developing it and using the values of b_0 and b_1. We then have a second expression of this R^2:

$$R^2 = \frac{\sum_i (\hat{y}_i - \bar{y})^2}{\sum_i (y_i - \bar{y})^2}$$

which is generally the way the R^2 is defined. Notice that this presentation is only valid if the linear regression (OLS) was used for computing the values of $\hat{y}$ (otherwise the square term of the preceding expression does not vanish).

A third expression can be found in developing it. We find:

$$R^2 = \frac{\left[\sum_i (x_i - \bar{x})(y_i - \bar{y})\right]^2}{\sum_i (x_i - \bar{x})^2 \sum_i (y_i - \bar{y})^2}$$

which is precisely the square of the correlation coefficient between x and y (see Chapter 5).

An Easy Interpretation of the R^2

All the formulae presented upwards do not give an intuitive interpretation of the coefficient of determination R^2. The following paragraph, even if it lacks mathematical rigor, allows for such an easy interpretation.

Statisticians generally consider that the average value $\bar{y}$ does not convey any information for the simple reason that, if all the data points have the same value of the dependent variable whatever the value of the causal variables, no correlation can be established with any other variable: they consider they know nothing (about possible relationships between variables) and they are unable to "explain" the value of this dependent variable.

Let us then define the level of information which is available in the sample as[4] $\Sigma_i(y_i - \bar{y})^2$: it is the square of the differences between the observed values and their average; if all observed values are identical to the average, then this level of information is 0, as mentioned in the previous paragraph. The level of information which is lost when the set of observed values is replaced by the dynamic center of the distribution is therefore given by:

$$\sum_i e_{+i}^2 \text{ or } \sum_i (y_i - \hat{y}_i)^2$$

and the level of information which is saved in the dynamic center by $\Sigma_i(\hat{y}_i - \bar{y})^2$. According to the formula given upwards, we can write:

information available in the sample = information saved in the dynamic center + information lost in the residuals

which gives a convenient interpretation of the formula:

$$R^2 = 1 - \frac{\sum_i e_{+i}^2}{\sum_i (y_i - \bar{y})^2}$$

Consequently if you multiply R^2 by 100, you may say that **it represents the percentage of available information which is incorporated in the dynamic center**, or that $1 - R^2$ represents the ratio of the information lost on the available information. The obvious consequence is that, as we want to incorporate as much as possible of the available information as possible in the model, we wish to get R^2 as close as possible to 1 for this model, whatever the model or the metric.

Application to the Population

The value of R^2 is of course computed for the sample. Is it reasonable to apply its value to the whole population? Let us call[5] $\hat{R}^2$ the estimated value of the coefficient for the population.

---

[4] We know we depart from Shannon definition of information. Our purpose is not to build a theory of information, but to give some perspective in the real meaning of the coefficient of determination.

[5] The purpose is always to distinguish between the sample and the population.

As explained in Chapter 15 we cannot respond directly to this question. The only thing we can do is to test the H_0 hypothesis $\hat{R}^2 = 0$ (Y and V_1 are independent in the population). If it is the case, what can be the distribution of the R^2 observed in a sample of size I? If the distribution of Y and V_1 are both normal[6] (which is a strong hypothesis!), then:

$$\frac{R}{\sqrt{1-R^2}}\sqrt{I-2}$$

follows a Student distribution with $I - 2$ degrees of freedom.

The center of this distribution is of course equal to 0. If we accept a level of confidence of 90%, it can easily be computed that:

- for $I = 10$, we will reject the hypothesis if $R^2 > 0.3$;
- for $I = 20$, we will reject the hypothesis if $R^2 > 0.27$.

Do not forget that says nothing about the true value of R^2 in the population.

A Correction

A correction can be used to estimate the value of $\hat{R}^2$ for the population. It can be built from the same formula as the R^2 taking into account the following modifications (based on the average information per product):

- An estimated value of the variance of the deviations in the population is given by (where $I - J - 1$ becomes $I - J$ if the intercept is forced to be 0):

$$\frac{1}{I-J-1}\sum_i e_{+i}^2$$

This represents the average amount of information lost per product.

- The average amount of information per product is given by:

$$\frac{1}{I-1}\sum_i (y_i - \bar{y})^2$$

The term $I - 1$ comes from the lost of one degree of freedom due to the computation of $\bar{y}$.

Then we can write:

$$\hat{R}^2 = 1 - \frac{\dfrac{1}{I-J-1}\sum_i e_{+i}^2}{\dfrac{1}{I-1}\sum_i (y_i - \bar{y})^2}$$

If $I = J + 1$, then necessarily $\Sigma_i e_{+i}^2 = 0$ and $\hat{R}^2$ is not definite, which is normal (the division 0/0 is not defined).

[6] As usual, in order to be able to do some computation, we need to start with some hypothesis.

What Does R^2 Says About the Values of B_0 and B_1?

Not that much. Malinvaud ([38], p. 91) wrote that "this coefficient is not particularly interesting" adding that for small samples, which is often the case in cost estimating, R^2 can be high although the variances of $\hat{B}_0$ and $\hat{B}_1$ may be high, whereas for a large sample $\hat{B}_0$ and $\hat{B}_1$ may give exact values of B_0 and B_1, although R^2 remains fairly low. Consequently R^2 must really be seen as the information contained in the formula, not a test about the coefficients.

Nevertheless one can easily demonstrate that:

$$\frac{\mathrm{var}(b_1)}{b_1^2} = \frac{1}{I-2}\frac{R^2-1}{R^2}$$

which means that the closer R^2 is to 1, the less is the variance of the slope (this is logic).

For Any Formula Not Using the Standard Regression Analysis

When the standard regression analysis is not used, the equality:

$$R^2 = \frac{\sum_i (\hat{y}_i - \bar{y})^2}{\sum_i (y_i - \bar{y})^2}$$

is not valid anymore because the cross term:

$$\sum_i (y_i - \bar{y})(\hat{y}_i - \bar{y})$$

does not vanish: using this formula may lead to values of R^2 higher than 1, which is rather embarrassing!

However, the formula:

$$R^2 = 1 - \frac{\sum_i e_{i+}^?}{\sum_i (y_i - \bar{y})^2}$$

which uses only the residuals and the level of information which is available in the sample, is still valid and can be used as an image of the information which is incorporated in the model. For this reason we always define the R^2 with this expression, whatever the formula type and the metric used.

16.5.4 The Fisher Test

The Fisher test (F-test) has a very limited interest for the cost analyst. But it is part of the panoply of statistician's tools and therefore deserves a comment.

For a Linear Formula Using the Standard Regression Analysis

The F-test is a test of the steepness of the slope b_1 in the relationship between Y and V_1.

The interest of this test comes from the fact that the statisticians do not know what they are looking for: they get a lot of data, with a lot of variables, and they try to find out some correlation between these variables, whatever they are. So they compute the dynamic center of the distribution of their data and find some slope b_1 (if only one parameter is used).

If this slope is large, they conclude that the causal variable they are investigating has an important influence on the dependent variable: they may be on the path of a discovery.

If, however, this slope is small, they consider that this causal variable has very little influence on the dependent variable and can be discarded: the dependent variable is more or less scattered around the average value $\bar{y}$: the sample does not contain for them, as we mentioned it earlier, any really usable information.

Therefore they decided to prepare a test for checking the importance of this slope. This test is based on the H_0 hypothesis of $b_1 = 0$ in the population. The sample is then used to validate or invalidate it.

The construction of the F-test follows about the same logic as the one indicated for the construction of the coefficient of determination R^2. It starts from the level of information which is available in the sample and decompose it:

$$\sum_i (y_i - \bar{y})^2 = \sum_i \left[(y_i - \hat{y}_i) - (\bar{y} - \hat{y})\right]^2$$

which can be simplified, due to the fact that the square term is equal, for a linear formula built from the standard regression analysis, to 0; therefore we can write:

$$\sum_i (y_i - \bar{y})^2 = \sum_i (y_i - \hat{y}_i)^2 + \sum_i (\bar{y} - \hat{y}_i)^2$$

which can be read the following way: the information available in the sample is equal to the information incorporated in the formula, plus the information lost.

An interesting figure to be computed in then the ratio of these two figures:

$$\frac{\sum_i (\hat{y}_i - \bar{y})^2}{\sum_i (y_i - \hat{y}_i)^2} = \frac{\text{information incorporated in the dynamic center}}{\text{information lost}}$$

The higher this quotient, the more pleased we are. It would then be interesting to use this quotient to test the H_0 hypothesis of $b_1 = 0$. In order to do that, we must make a small modification to it in order to use a well-known distribution:

- First it can be shown that:

$$\frac{\sum_i (\hat{y}_i - \bar{y})^2}{S^2} \times 1$$

(where 1 is the number of degrees of freedom of the numerator and S^2 the variance of the deviations inside the population) follows a χ^2 distribution with 1 degree of freedom.

- Second it was already shown that:

$$\frac{\sum_i (\hat{y}_i - y_i)^2}{S^2} \times (I - J - 1)$$

also follows a χ^2 distribution with $I - J - 1$ degrees of freedom.
These two variables being independent, their ratio follows a $F(1, I - J - 1)$ distribution if $b_1 = 0$. The F distribution is defined in Chapter 3.

To test the hypothesis we compute the value of this ratio: if the value is higher than a threshold, then we conclude, with a given level of confidence, that this value has little chance to be found in a sample drawn from a population with $b_1 = 0$ and we reject this hypothesis. For example, with a confidence level of 90%:

- if $I = 10$, we will reject the null hypothesis if the ratio exceeds 3.46,
- if $I = 20$, if the ratio exceeds 3.01.

For a similar scattering of the y_i around the dynamic center $\hat{y}_i$, the value of the quotient will be greater if the "distances" between the dynamic center and the cost average is high: this is the reason why the F-test can be called a test of the "steepness of the formula".

This steepness has not a great interest in the domain of cost, especially when we try to correlate the specific cost (the cost per unit of size) to the size: this specific cost generally has a limited steepness. Consequently the F-test will always appear poor in such situations.

A last word about this test: obviously there is a correlation between the distribution of the R^2 and the F-test, as it can easily be shown that the quotient we used for defining the F-test is equal ([50], p. 368) to:

$$\frac{R^2}{1 - R^2} \times (I - 2)$$

For Any Formula Not Using the Standard Regression Analysis

A similar discussion can be done: the F-test cannot anymore be interpreted as the amount of available information divided by the lost information, due to the fact that the cross term does not vanish in the computation, but nevertheless the same test can be used: after all it gives an idea about it.

Bibliography

1. AFITEP. *Estimation des coûts d'un projet industriel*. AFNOR, 1995.
2. George Anderlohr. What production breaks cost? *Industrial Engineering*, September, 1969.
3. Claude Andrieux. Normes de temps. *Lavoisier*, 1983.
4. Robert N. Anthony. *Management Accounting. Text and Cases*. Richard D. Irwin, Inc., 1964.
5. P. W. Atkins. *Physical Chemistry*. Oxford University Press, 1990.
6. A. R. Ballman. *A tutorial application of the Anderlohr break-in production technique*. Westinghouse Electric Corporation, ILSD Engineering, Hunt Valley, MD.
7. Ralph M. Barnes. Motion and time study. *Design and Measurement of Work*. John Wiley & Sons, 1980.
8. Barry W. Boehm. *Software Engineering Economics*. Prentice-Hall, Inc., Englewood Cliffs, NJ, 1981.
9. S. A. Book. *The Aerospace Corporation's Capabilities in Cost and Risk Analysis*. June, 1994.
10. Carl de Boor. *A Practical Guide to Splines*. Springer, 2000.
11. Pierrine Bouchard (born Foussier). Parametrics for underground construction. *Proceedings of the International Society of Parametric Analysts, 22nd Annual Conference*, Noordwijk, 2000.
12. John Bowers. Assessing risk in major projects. *Proceedings of the International Society of Parametric Analysts, 10th Annual Conference*, 1988.
13. George Bozoki. An expert judgment based software sizing model. *Journal of Parametrics*, XIII(1), May, 1993.
14. David A. Belsley, Edwin Kuh and Roy E. Welsh. *Regression Diagnostics*. John Wiley & Sons, 1980.
15. Barry J. Brinker (ed.). *Emerging Practices in Cost Management*. Warren, Gorham & Lamont, 1990.
16. C. Burnet. *The effect of aircraft size and complexity on the production learning curve*. British Aerospace Public Limited Co., About 1975.
17. Jean-Marie Chauveau. *Valoriser l'imprécision des coûts en gestion de projet*. ESCP, 1991.
18. A. Chauvel, G. Fournier and C. Raimbault. *Manuel d'Évaluation Economique Des Procédés*. Editions Technip, 2001.
19. *Defense System Management College Risk Assessment Techniques*, 1st edition. July, 1983.
20. Norman Draper and Harry Smith. *Applied Regression Analysis*. John Wiley & Sons, 1981.
21. D. Dubois. *Modèles mathématiques de l'imprécision et de l'incertitude en vue d'applications aux techniques d'aide à la décision*. Thèse d'état. Université de Grenoble, 1983.
22. D. Dubois and H. Prade. *Théorie des possibilités. Application à la représentation des connaissances en informatique*. Masson, 1988.
23. D. Dubois and H. Prade. Fuzzy real algebra: some results. *Fuzzy Sets and Systems*, 1979.
24. Bradley Efron and Robert J. Tibshirani. *An Introduction to the Bootstrap*. Chapman & Hall/CRC, 1993.
25. Merran Evans, Nicolas Hastings and Brian Peacock. *Statistical Distributions*. John Wiley & Sons, 2000.
26. R. E. Fairbairn and L. T. Twigg. Obtaining probabilities distributions of cost with approximations of general cost model functions. *Proceedings of the International Society of Parametric Analysts, 13th Annual Conference*, 1991.
27. Pierre Foussier and Jean-Marie Chauveau. Risk analysis. Are probabilities the right tool? *Proceedings of the International Society of Parametric Analysts, 16th Annual Conference*, Boston, 1994.
28. Larry D. Gahagan. A practical approach to conducting a cost risk analysis. *Proceedings of the International Society of Parametric Analysts, 13th Annual Conference*, 1991.
29. Paul F. Gallagher. *Parametric Estimating for Executives and Estimators*. Van Nostrand Reinhold Company, 1982.
30. Paul R. Garvey. *Probability Methods for Cost Estimating Analysis*. Marcel Dekker, Inc., 1999.
31. Gene H. Golub and Charles F. Van Loan. *Matrix Computations*. The John Hopkins University Press, 1996.
32. Glyn James. *Modern Engineering Mathematics*, 3rd edition. Prentice Hall, 2001.

33. Douglas T. Hicks. *Activity-Based Costing for Small and Mid-Sized Businesses. An Implementation Guide*. John Wiley & Sons, 1992.
34. J. Johnston. *Econometric Methods*. McGraw-Hill Book Company, 1960.
35. M. Lambert Joseph. Software estimation and prioritized hierarchies. *Proceedings of the International Society of Parametric Analysts*, V438–V445, 1986.
36. A. Lichnerowicz. *Algèbre et analyse linéaire*. Masson, 1955.
37. Donald W. Mackenzie. Price-based estimating risk model. *Proceedings of the International Society of Parametric Analysts, 13th Annual Conference*, 1991.
38. E. Malinvaud. *Statistical Methods of Econometrics*. North-Holland Publishing Company, 1980.
39. John Mandell. *The Statistical Analysis of Experimental Data*. Dover Publications, 1964.
40. Harry F. Martz and Ray A. Waller. *Bayesian Reliability Analysis*. John Wiley & Sons, 1982.
41. Matz, Curry and Frank. *Cost Accounting*. South-Western Publishing Company, 1967.
42. Gerald R. McNichols. An historical perspective on risk analysis. *Proceedings of the International Society of Parametric Analysts, 5th Annual Conference*, 1983.
43. Frederick Mosteller and John W. Tukey. Data analysis and regression. *A Second Course in Statistics*. Addison-Wesley Publishing Company, 1977.
44. National Association of Accountants. *Current application of direct costing*. Research Report, 1961.
45. Anthony J. Pettofrezzo. *Matrices and Transformations*. Dover Publications, 1966.
46. C. Radhakrishna Rao. *Linear Statistical Inference and Its Applications*. John Wiley & Sons, 2002.
47. David A. Ratkowsky. *Nonlinear Regression Modelling: A Unified Practical Approach*. UMI, 2002.
48. T. L. Saaty. *The Analytic Hierarchy Process*. McGraw-Hill, New York, NY, 1980.
49. Lothar Sachs. *Applied Statistics*. Springer-Verlag, 1984.
50. G. Saporta. *Probabilités, Analyse des données et Statistique*. Editions Technip, 1990.
51. Jun Shao and Dongsheng Tu. *The Jacknife and Bootstrap*. Springer, 1985.
52. Society of Manufacturing Engineers. *Tool and Manufacturing Engineers Handbook*, 3rd edition. McGraw-Hill Book Company, 1976.
53. P. Sprent and N. C. Smeeton. *Applied Nonparametric Statistical Methods*. Chapman & Hall/CRC, 2001.
54. Rodney D. Stewart. *Cost Estimating*. John Wiley & Sons, 1982.
55. Charles R. Symons. *Software Sizing and Estimating. Mk II FPA*.
56. Henri Theil. *Principles of Econometrics*. John Wiley & Sons, 1971.
57. US Department of Defense. *Armed Services Procurement Regulation Manual*, 1975 edition.
58. Michel Volle. *Analyse des données*. Economica, 1985.
59. Charles Winklehaus and Robert Michel. *Cost Engineering*, August, 1982.
60. Thomas H. Wonnacott and Ronald J. Wonnacot. *Statistique*. Economica, 1990.
61. L. Zadeh. Fuzzy sets as a basis for a theory of possibility. *Fuzzy Sets and Systems*, 1, 3–28, 1978.

Index

Printing: Krips bv, Meppel
Binding: Stürtz, Würzburg